CHEMICAL AND BIOCHEMICAL APPLICATIONS OF LASERS

VOLUME IV

Contributors

Alessandra Andreoni

J. L. Beauchamp

John C. Bellum

D. S. Bomse

Paul L. DeVries

Sylvie Druet

Thomas F. George

John E. Hearst

Bryan E. Kohler

Kai-Shue Lam

Hai-Woong Lee

Jui-teng Lin

Richard Mathies

Carlo A. Sacchi

Mark S. Slutsky

Orazio Svelto

Jean-Pierre Taran

Richard P. Van Duyne

R. L. Woodin

Jian-Min Yuan

I. Harold Zimmerman

CHEMICAL AND BIOCHEMICAL APPLICATIONS OF LASERS

edited by C. BRADLEY MOORE

Department of Chemistry
University of California
and
Materials and Molecular Research Division
of the Lawrence Berkeley Laboratory
Berkeley, California

VOLUME IV

ACADEMIC PRESS   New York   San Francisco   London   1979
A Subsidiary of Harcourt Brace Jovanovich, Publishers

Academic Press Rapid Manuscript Reproduction

ACADEMIC PRESS, INC.
111 Fifth Avenue, New York, New York 10003

United Kingdom Edition published by
ACADEMIC PRESS, INC. (LONDON) LTD.
24/28 Oval Road, London NW1 7DX

LIBRARY OF CONGRESS NUMBER 79–398

ISBN 0–12–5054–04–1

PRINTED IN THE UNITED STATES OF AMERICA

79 80 81 82   9 8 7 6 5 4 3 2 1

Contents

4. Laser Excitation of Raman Scattering from Adsorbed
    Molecules on Electrode Surfaces

    Richard P. Van Duyne

5. Coherent Anti-Stokes Raman Spectroscopy

    Sylvie Druet and Jean-Pierre Taran

6. Theory of Molecular Rate Processes in the Presence of Intense
    Laser Radiation

    *Thomas F. George, I. Harold Zimmerman, Paul L. DeVries,
    Jian-Min Yuan, Kai-Shue Lam, John C. Bellum, Hai-Woong
    Lee, Mark S. Slutsky, and Jui-teng Lin*

7. Multiphoton Dissociation of Gas Phase Ions Using Low
    Intensity cw Laser Radiation

    R. L. Woodin, D. S. Bomse, and J. L. Beauchamp

8. Photochemical Fixation of the Nucleic Acid Double Helix
    Utilizing Psoralens

    John E. Hearst

List of Contributors

(Numbers in parentheses indicate the pages on which the authors' contributions begin.)

Alessandra Andreoni (1), Centro di Elettronica Quantistica e Strumentazione Elettronica, Istituto di Fisica del Politecnico, Piazzale Leonardo da Vinci, 32, 20133 Milan, Italy

J. L. Beauchamp (355), Division of Chemistry and Chemical Engineering, California Institute of Technology, Pasadena, California 91125

John C. Bellum (253), Fachbereich Physik, Postfach 3049, Universität Kaiserslautern, D-6750 Kaiserslautern, West Germany

D. S. Bomse (355), Division of Chemistry and Chemical Engineering, California Institute of Technology, Pasadena, California 91125

Paul L. DeVries (253), Department of Chemistry, University of Rochester, River Station, Rochester, New York 14627

Sylvie Druet (187), Office National d'Etudes et de Recherches Aérospatiales (ONERA), 29 Avenue de la Division Leclerc, 92320 Châtillon, France

Thomas F. George (253), Department of Chemistry, University of Rochester, River Station, Rochester, New York 14627

John E. Hearst (389), Department of Chemistry, University of California, Berkeley, California 94720

Bryan E. Kohler (31), Department of Chemistry, Wesleyan University, Middletown, Connecticut 06457

Kai-Shue Lam (253), Department of Chemistry, University of Rochester, River Station, Rochester, New York 14627

Hai-Woong Lee (253), Department of Chemistry, University of Rochester, River Station, Rochester, New York 14627

Jui-teng Lin (253), Department of Chemistry, University of Rochester, River Station, Rochester, New York 14627

Richard Mathies (55), Department of Chemistry, University of California, Berkeley, California 94720

Carlo A. Sacchi (1), Centro di Elettronica Quantistica e Strumentazione Elettronica, Istituto di Fisica del Politecnico, Piazzale Leonardo da Vinci, 32, 20133 Milan, Italy

Mark S. Slutsky (253), Department of Chemistry, University of Rochester, River Station, Rochester, New York 14627

Orazio Svelto (1), Centro di Elettronica Quantistica e Strumentazione Elettronica, Istituto di Fisica del Politecnico, Piazzale Leonardo da Vinci 32, 20133 Milan, Italy

Jean-Pierre Taran (187), Office National d'Etudes et de Recherches Aérospatiales (ONERA), 29 Avenue de la Division Leclerc, 92320 Châtillon, France

Richard P. Van Duyne (101), Department of Chemistry, Northwestern University, 2145 Sheridan Road, Evanston, Illinois 60201

R. L. Woodin (355), Exxon Research and Engineering Company, Corporate Research Laboratories, P.O. Box 45, Linden, New Jersey 07036

Jian-Min Yuan (253), Physics Department, Drexel University, 32nd and Chestnut Streets, Philadelphia, Pennsylvania 19104

I. Harold Zimmerman (253), Physics Department, Clarkson Memorial College, Potsdam, New York 13676

Preface

Since the publication of the first volume of this series, many new kinds of lasers have been discovered and brought into commercial production. The development of reliable, push-button tunable lasers in the ultraviolet, visible and infrared is just now removing technological barriers to many new research areas in chemistry and biochemistry. Relatively little of a researcher's effort need be expended on the laser itself. Most of the experiments described in this volume may be carried out with standard, commercial laser systems. Structural spectroscopy and photochemistry is done with ions in a high vacuum, with DNA in solution, and with molecules adsorbed on surfaces.

Structural studies of DNA by fluorescence microscopy are described in Chapter 1. Sharp fluorescence spectra of complex molecules in solids may be obtained when a narrow-band laser selectively excites molecules in specific sites, Chapter 2. Chapter 3 describes the theory and application of resonance Raman spectroscopy to several biological systems. It has recently been discovered that the intensity of Raman scattering by molecules adsorbed on electrode surfaces can be enhanced by many orders of magnitude. Chapter 4 relates the state of this art and its applications. Chapter 5 gives a thorough treatment of coherent anti-Stokes Raman spectroscopy and its applications in analytical chemistry and combustion diagnostics. Chapter 6 is a theoretical treatise on a new kind of photochemistry in which potential surfaces are forced to cross in extremely high laser fields. Multiphoton infrared excitation of molecules by very intense lasers has considerably broadened our ideas concerning the interaction of radiation and matter (Vol. III, Chapter 2). Chapter 7 reveals that dissociation of molecular ions by low power continuous infrared lasers is possible. The final chapter describes laser control of the sequential photochemical reaction of the drug psoralen with the two strands of the DNA double helix.

Since this series is intended to preview rather than to review research using lasers, the editor has imposed greatly upon the authors and their secretaries to produce their work in camera-ready form. The copy was revised and completed with great skill and patience by Luce J. Denney, Margaret Knight, and Nancy Monroe. The editor is most grateful to all of the authors, typists, and draftsmen whose very professional and timely work make up this volume. He is likewise grateful to Penny Percival who has continued the thorough and thoughtful indexing of this series.

STRUCTURAL STUDIES OF BIOLOGICAL MOLECULES
VIA LASER-INDUCED FLUORESCENCE:
ACRIDINE-DNA COMPLEXES[1]

Alessandra Andreoni
Carlo A. Sacchi
Orazio Svelto

Centro di Elettronica Quantistica
e Strumentazione Elettronica
Istituto di Fisica del Politecnico
Milano, Italy

I. INTRODUCTION

Measurements based on fluorescence emission are particular
ly valuable and sensitive for structural studies of biological
molecules (1-4). The emission may be due to either the biomole
cule itself (primary fluorescence) or to a suitable dye bound
to a specific position of the biomolecule (secondary fluorescen
ce). In the former case, UV excitation is usually required and
the information to be gained is related to the biological mole
cule either alone (e.g. purine and pirimidine bases, which are
the constituents of the deoxyribonucleic acid, DNA) or interact
ing with the other biomolecules in the sample (as in the DNA
macromolecule). In the latter case, the fluorescence properties
depend, in general, on the mechanism of interaction between the
dye and the macromolecules. In both cases, the usual fluorescen
ce parameters are of interest, namely, the emission and excita
tion spectra, the decay time, the quantum yield and the polari
zation. In particular, the measurement of the decay time is an
especially valuable technique for a large variety of biophysical processes (4).
For structural studies in the field of cell biology, the

[1]*Work supported by the Italian National Research Council in
part through the special program "Tecnologie Biomediche".*

technique of fluorescence microscopy has been increasingly used to investigate biophysical and biochemical processes in sin
gle cells or cellular organelles, eventually in the living sta
te (5). The aim of such an investigation is to obtain a better
knowledge of the relations between the structure of the biological macromolecule and the functional properties of the cell.
Among the current trends in this field are, for example, (i)
the biophysical study of the secondary structure of nucleic
acids, using acridine derivatives as fluorescent probes, (ii)
the biochemical study of enzymatic reactions which either produce fluorescent emission or occur on a fluorogenic substrate
(6,7); (iii) the study of structural and functional properties
of membranes related, for example, to the potential-dependent
release and uptake of fluorescent dyes (8).

The use of laser beams to excite the fluorescence emission
of biological molecules and cell constituents appears to be a
very promising technique for several reasons: (i) Short light
pulses with duration well below the usual limit of conventional
light flashes ($\sim$ 1 ns) are nowadays available with a large variety of lasers (9). (ii) Since laser beams are usually diffrac
tion limited the beam can be focused down to a spot with a radius approximately equal to the beam wavelength. This allows
the possibility of selective excitation in space of single
cells and cell constituents. (iii) Since a few of these lasers
can be made tunable over a sufficiently wide wavelength range,
selective excitation at the desired wavelength is also possible.

The technique of laser induced fluorescence has already
been applied to a few cases of biological interest and, in parti
cular, a great deal of work has been devoted to the study of
the primary events of photosynthesis (10). In this work we apply ourselves to what we believe to be another very interesting example of biological molecules which can be profitably
studied by laser techniques: namely the complex which is formed when a dye, belonging to the acridine family, is bound to
both synthetic polynucleotides and native DNAs. These complexes have many interesting properties with implications of relevance in several fields from chemical physics to biophysics,
pharmacology and medicine. Their study may, in particular, be
useful for understanding the origin of the fluorescence patterns
of chromosomes when stained with some of the acridine dyes. For
these works, a special microfluorometer, based on a pulsed dye
laser and on a home designed signal averager, has been construct
ed in our laboratory. At present, the system is capable of a
spatial resolution of $\sim$ 0.3 μm, a temporal resolution of $\sim$ 0.3
ns, and the excitation beam can be tuned over the whole visible
and UV spectrum. A modified version of the system, which is now
being developed, is expected to allow measurements of lifetimes
below 0.1 ns.

II. THE LASER MICROFLUOROMETER

The block diagram of the experimental apparatus is shown in Fig. 1. It is based on (i) a subnanosecond-pulsed tunable dye-laser, (ii) a microscope system for fluorescence microscopy (Leitz MPV) and (iii) a digital signal averager (11-16).

The dye laser, described in detail in Ref. 17, is pumped by a nitrogen laser that generates pulses with peak power up to 500 kW, duration of 2.5 ns and repetition rate up to 40 Hz. Due to a special design of the dye cell, the dye laser can generate pulses with a duration of $\sim$ 0.4 ns and peak power up to 10 kW. Since the beam is diffraction limited, it can be focused down to the resolving spot of the microscope, i.e. to $\sim$ 0.3 μm. In most of our experiments a 1.5 x 10^{-3}M solution of POPOP in Toluene was used as lasing medium. As shown in Fig. 1, the excitation beam enters the fluorescence microscope through two lenses and two diaphragms which select the central part of the beam. Both the lenses and the diaphragms determine, together with the microscope optics, the spot size on the plane of the sample. Part of the beam is sent, through a beam-splitter, on a fast photodiode (Hadron TRG mod. 105 C, risetime 0.3 ns) whose output signal is used to monitor the beam power (on a 519 Tektronix oscilloscope) and to provide for an external trigger signal for the digital averager.

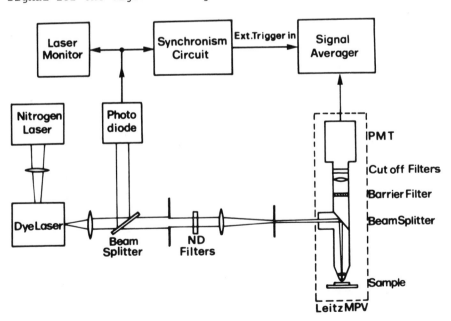

FIGURE 1. Block diagram of the experimental apparatus.

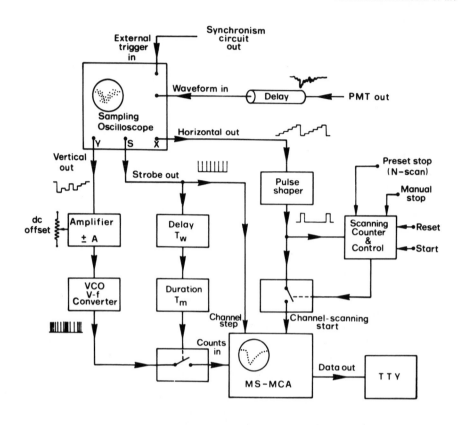

FIGURE 2.   Block diagram of the digital averager.

The Leitz MPV apparatus is a commercial instrument for flu
orescence microscopy. It is made mainly of a microscope with
side windows for both the entrance of the excitation beam and
sample illumination by conventional lamps. The fluorescence
light is collected by the microscope objective itself and focused on the photocathode of a photomultiplier mounted at the
top of the Leitz MPV instrument. Suitable barrier filters are
inserted in the path of the fluorescence light to select the
desired spectral interval. In our experiment a fast photomulti
plier (RCA, mod. 70045 D), with a risetime of ~ 0.7 ns has
been used. Due to several reasons (mainly the smallness of
the irradiated volume and the fluorescence and detection efficiencies) the fluorescence power may be very weak, typically a
few photons per nanosecond. In these conditions, due to radiation emission statistics, the detected signal presents fluc-

tuations comparable to the average value. The signal must then
be averaged over a high number N of repetitions in order to re
duce (by the factor $N^{\frac{1}{2}}$) the relative fluctuations.

The averager has been developed in our laboratory, and
makes use of the internal facilities of a sampling oscillosco
pe (Tektronix 564 with S-2 sampling head) and of a multichannel pulse analyzer (MCA), with suitable interconnections. As
shown in the block diagram in Fig. 2, the vertical output of
the oscilloscope is converted into the frequency of a train of
pulses by means of a free-running voltage-controlled oscillator (VCO). The frequency is then digitally measured by counting the VCO pulses falling in a prefixed time-gate of duration
T_m (measuring time). This is done by the MCA which operates as
a multiscaler. The strobe output of the oscilloscope (blanking
signal) provides for the channel stepping and ensures the correspondence between the temporal position of the samples and
the channel address in the MCA. The horizontal output of the
oscilloscope, which is operated in the sequential-scanning mode, is used to sweep, in synchronism and sequentially, the
channel address in the MCA. The gate of the VCO pulses is open
ed after a waiting time T_W with respect to the sampling action,
to let the vertical output of the oscilloscope reach its proper value. The number of averaging scans is externally preset
and counted. It is worth noting that our averager has the following important properties, which are not easily found in com
mercial systems: (i) The same time resolution as the sampling
head, (ii) a high degree of accuracy due to the method used for
analog-to-digital conversion, (iii) a substantial reduction of
drift due to the fast sequential scanning.

Before ending this section, we wish to summarize the per-
formances of our laser microfluorometer: (i) We verified pho-
tographically that the laser beam can be focused down to a
spot diameter equal to the resolving diameter of the microscope ($\sim$ 0.3 μm). (ii) We verified that the prompt response of
the system has a time width (FWHM) of $\sim$ 1 ns. This width arises from the combination of the laser pulse duration, the response time of the PMT and that of the signal averager. A typi
cal example of this response is shown in Fig. 3. (iii) We notic
ed that our laser beam during a typical measuring cycle,produc
es little, if any, photodecomposition of the dye to be studied.
This is shown in Fig. 4 in the case of frog's erythrocytes
stained by the conventional Feulgen reaction (pararosaniline-
-SO$_2$) for DNA demonstration (11). Figure 4a shows the time-re
solved fluorescence of a nucleus after a few laser shots,while
Fig. 4b shows the fluorescence of the same nucleus after irradiating the sample for about 30 minutes (corresponding to
approximately 50,000 laser shots). Since the two figures are
nearly identical, we conclude that dye photodecomposition did

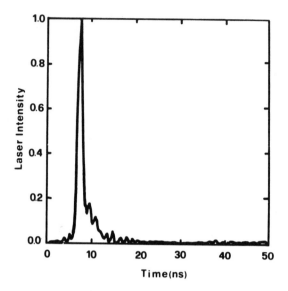

FIGURE 3. Prompt response of the whole apparatus.

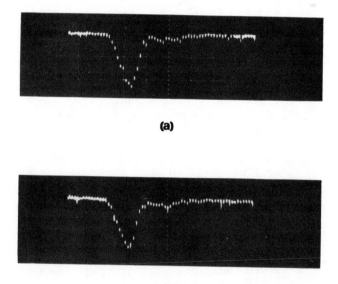

FIGURE 4. Test to show the lack of dye photodecomposition by pulsed laser excitation: (a) fluorescence signal of the specimen at the beginning of the test, (b) fluorescence signal of the same sample after 30 minutes of laser irradiation. Time scale is 2 ns/div.

not appreciably occur. Similar results were obtained in all
other cases considered. This is a particularly relevant result,
to be contrasted with what happens in ordinary fluorescence
microscopy experiments, where a UV lamp is used as the excitation source. In this case dye photodecomposition (as evidenced
by the fading of the fluorescence) usually occurs in a few seconds, so that reliable measurements are often difficult to
make. The lack of dye photodecomposition in the case of laser
excitation is believed to be due to the small amount of energy density which is normally used ($\sim$ 1 mJ/cm^2). This energy is
however still able to give a good signal-to-noise ratio becau
se it is concentrated in a small time interval.

Finally, we should note that an improved version of the mi
crofluorometer is now being built. It makes use of a nitrogen
pumped dye laser which gives laser pulses of $\sim$ 150 ps duration
(FWHM) with a repetition rate up to 150 Hz (18). The detecting
system is made of a Varian type 154 M crossed-field photomulti
plier whose measured risetime is 100 ps and falltime is 150 ps.
With this system, decay times of the order of 100 ps or perhaps
(by appropriate deconvolution) somewhat shorter than that
should be measured.

III. THE ACRIDINE-DNA COMPLEX

The acridines form a large family of fluorescent dyes which
absorb in the blue and emit in the green region of the spectrum
(19,20). The chemical structure of a few of them is shown in
Fig. 5, where the cationic form is indicated for Acriflavine
(AF), Quinacrine (QAC), and Quinacrine Mustard (QM). The absorption and fluorescence spectra of the dye of greater interest to us, QM, are shown in Fig. 6. It should be noted that the
spectra of the other acridine dyes are all very similar to that
of QM, since they are mainly due to transitions of the π-elec-
trons of the acridine ring. The transition involved is believed to be of π-π^* type (21). For most acridines the dipole moment of this blue transition is polarized parallel to the major
geometrical axis of the molecule (22).

A. *Survey of the Binding Processes*

When an acridine dye is added to a solution of DNA, an
acridine-DNA complex is formed. There is general agreement that
the binding process can be divided into a strong process (I)
and a weak process (II) (23). The strong binding process accord
ing to the original model of Lerman (24) as modified by Peacocke and co-workers (25) is depicted in Fig. 7. The binding oc-

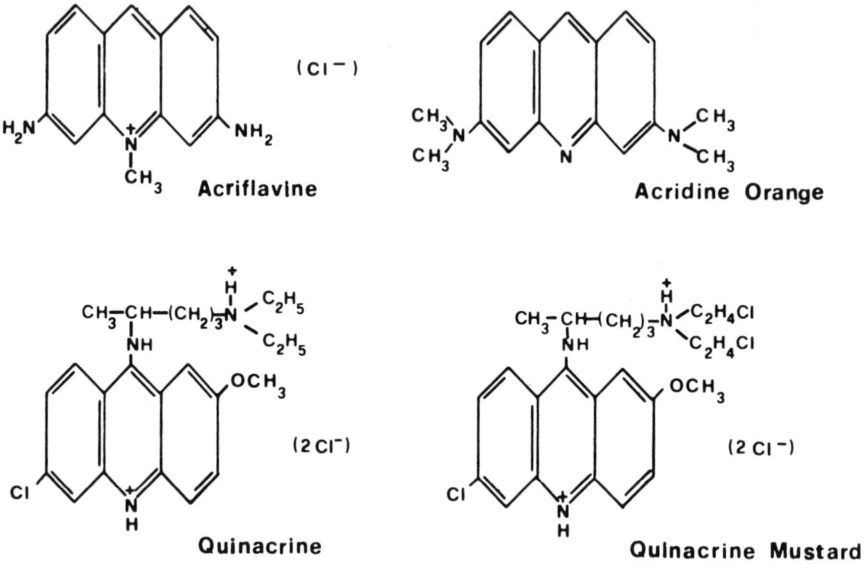

Alessandra Andreoni et al.

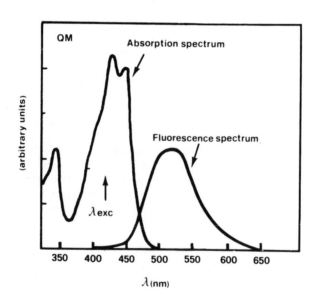

FIGURE 5.   Chemical structure of a few acridine dyes.

FIGURE 6.   Absorption and fluorescence spectra of Quinacri_
ne Mustard (QM) in a 0.2 M Acetate buffer solution (pH= 4.6).

curs mainly by intercalation of the acridine ring between two
neighboring bases of the same DNA strand and enlarges the distance between these bases twofold. The negatively charged Oxy
gen atom of the phosphate group, which lies between the separat
ed bases, swings into the inside of the polynucleotide chain.
The main binding force thus arises from Van der Waals interactions between the π-electrons of the acridine ring and the π-
-electrons of the bases. The intercalation is further strength-
ened by the ionic binding between the basic heterocyclic Nitro
gen atom of the acridine ring (bearing a positive charge) and
the Oxygen atom of the phosphoric acid of the DNA backbone.
When the QM dye is used and at least one of the two adjacent
bases is Guanine, a covalent bond between the alkylating group
of the QM side chain and the 7-amino group of Guanine is also
formed (26). Although the binding constant for this strong
binding process is quite large, the DNA molecule must clearly
undergo strong conformational changes (increased separation
between the base planes at the intercalation site followed by

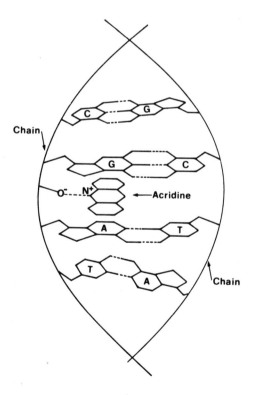

FIGURE 7. *Modified Lerman's model of acridine-DNA complex.*

an unwinding of the DNA double helix) to accomodate an acridine dye. Consequently not all available positions can be utiliz
ed by the strong binding process. If we call   r   the ratio of
acridine bound per DNA phosphate group, this amounts to saying
that   r   can reach a limiting value   n   which is appreciably
smaller than 1. The actual value of   n   depends on the acridine dye which is used but it is always smaller than ~ 0.2 .
This leads to the so-called excluded site model which assumes
that, owing to the increase of base pair distance upon intercalation, no two acridine rings can intercalate adjacent bind
ing sites (27). Indeed such a model would give a limit value
of one acridine dye bound every four phosphorus nucleotides
(i.e. $n = 0.25$).

When   r   reaches the limit value   n   any other acridine dye
which may be further added cannot intercalate the DNA molecule. At this stage a new and weaker binding process takes over
(process II). In this process the acridine cation is believed
to be bound approximately edgewise and externally to the double helix of DNA with its positive ring nitrogen-atom close to
the phosphate group of the DNA backbone. By such a process the
value of   r   can keep increasing up to the limit value of 1
(electrical neutrality). When   r   is large enough, the acridine rings can stack upon each other in a direction parallel to
the helix axis. In this weaker binding process, the binding
forces thus arise from both electrostatic interaction with the
phosphate groups and Van der Waals interaction between stacked
dyes.

All previous qualitative discussion about process I and II
can be put in a quantitative form (23). If only one binding
mechanism were present, the equilibrium condition of the association reaction

DNA + acridine $\rightleftarrows$ acridine-DNA complex

would lead to the equation

$$r = k\ (n-r)\ C, \tag{1}$$

where   C   is the molar concentration of the unbound acridine
in the bulk solution and the constant   k,   which is called the
association constant and measured in M^{-1}, gives the strength
of the binding process. From Eq. (1) we get

$$r = \frac{n\ kC}{1+kC} \tag{2}$$

When two binding processes are effective, Eq. (2) generalizes
to

$$r = \frac{n_I k_I C}{1+k_I C} + \frac{n_{II} k_{II} C}{1+k_{II} C} \cdot \qquad (3)$$

If only one binding process is present, a plot of r/C vs r , which is called Scatchard plot, gives a straight line with an intercept at $r = n$ on the r axis and with slope $-k$ (see Eq. 1). When more than one binding mechanism is effective, the plot gives a curve which is not a straight line any more. However, it is often possible to discern distinct linear portions of this plot from which the individual n_j and k_j can be determined. This is especially true when the (n,k) values of each binding site are considerably different. For the strong and weak processes considered before (process I and II respectively) one has:

$$n_I \simeq 0.2; \; k_I \simeq 10^7 \div 10^8 \; \text{M}^{-1}; \; n_{II} \simeq 0.8; \; k_{II} \simeq 10^5 \div 10^6 \text{M}^{-1}.$$

The relatively high value of the binding constant k_I explains why acridine dyes exhibit a pronounced specificity to nucleic acids even in the presence of substantial amounts of proteins. Thus an acridine dye, when used to stain a cell, will only stain the nucleus and not the extranuclear proteins.

B. Fluorescence Properties of the Complex

A first important property of the acridine-DNA complexes (when process I, i.e. intercalation, is mainly taking place) is that their fluorescence quantum yield depends strongly on the type of binding site (28-30). This is particularly true for dyes such as Quinacrine (QAC), Quinacrine Mustard (QM), Proflavine (PF), Acriflavine (AF). Indeed the fluorescence quantum yield increases (as compared to the case of the same dye in buffer solution) when these molecules intercalate two Adenine-Thymine(AT) base pairs. The increase is moderate for AF and PF($\sim$1.3) while it is somewhat higher for QAC and QM ($\sim$ 4). What is even more relevant, however, is that the quantum yield decreases (by a factor 4 to 10 for QAC and QM) when the molecule intercalates either two Guanine-Cytosine (GC) base pairs or a GC-AT base-pair sequence. By comparison, the absorption band undergoes a limited red-shift ($\sim$ 10 nm) when the dye goes from the unbound to the bound state. Furthermore the absorption spectrum remains approximately the same irrespective of the type of binding site (i.e. of the DNA base pair composition). Similarly the fluorescence spectrum undergoes a blue--shift by the same amount when the dye goes from free to bound state. These spectral shifts are somewhat moderate and perhaps

easier to understand. Indeed, shift of opposite directions in
absorption and fluorescence spectra, which bring the maxima
closer, strongly suggest an alteration of the vibrational pattern with a higher probability of population of the low vibrational levels (31). The change in quantum yield upon the type
of binding site is however more dramatic and difficult to explain. The increase of quantum yield when the acridine intercalates an AT-AT sequence can perhaps be understood in terms
of an increase of viscosity and decrease of dielectric constant
of the enviroment (32). Indeed a similar increase is observed
for acridine dyes in solution upon increasing the viscosity of
the solvent. The decrease of quantum yield upon binding to a
GC-AT or GC-GC sequence is however more difficult to explain.
On the basis of theoretical calculations of the electron affinity of these dyes and of the purine and pyrimidine bases (33),
the quenching has been attributed to the formation of a charge-
-transfer complex between the acridine dye (acting as electron
acceptor) and the Guanine residue (acting as electron donor)
(34,35). However Schreiber and Daune (30) have not been able
to detect any additional absorption band which should accompany such a complex (36). Furthermore one should explain why
Acridine Orange does not undergo any quantum yield reduction
upon binding to a Guanine residue (actually its quantum yield
increases by a factor of $\sim$ 2.5 upon binding to both AT-AT
and GC-GC sequences (30)).

C. Chromosome Fluorescence Banding Pattern

Another relevant property of the acridine-DNA complex is
observed when some of the acridine dyes (e.g. QM, QAC, PF,AF)
are used to stain the DNA of methaphase chromosomes of both
mammalians and plants. When the chromosomes are illuminated as
a whole by a UV lamp, some specific and narrow ($\sim$ 1 µm wide)
portions of them (bands) fluoresce more intensely than others
(37,38). A schematic fluorescence pattern of the M-chromosome
of Vicia Faba is sketched in Fig. 8, while a picture of what
actually is seen under a fluorescence microscope is shown in
Fig. 9. This is the well known phenomenon of chromosome fluorescence bands which is now widely used for chromosome recognition and characterization and it is now the primary internatio
nal standard for human methaphase chromosomes (39). Furthermore, the presence of new additional bands or the disappearance
of some already existing ones in human chromosomes have been
consistently observed in connection with certain diseases (40-
-42). For instance, an additional fluorescence band in a chromosome N.14 has been observed in 10 out of 12 patients which
were affected by Burkitt Lymphomas (40). Despite the fact that

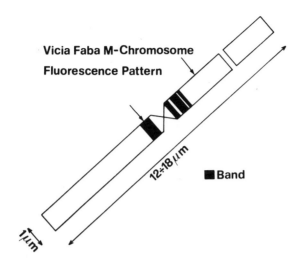

FIGURE 8. Schematic drawing of Vicia Faba M-chromosome banding pattern.

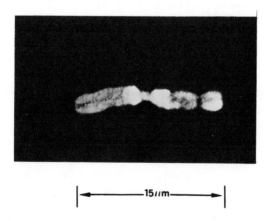

FIGURE 9. Picture of a Vicia Faba M-chromosome fluorescence banding pattern as obtained when staining with QM.

standard banding patterns of chromosomes are widely used for practical purposes, their origin is still unclear. The most appealing possibility is related to the base dependent behavior of the acridine-DNA complex which was previously discussed. According to this possibility the more intensely fluorescing regions of the chromosomal DNA should be richer in AT-AT base se

quences. The other possibility is however that the banding regions could be due to an increased binding of acridine to
parts of DNA where interaction between DNA and surrounding pro
teins is less strong. Although the above possibility related
to the base dependent behavior seems to be favored, no defini-
tive evidence is available in favour to this assumption (43-45).

D. *Other DNA Intercalating Dyes*

Before ending this section we wish to mention that some of
the properties of the acridine-DNA complexes (in particular
base specificity and chromosome fluorescent bands) are also
shown by other dyes which, although not belonging to the acridine family, are also able to intercalate the DNA double helix.
Most of these compounds consist of three coplanar fused aromatic rings and are therefore similar to the acridines in their
configuration. Among these intercalating dyes we like to mention: (i) the Ethidium Bromide molecule (46) which gives, in
some chromosomes, fluorescent bands which are the reverse of
those observed with acridines (i.e. a fluorescence decrease
where acridines give a fluorescence increase, a phenomenon call
ed *reverse banding* ; see ref. 44 and 45), (ii) the Actinomycin
D molecule which shows antibiotic action (39), (iii) diacridi-
ne derivatives which bisintercalate the DNA chain (47), (iv)
the Adriamycin and Daunomycin molecules (48) which are among
the strongest antitumor agents so far available.

IV.   LASER-INDUCED FLUORESCENCE EXPERIMENTS ON ACRIDINE-DNA
      COMPLEXES

A. *Possible Experiments Via Laser Excitation*

After the previous description of the properties of the
acridine-DNA complexes, we can ask ourselves how a laser beam
can be used to provide information which would be hard if not
impossible to obtain by conventional methods. According to
what is described in section II, we can say that lasers can be
profitably used for the following experiments: (i) Acridine
bound to a GC-GC or GC-AT sequence. In this case, on account
of the low quantum yield, the fluorescence decay time is expected to be rather short (less than 1 ns) and hence difficult
to measure by conventional pulsed excitation sources. The study of such complexes would help to elucidate the mechanisms
which are responsible for the fluorescence quenching. (ii)
Acridine bound to either AT-AT, or AT-GC, or GC-GC

(as in native DNAs). In this case, the fluorescence decay curve is expected to result from the superposition of fluorescence with long lifetime (dye bound to AT-AT) and fluorescence with short lifetime (dye bound to AT-GC or GC-GC). Then the shape of the decay should give information about the amount of repeated AT-AT sequences which is present. (iii) In native DNAs and at sufficiently high dye concentration, energy transfer might occur between high quantum yield sites and low quantum yield sites (49). Since the structure of DNA is quite well known, this is certainly a rather simple and appealing case of studying energy transfer and migration in biological substances. (iv) Many synthetic mononucleotides and polynucleotides of known base sequences can be used to provide known binding sites for the acridine molecules. Some of them do show fluorescence quenching thus being more easily studied in pulsed laser excitation. (v) Since the laser beam can be focused to a spot which is smaller than a typical fluorescent band of a chromosome, the fluorescence properties of individual bands can be studied in order to shed light on the origin of the banding pattern and on the corresponding genetic implications. (iv) The precise location of some of the intercalating drugs (i.e. the antitumor agents Adriamycin and Daunomycin) within the cell can be studied with our apparatus even at the low concentrations which occur *in vivo*.

B. *Quinacrine Mustard in Synthetic Polynucleotides*

With some of these possibilities already in mind, a few years ago we started an extensive program with our apparatus. Most of our experiments have been done on Quinacrine Mustard because it gives the brightest and most stable chromosome fluorescent bands (44,45). The laser wavelength was tuned to 419 nm which corresponds to the absorption peak of QM (see Fig. 6).

In a first set of experiments, the decay curves of QM which is either free in solution or bound to some synthetic polynucleotides were studied (14-16). The decay curve of QM in a 0.2 M Acetate buffer solution (pH = 4.6) is shown in Fig.10. The dye concentration is 1.6 x 10^{-5}M, which is comparable to that normally used to stain the cytological preparations. The curve can be fitted to a single exponential with decay time $\tau = 4 \pm 0.2$ ns. This value turns out to be the same as that recently reported for QAC (50,51). The synthetic polynucleotides Poly dA-Poly dT and Poly dG-Poly dC,in which all the base pairs are AT or GC respectively,were studied. When QM is bound to Poly dA-Poly dT, the fluorescence decay curve (Fig. 11) is, at a first approximation, exponential with decay time

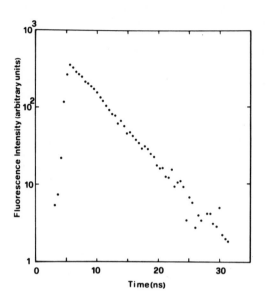

FIGURE 10. *Fluorescence decay curve of a 1.6 x 10^{-5} M solution of QM in 0.2 M Acetate buffer (pH = 4.6).*

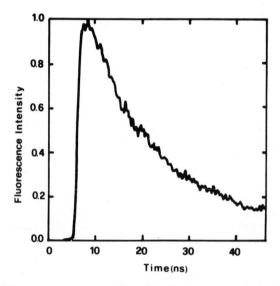

FIGURE 11. *Fluorescence decay curve of QM bound to Poly dA - Poly dT.*

$\tau \simeq$ 18 ns. This value compares well with that reported for QAC (50,51). The increase of decay time in respect to the case of free QM is by a factor ($\sim$ 4.5) which is in agreement with, although somewhat larger than, the corresponding increase in quantum yield for QM bound to Poly d(AT) (28). In contrast to the rather simple behavior of the fluorescence decay of QM both free or bound to Poly dA-Poly dT, the decay curve of QM bound to Poly dG-Poly dC is markedly different (Fig. 12). The curve is clearly not exponential but rather formed by the superposition of a fast initial decay with time constant τ_1 = 0.5 $\pm$ 0.2 ns and a slower tail with time constant $\tau_2 \simeq$ 5 ns. Note that, according to the quantum yield data (28), one would have expected a decay time of $\sim$ 0.45 ns for a pure exponential decay. Although a detailed study of this case is still to be made, we wish to point out that: (i) The decay curve remains unchanged when the laser intensity is changed by more than one order of magnitude. This eliminates the possibility that the fast initial decay is due to a stimulated effect. (ii) The decay curve, and, in particular, the slower tail, remains unaffected when KI is added to the solution. Potassium iodide is known to quench strongly the fluorescence of free acridines (a circumstance which was verified by us to occur for free QM in solution) while not affecting an intercalated dye (52). On account

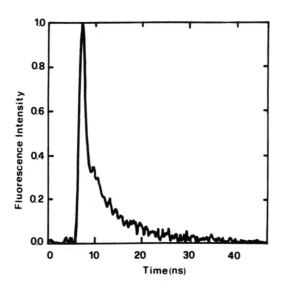

FIGURE 12. Fluorescence decay curve of QM bound to Poly dG-Poly dC.

of this we conclude that the slower tail is not due to free
dye molecules which were left in the solution. By the same
argument we are led to believe that this tail is not due to
QM molecules which are bound to the DNA but not intercalated
(i.e. bound according to the weaker binding process described
in section III). We therefore conclude that the double-exponen
tial curve of Fig. 12 is due to a single type of complex and
suggest that this complex is an excimer formed by the excited
QM molecule with the Guanine residue. Excimer fluorescence is
indeed well known to lead to a double-exponential fluorescence
decay (53). On account of the good electron donor property of
QM and of the good electron acceptor property of Guanine (33),
we also suggest that the excimer is, predominantly, of the
charge-transfer type. Time resolved spectra of the emission
would be helpful to substantiate this assumption.

C. *Quinacrine Mustard in Bacteria, Eukaryotic Cells and Chromosomes*

As a next step, the fluorescence of bacteria (i.e. of rather simple organisms) with different AT percentages when
stained with QM has been studied. To this purpose, the laser
light was focused by the microscope to a spot smaller than the
bacteria themselves. The original idea which led to this experiment was as follows. Suppose that, in a given bacterium, the
decay curve of Fig. 11 gives the decay of QM dyes intercalating AT-AT sequences and the curve of Fig. 12 gives the decay
of QM dyes intercalating either AT-GC or GC-GC sequences. Then,
if the whole fluorescent process were additive, the overall de
cay would merely be the weighted sum of the two curves shown
in Fig. 11 and 12, the weighting factor being given by the ave
rage percentage of AT-AT and (AT-GC + GC-GC) sequences in the
bacterium. The decay curves obtained with two bacteria of widely different AT percentages are shown in Fig. 13. The general aspect of these curves is of the kind one would have expect
ed on account of a superposition of a fast decay (QM bound to
AT-GC or GC-GC) and a slower tail (QM bound to AT-AT). A more
careful study of the curves, however, shows that the (average)
decay time of the slower tail is not quite that of QM bound to
AT-AT sequences. Actually this decay time was found to increase
linearly with the square of the AT percentage (α) of the bacte
rium as shown in Fig. 14 (54). To understand these results we
checked to see whether the presence of other cellular components in the bacterium (e.g. proteins etc.) were heavily affect
ing the fluorescence decay. To this purpose, a set of measur –
ements was carried out to compare the fluorescence response of
cytological preparations to that of the corresponding purified

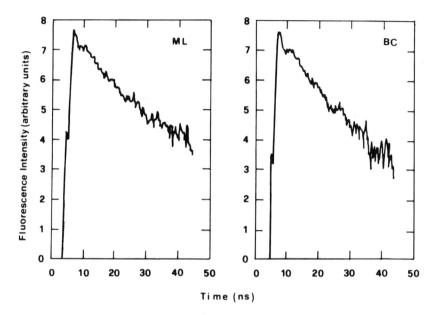

FIGURE 13. *Semilogarithmic plot (base e) of the fluorescen*
ce decay curve of QM bound to cytological preparations of Micro
coccus Lysodeikticus *(ML) (28% AT) and* Bacillus Cereus *(BC)*
(62.5% AT) .

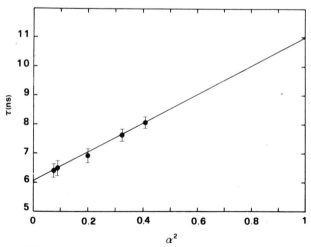

FIGURE 14. *Average decay time of the slower tail of seve-*
ral bacteria as a function of the square of the AT percentage
(α) .

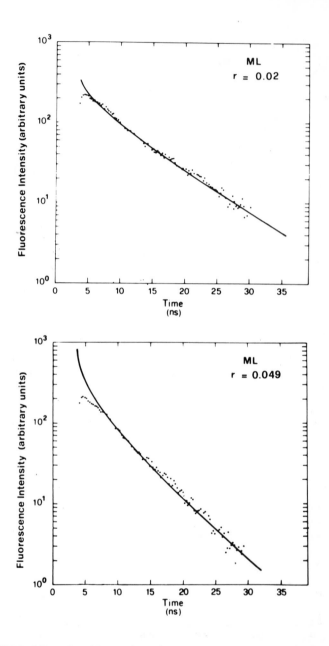

FIGURE 15.   *Semilogarithmic plot of the fluorescence decay for QM bound to* Micrococcus Lysodeikticus   *purified DNA at two different values* r_1.

DNAs. The case of *Micrococcus Lysodekticus* (ML) was particularly studied. It was found that: (i) The slower tail of the fluorescence curve for the purified DNA depends on the fraction of DNA base pairs which are intercalated by the QM molecules (55). The corresponding decay time decreases by increasing this fraction. Examples of two decay curves obtained in this way are reported in Fig. 15[1], where the fraction of intercalated base pairs is indicated by the quantity r_1. (ii) This should be contrasted with the case of the smeared bacterium (i.e. with the case of the cytological preparation) wherein the fluorescence curve does not depend on the staining conditions (i.e. on the concentration of the QM solution used to stain the cytological preparation). (iii) A specific value of r_1

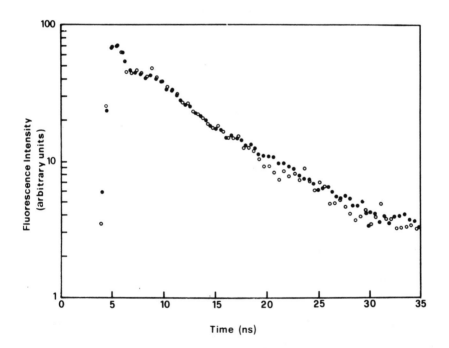

FIGURE 16. Semilogarithmic plot of the fluorescence decay of QM bound to Micrococcus Lysodeikticus: o-cytological preparation; •-purified DNA solution stained at $r_1 = 0.02$.

[1]Note that r_1 in Fig. 15 and 16 is defined as the fraction of intercalated QM molecules per couple of phosphate groups (i.e. $r_1 = r_I/2$, where r_I was defined in section III).

could be chosen at which the fluorescence curve of the purified
DNA was identical to that of the bacterial smear. This is shown
in Fig. 16. The result discussed at point (iii) of above shows
that the fluorescence of the cytological preparations is to be
attributed to the dye bound to the DNA content of the bacteria,
while the other cellular components seem to play a minor role.
The result at point (ii) of the above can be interpreted to
say that, in the bacterium, all the available binding sites are
occupied by the QM molecules. If this is true, then from Fig.
16 we can say that the limit value of the available binding si
tes in the cytological preparation of ML is r_1 = 0.02 (i.e.
only 2% of the intercalation sites can be occupied by the QM
molecules). Lastly, we faced the problem of explaining the observed dependence of the decay time of the slower tail on the
AT percentage and on the staining conditions. Quantitative
studies of this problem showed that the effect could be interpreted in terms of a Förster type energy transfer between QM
molecules bound to AT-AT (which act as donors) and QM molecules bound to either AT-GC or GC-GC (which are supposed to be
nonfluorescent acceptors) (54,55). Indeed, in such a case, the
decay time of the slower tail is expected to depend linearly
on the square of the AT percentage (α) since the probability
of a donor being followed by an acceptor along the DNA chain
is linearly dependent on α^2. Furthermore the decrease in decay time with increased staining of the DNA is readily explain
ed since when staining increases, the average donor-to-accep-
tor distance decreases, and the energy transfer rate increases.

Although most of the results obtained in single bacteria
were understood reasonably well, the complexity of the phenomena which are occuring and the lack of a quantitative study
of the faster part of the curve of Fig. 13 make us very cau-
tious about the overall picture. The relevance of studies of
this sort for understanding some structural properties of biological samples can however be further evidenced by the two
following examples. In Fig. 17, the fluorescence curves obtain
ed when QM stains the nuclei of two different plant cells are
shown. The general appearance is somewhat similar to that of
bacteria (Fig. 13), although the biological complexity of these
plant cells is certainly much greater. Note however that, although the AT percentage is not much different in the two cases, the fluorescence decay curves are radically different.
One possibility could be that the two cells, although having
similar AT percentages, do have drastically different percentages of repeated AT-AT sequences. As the second example, Fig. 18
shows two fluorescence curves of the *Vicia Faba* M-chromosome
The solid curve and the dots give the fluorescence response
after laser excitation outside and inside a fluorescent band,
respectively. The exact locations of the laser excitation are

Allium Coepa: 63%AT

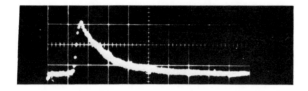

Ginco Biloba: 65.2%AT

FIGURE 17. Comparison of fluorescence decay curves of two plant cells with similar AT content. The time scale is 5 ns/div.

indicated by the two arrows in Fig. 7. For purposes of compari son in Fig. 18 the intensity of the in-band fluorescence response has been reduced by a factor 1.54 . Since the two responses are almost identical,the ratio of the two quantum yields is also equal to the same scale factor introduced before (i.e. 1.54). On account of what was said before, one would expect that different AT percentages at the two points of the chromosome would have given different shapes of fluorescence curves. The results of Fig. 18 show however that, at least in the case considered, the quantum yield difference is only accounted for by a simple scale factor, i.e. it seems to be due essentially to a different dye concentration.

D. Adriamycin

As a final example of the possibilities offered by these techniques for structural studies of biological molecules, we wish to mention a few results which have recently been obtained on the antitumor drug Adriamycin. Figure 19 shows the fluorescence decay of a 1.7×10^{-3} M solution of Adriamycin in phy siological solution when excited at 476 nm which corresponds to the peak of the absorption spectrum of Adriamycin Hydrochlo

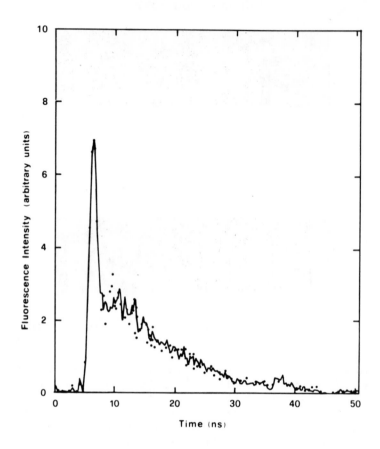

FIGURE 18.   *Comparison of the fluorescence decay curves of*
Vicia Faba *M-chromosome stained with Quinacrine Mustard. The
solid line and the dots refer to laser excitation outside and
inside a fluorescent band, respectively. For purposes of comparison the intensity of the in-band fluorescence has been reduced by a factor 1.54.*

ride. We see that the curve is highly non exponential which is
indicative of a rather complex behavior of this drug even in a
simple solution. Note that this behavior remains unchanged even
when the drug concentration is reduced by several orders of mag
nitude. Actually the minimum concentration which could be measured is 10^{-9}M (56) which is two orders of magnitude lower than
what can presently be measured by the best available convention
al fluorometers (i.e. $\sim$ 100 ng/cm^3). Hence one could hope that
in situ measurements of the drug distribution inside tissues

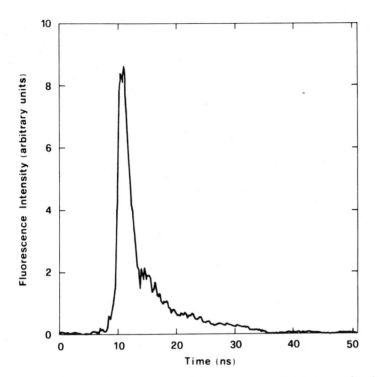

*FIGURE 19. Fluorescence decay curve of Adriamycin in phy-
siological solution.*

and single cells (eventually in living conditions) could be ma̲
de for the first time. This has actually been proved in frozen
slices of rat liver and in human granulocytes (57).

V. CONCLUSIONS AND PROSPECTS

In this work a few cases of laser-induced fluorescence stu̲
dies of biological molecules have been considered. The acridine̲
dye Quinacrine Mustard (QM) and the antitumor agent Adriamycin
bound to DNA have been selected. The following cases have been
particularly studied: (i) QM bound to some synthetic polynucleo̲
tides, (ii) QM bound to the DNA of bacteria, (iii) QM bound to
the DNA of eukaryotic cells, (iv) QM bound to *Vicia Faba* M-
-chromosomes, (v) Adriamycin bound to rat liver tissue and human granulocytes. In all these cases, the laser beam has prov-
en to be very useful in getting new information, mainly for the
following reasons: (i) Possibility of generating short light

pulses of duration well below the limit value presently obtainable with conventional lamps (∿ 1 ns). (ii) Tunability of the laser wavelength over the whole visible and near UV range. (iii) Possibility of focusing the beam down to the resolving spot of the microscope objective which is used. (iv) Absence of photodecomposition of the sample with an overall (i.e. summed over all laser pulses) energy density on the sample which is enough to obtain precise fluorescence measurements. Although, in all the cases considered, valuable pieces of information have already been obtained, we can say that often we have only scratched the surface of the problems which are still to be understood and which can be studied by these techniques. In particular we wish to mention: (i) Nature of the binding process between QM and the Guanine residue, (ii) Initial fast decay in bacteria, (iii) Systematic studies of eukaryotic cells of different AT percentages, (iv) Systematic study of fluorescence decay curves along a given chromosome, (v) Extension to other dyes of the acridine family, (vi) Nature of the binding and drug location with some of the antitumor drugs. Besides these examples, which more directly follow from what was said in this chapter, we wish to mention other possibilities: (i) Studies of energy transfer between purine and pyrimidine bases in the DNA (58) and between these bases and a fluorescent label (59). (ii) Studies on the decay of the fluorescence anisotropy which can give information about the overall shape of the macromolecule and on its orientational relaxation (60,61). (iii) Use of laser excitation to act on rather than to probe the biological molecule. In this respect, we wish to mention the possibility of selective excitation of a given binding site by two-step selective laser excitation using picosecond pulses (62). Another possibility, more easily implemented, is related to the wide difference in decay curve which is expected when an acridine dye binds either an AT-AT or a GC-GC sequence (see Fig. 11 and 12). Accordingly, if a second laser pulse is sent with a suitable delay and with a suitable wavelength, selective excitation to a higher upper state of the acridine bound to AT-AT (which is still in the upper vibronic state, while the corresponding binding to GC-GC has already relaxed to the ground state) is possible. Of course, many other examples can be given. We believe however that the results and the discussion presented in this paper provide for enough evidence to say that lasers will have a bright future for structural studies of biological molecules.

ACKNOWLEDGMENTS

We are grateful to Dr. G.Bottiroli, Prof. S.Cova and Dr.A. Longoni for very helpful discussions. We thank also the Istitu to Ricerche Farmitalia and the Istituto di Ricerche Farmacologiche "Mario Negri" for kindly supplying the samples of Adriamycin.

REFERENCES

1. Konev, S. V., "Fluorescence and Phosphorescence of Proteins and Nucleic Acids". Plenum Press, New York, (1967).
2. Udenfried, S., "Fluorescence Assays in Biology and Medicine", Vol. II. Academic Press, London, New York, (1969).
3. "Physico-Chemical Properties of Nucleic Acids", Vol. I (J. Duchesne, ed.). Academic Press, London, New York, (1973).
4. Brand, L., and Gohlke, J. R., *Ann. Rev. Biochem. 41*, 843 (1972).
5. Böhm, N., *in* "Techniques of Biochemical and Biophysical Morphology", Vol. I (D. Glick and R. M. Rosenbaum, ed.),p. 89. Wiley-Interscience, London, New York, (1972).
6. Kohen, E., Kohen, C., Hirschberg, J. G., Wouters, A., and Thorell, B., *Photochem. Photobiol. 27*, 259 (1978).
7. Prenna, G., Bottiroli, G., and Mazzini, G., *Hystochem. J. 9*, 15 (1977).
8. Hoffmann, J. F., and Laris, P. C., *J. Physiol. 239*, 519 (1974).
9. "Topics in Applied Physics. Vol. XVIII: Ultrashort Light Pulses" (S. L. Shapiro, ed.). Springer-Verlag, Berlin,Heidelberg, New York, (1977).
10. Campillo, A. J., and Shapiro, S. L., *in* "Topics in Applied Physics. Vol. XVIII: Ultrashort Light Pulses" (S. L. Shapiro, ed.), p. 317. Springer-Verlag, Berlin, Heidelberg, New York, (1977).
11. Sacchi, C. A., Svelto, O., and Prenna, G., *Histochem. J. 6*, 251 (1974).
12. Sacchi, C. A., and Andreoni, A., *in* "Proceedings of the Second European Electro-Optics Markets and Technology Conference" (H. A. Elion, ed.), p. 67. Mack Brooks Exh. Ltd, London, (1975).
13. Andreoni, A., Sacchi, C. A., Cova, S., Bottiroli, G., and Prenna, G., *in* "Lasers in Physical Chemistry and Biophysics" (J. Joussot-Dubien, ed.), p. 413. Elsevier Scientific Publishing Co., Amsterdam, (1975).

14. Andreoni, A., Sacchi, C. A., Svelto, O., Longoni, A., Bot tiroli, G., and Prenna, G., *in* "Proceedings of the Third European Electro-Optics Conference" (H. A. Elion, ed.), Vol. 99, p. 258. SPIE, Washington, (1976).

15. Andreoni, A., Longoni, A., Sacchi, C. A., Svelto, O., and Bottiroli, G., *in* "Optical Sciences. Vol. III: Tunable La sers and Applications" (A. Mooradian, T. Jaeger and P. Stokseth, ed.), p. 303. Springer-Verlag, Berlin, Heidelberg, New York, (1976).

16. Andreoni, A., Sacchi, C. A., Svelto, O., Bottiroli, G., and Prenna, G., *Sov. J. Quantum Electron.*, in press.

17. Andreoni, A., Benetti, P., and Sacchi, C. A., *Appl.Phys.* 7, 61 (1975).

18. Cubeddu, R., De Silvestri, S., and Svelto, O., to be published.

19. Albert, A., "The Acridines". E. Arnold Ltd, London,(1966).

20. "Heterocyclic Compounds: Acridines (Vol. IX)" (R. M. Ache son, ed.). Interscience, New York, (1973).

21. Zanker, V., *Z. Phys. Chem., N.F. 2* (1954).

22. Kelly, G. R., and Kurucsev, T., *Biopolymers 15*, 1481 (1976).

23. Peacocke, A. R., *in* "Heterocyclic Compounds: Acridines (Vol. IX)" (R. M. Acheson, ed.), p. 723. Interscience, New York, (1973).

24. Lerman, L. S., *J. Mol. Biol. 3*, 18 (1961).

25. Pritchard, N. J., Blake, A., and Peacocke, A. R., *Nature 212*, 1360 (1966).

26. Selander, R.-K., *Biochem. J. 131*, 749 (1973).

27. Crothers, D. M., *Biopolymers 6*, 575 (1968).

28. Michelson, A. M., Monny, C., and Kovoor, A., *Biochimie 54*, 1129 (1972).

29. Latt, S. A., and Brodie, S., *in* "Excited States of Biological Molecules" (J. B. Birks, ed.), p. 178. Wiley-Interscience, London, New York, (1976).

30. Schreiber, J. P., and Daune, M. P., *J. Mol. Biol. 83*, 487 (1974).

31. Weill, G., and Calvin, M., *Biopolymers 1*, 401 (1963).

32. Weill, G., *Biopolymers 3*, 567 (1965).

33. Pullman, B., and Pullman, A., *Rev. Mod. Phys. 32*, 428 (1960).

34. Tubbs, R. K., Ditmars, W. E., jr., and Van Winkle, Q., *J. Mol. Biol. 9*, 545 (1964).

35. Thomes, J. C., Weill, G., and Daune, M., *Biopolymers 8*, 647 (1969).

36. Birks, J. B. "Photophysics of Aromatic Molecules", p. 403. Wiley-Interscience, London, New York, (1970).

37. Caspersson, T., Farber, S., Foley, G. E., Kudynowski, J., Modest, E. J., Simonsson, E., Wagh, U., and Zech, L., *Exptl. Cell Res. 49*, 219 (1968).
38. Caspersson, T., Zech, L., Johansson, C., and Modest, E. J., *Chromosoma 30*, 215 (1970).
39. Modest, E. J., and Sengupta, S. K., *in* "Fluorescence Techniques in Cell Biology" (A. A. Thaer and M. Sernetz, ed.), p. 125. Springer-Verlag, Berlin, Heidelberg, New York, (1973).
40. Manolov, G., and Manolova, Y., *Nature 237*, 33 (1972).
41. Rowley, J. D., *Nature 243*, 290 (1973).
42. Zankl, H., and Kang, K. D., *Humangenetik 14*, 168 (1972).
43. Weisblum, B., and De Haseth, P. L., *Proc. Natl. Acad. Sci. US 69*, 629 (1972).
44. Rigler, R., *in* "Chromosome Identification. Nobel Symposium XXIII" (T. Caspersson and L. Zech, ed.), p. 335. Academic Press, New York, (1973).
45. Modest, E. J., and Sengupta, S. K., *in* "Chromosome Identification. Nobel Symposium XXIII" (T. Caspersson and L. Zech, ed.), p. 327. Academic Press, New York, (1973).
46. Le Bret, M., Le Pecq, J.-B., Barbet, J., and Roques, B.P., *Nucleic Acid Res. 4*, 1361 (1977).
47. Le Pecq, J.-B., Le Bret, M., Barbet, J., and Roques, B.P., *Proc. Natl. Acad. Sci. US 72*, 2915 (1975).
48. Di Marco, A., Arcamone, F., and Zumino, F., *in* "Antibiotics. Vol. III: Mechanism of Action of Antimicrobial and Antitumor Agents" (J. W. Corcoran and F. E. Hahn, ed.), p. 101. Springer-Verlag, Berlin, Heidelberg, New York, (1974).
49. Latt, S. A., Brodie, S., and Munroe, S. H., *Chromosoma 49*, 17 (1974).
50. Seligy, V. L., Szabo, A. G., and Williams, R. E., *in* " Excited States of Biological Molecules" (J. B. Birks, ed.), p. 233. Wiley-Interscience, London, New York, (1976).
51. Duportail, G., Mauss, Y., and Chambron, J., *Biopolymers 16*, 1397 (1977).
52. Fröhlich, P., and Wehry, E. L., *in* "Modern Fluorescence Spectroscopy, Vol. II" (E. L. Wehry, ed.), p. 319. Heyden, London, New York, Rheine, (1976).
53. Birks, J. B., "Photophysics of Aromatic Molecules", p.301. Wiley-Interscience, London, New York, (1970).
54. Bottiroli, G., Prenna, G., Andreoni, A., Sacchi, C. A., and Svelto, O., *Photochem. Photobiol.*, in press.
55. Andreoni, A., Cova, S., Bottiroli, G., and Prenna, G., submitted to *Photochem. Photobiol.*
56. Andreoni, A., to be published.
57. Andreoni, A., Sacchi, C. A., Bottiroli, G., and Salmona, M., to be published.

58. Shapiro, S. L., Campillo, A. J., Kollman, V. H., and Goad, W. B., *Opt. Commun. 15*, 308 (1975).
59. Anders, A., "Spectroscopic Investigations of DNA-dye Complexes with Tunable Lasers" (in German) Thesis. University of Bielefeld (West Germany), 1977.
60. "Biochemical Fluorescence: Concepts" (R. F. Chen and H. Edelhock, ed.). Marcel Dekker, Inc., New York, (1975).
61. Rigler, R., and Ehrenberg, M., *Q. Rev. Biophys. 9*,1 (1976).
62. Letokhov, V. S., *in* "Optical Sciences. Vol. III: Tunable Lasers and Applications" (A. Mooradian, T. Jaeger and P. Stokseth, ed.), p. 122. Springer-Verlag, Berlin, Heidelberg, New York, (1976).

SITE SELECTION SPECTROSCOPY

Bryan E. Kohler[1]

Joint Institute for Laboratory Astrophysics
University of Colorado and National Bureau of Standards
Boulder, Colorado

I.   INTRODUCTION

To physicists and chemists the spectrum of a physical
system is just a plot of the probability of that system interacting with light versus frequency. As laser technology
provides increasingly sophisticated methods for manipulating
the radiation field part of the spectroscopic process, new
types of spectra constantly appear (ODMR, CARS, RIKE, etc.).
Here we limit the discussion to conventional optical absorption and emission spectra of molecules in condensed phases and
focus on simple laser based methods for enhancing the information content of these spectra.

The need for improving resolution well beyond that attainable with classical methods can be highlighted by discussing a
specific case, for example, the connection between optical
absorption and emission spectra and photochemical properties.
To account for observed photochemistry in terms of fundamental
molecular properties several pieces of information are needed.
Most important are the energy, multiplicity, and symmetry of
the excited state and information on how the excited state
energy varies with nuclear displacement.   This information is,
in principle, contained in the absorption and emission spectra.

The frequencies and intensities of the vibrational fine
structure observed in absorption to and emission from the
lowest energy excited state of given multiplicity (the lowest
energy excited state is the most important one because the

---

[1]*JILA Visiting Fellow, 1978-79, on leave from Wesleyan
University, Middletown, CT 06457.*

rate of relaxation from higher electronic states is usually
much faster than the rate of photochemistry in condensed
phases) are related to the changes in the potential energy
surface that accompany electronic excitation. This is schematically indicated in Figure 1. A more detailed description
of the connection between the observed spectra and chemical
properties is implicit in the qualitative description of
vibronic spectra given in Section II. Unfortunately, in most
cases the information contained in the condensed phase optical
spectrum of a large molecule is quite inadequate for such an
analysis. For example, the fluorescence spectrum obtained
using standard techniques for perylene, a polycyclic aromatic
hydrocarbon, in ethyl alcohol at 77°K is shown in Figure 2.
Perylene has 90 normal vibrational modes. Even for the
highest possible symmetry (D_{2h}), there are 16 totally sym-

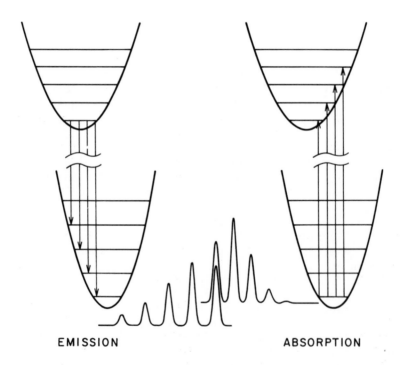

EMISSION   ABSORPTION

*FIGURE 1. The connection between absorption and emission
spectra in condensed phases and molecular energy levels. The
absorption spectrum contains vibrational intervals corresponding to the upper state while the emission spectrum contains
intervals for the ground state.*

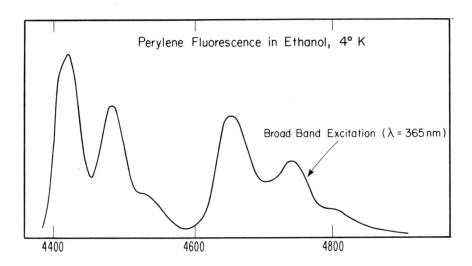

*FIGURE 2. Fluorescence of perylene in ethanol at 4.2°K.
This spectrum was excited with the 365 nm line from a pressure
broadened Hg arc. Fluorescence wavelength in Å.*

metric modes that could couple to an allowed transition. To
attempt to deduce the frequency changes and displacements for
these 16 modes from the three or four ill-resolved features
seen in the spectrum shown in Figure 2 could charitably be
called ambitious and realistically is quite impossible. Of
course, if the guest molecules have lower than D_{2h} symmetry,
as is likely, the scale of the problem goes up dramatically.
The question then is, what makes these spectra so bad and what
can be done to improve matters.

For the width there are two extreme possibilities: either
every molecule in the experimentally addressed ensemble has
the same broad spectrum and the observed average faithfully
represents the spectrum of an individual molecule (the homogeneous limit) or every molecule has a sharp but, in general,
different spectrum and the observed average is not representative of any one molecule (the inhomogeneous limit). This is
sketched in Figure 3. For the first case, homogeneous
broadening, nothing can be done short of appealing to different types of experiments with different selection rules in
order to simplify the spectrum. For the second case, the high
flux and monochromaticity of lasers can be used to interrogate
only a subset of the total experimental ensemble and greatly
enhance the information content of the measured spectra. For
the particular case of perylene an emission spectrum like the
one shown in Figure 4 can be easily realized. In the best

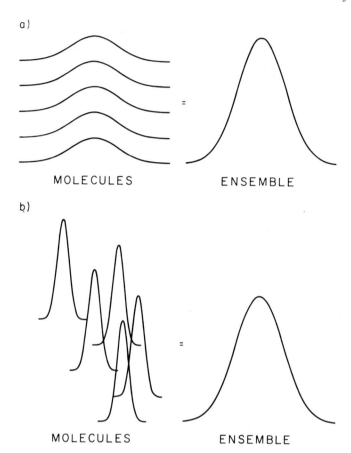

FIGURE 3. The limiting cases of homogeneous and inhomo-
geneous broadening. In the homogeneous case, a), the spectrum
summed over all molecules in the sample corresponds identical-
ly to that for any one of the individual molecules. In the
inhomogeneous case, b), the spectrum summed over all molecules
in the sample differs from the spectrum that would be seen
for any single molecule.

case for such a laser selection experiment on a perylene solution, approximately 160 vibronic lines of less than 1 cm^{-1}
width have been measured and fit by 16 fundamentals which are
found to have uncertainties of at most 0.3 cm^{-1} (1). With
such an improvement in the quality of the data the interpretation of these spectra in terms of fundamental chemical
properties becomes a much more realistic endeavor. In this
review we shall discuss how these selection experiments work,
what the range of applicability of these techniques is likely
to be and present some specific examples.

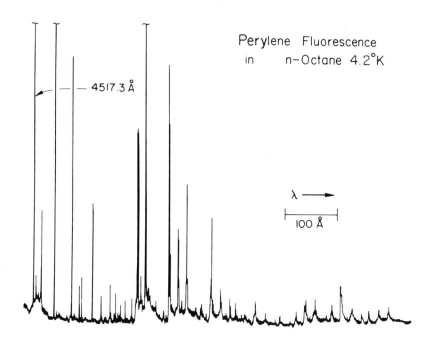

Perylene Fluorescence
in   n-Octane 4.2°K

— — 4517.3 Å

λ ⟶

⊢——— 100 Å ———⊣

FIGURE 4. *Laser selected fluorescence of perylene in*
n-octane at 4.2°K. This spectrum was excited with a narrow
band (FWHM 0.1 cm⁻¹) dye laser operating at 4517.3 Å.

II. OPTICAL SPECTRA IN CONDENSED PHASES

Before discussing laser selection spectroscopy in detail
it is useful to review the theory of vibrational structure in
optical absorption and emission spectra and discuss how the
local field and the additional degrees of freedom associated
with the solvent manifest themselves in the observed spectra.
This review is by no means comprehensive and, in fact, covers
qualitatively only the minimum number of concepts needed to
understand how laser selection experiments work.

A. *The Frozen Gas*

In discussing condensed phase optical spectra we begin by
ignoring the interactions of the solute molecule with its solvent environment except to acknowledge that in the case of
solid solutions these interactions quench rotation and translation. The energy of a non-rotating, non-translating frozen

gas molecule depends on both the electronic and nuclear coordinates as·summarized in the Hamiltonian

$$H = T_e(\underset{\sim}{r}) + T_N(\underset{\sim}{R}) + V_e(\underset{\sim}{r}) + V_N(\underset{\sim}{R}) + V_{eN}(\underset{\sim}{r},\underset{\sim}{R}) \tag{II.1}$$

where the individual terms represent respectively the kinetic
energy of the electrons, the kinetic energy of the nuclei, the
Coulomb repulsion of the electrons, the Coulomb repulsion of
the nuclei and the electron-nuclear Coulomb attraction. $\underset{\sim}{R}$
represents the coordinates of all the nuclei and $\underset{\sim}{r}$ represents
the coordinates of all the electrons.

In the adiabatic approximation (2) the wave function can
be written

$$\Psi_{m,\mu}(\underset{\sim}{r},\underset{\sim}{R}) = \Phi_m(\underset{\sim}{r},\underset{\sim}{R})\chi_\mu^m(\underset{\sim}{R}) \tag{II.2}$$

where

$$(T_e + V_e + V_{eN})\Phi_m = U_m(\underset{\sim}{R})\Phi_m \tag{II.3}$$

and

$$(T_n + V_n + U_m)\chi_\mu^m = E_\mu^m\chi_\mu^m. \tag{II.4}$$

The probability of an electronic transition is proportional to
the squared amplitude of the transition moment

$$\underset{\sim}{M} = \sum_i e\underset{\sim}{r}_i - \sum_k eZ\underset{\sim}{R}_k = \underset{\sim}{M}_e + \underset{\sim}{M}_N.$$

The electronic wave functions Φ_m can be expanded in a Taylor
series in the nuclear coordinates about the equilibrium
nuclear positions of the appropriate states (3). Taking only
the first term in this expansion [the Condon approximation (4)]
gives for the squared transition moment

$$\left|\underset{\sim}{M}_{m,\mu\to n,\nu}\right|^2 = \left|<\chi_\mu^m|\chi_\nu^n><\Phi_m^o|\underset{\sim}{M}_e|\Phi_n^o>\right|^2. \tag{II.5}$$

Thus, in the Condon approximation, a transition from one
electronic-vibrational level to another is proportional to a
factor that depends only on the electronic states involved
multiplied by the squared overlap integral of the vibrational
wave functions, the Franck-Condon factor (4,5). Proper symmetry labels may be deduced from the intensity and polarization of the observed transition and the accompanying
vibrational fine structure carries information about the

differences between the ground and excited state potential surfaces. In the simplest case, the vibrational degrees of freedom can be described as a collection of non-interacting harmonic oscillators. If all the electronic states are so described, then the normal coordinates of one state can be expressed in terms of those of another by a linear transformation

$$Q_\nu^n = \sum_\mu A_{\nu\mu}^{nm}(Q_\mu^m - X_\mu^{nm}),$$

where the Q_ν^n are the normal coordinates for the νth normal mode in the nth electronic state. When $\underset{\approx}{A}$ is the identity, the transformation corresponds to displacement of the equilibrium coordinates of state m with a transition to state n. If $\underset{\approx}{A}$ is diagonal but not the identity, then the force constants are different for the different electronic states. A non-diagonal $\underset{\approx}{A}$ corresponds to a mixing of normal modes in state n relative to m [the Duschinskii effect (6)]. By modeling coordinate changes to get a fit to the observed vibronic intensities, much can be learned about the chemical properties of excited states. Although relatively little has been attempted in this regard since the pioneering work of Craig on benzene (7), the increased availability of high quality spectra and efficient algorithms will hopefully lead to more effort in this area.

B. *Effects of the Solvent*

To understand site selection phenomena, it is necessary to go into a little more detail regarding the effects of the solvent environment on the observed spectrum. Following the established custom, these can be separated into a shifting of the transition energies from the values that would be assumed for the isolated molecule and the addition of lattice modes to the vibrational fine structure. The former is the key to analyzing inhomogeneous broadening in solid solutions and the latter rationalizes the experimentally observed importance of red-edge excitation and low temperatures.

1. Transition energy shift. Numerous equations have been advanced to describe the dependence of electronic transition energies on local solvent fields (8). The usual approach is to consider the interaction between a neutral solvent and a neutral solute molecule to second order in perturbation theory. With appropriate summing over an assumed solvent configuration, expressions that predict shifts in solute transition energies from the values that would obtain *in vacuo* result. While we needn't develop the theory in full detail, it is pertinent to review two properties of the final expressions: they depend strongly on solvent configuration and they are almost independent of solute vibrational level.

With the assumptions of no overlap between solvent and solute molecules and electrical neutrality, the perturbation Hamiltonian for the interaction of a molecule A with a molecule B may be written

$$H'_{AB} = \frac{1}{R^3_{AB}} \, \underset{\sim}{M}_A \cdot (\underset{\sim}{I} - 3R_{AB}R_{AB}) \cdot \underset{\sim}{M}_B \qquad (II.6)$$

where R_{AB} is a vector from A to B and M_A is the dipole moment operator for molecule A. The shift is computed by evaluating to second order H'_{AB} summed over all solvent molecules for each of the two solute states involved in the transition and then taking the difference.

As was the case with the transition moments considered above, within the adiabatic approximation the dipole moment expectation values will have the form of an electronic matrix element times a vibrational overlap factor. In the zero-order basis of product states this factor will be 1 for all the first order corrections and will again sum to 1 in the sums over states for the second order corrections since the vibrational states are a complete set. Thus, to this degree of approximation, the solvent shift of a given electronic transition is expected to be the same for all solute vibrational sublevels. While the shift will be different for different configurations of solute and solvent and different for different solute transitions, the result for a given configuration and solute transition will simply be a shift of every peak associated with that electronic transition in the isolated molecule by the same amount. Actually, a proper accounting of the mixing of solute electronic levels by the solvent perturbation would introduce a dependence of vibrational frequency on solvent configuration but these "vibrational solvent shifts" can be expected to be small relative to the vibrational frequencies. Even in very heterogeneous systems, infrared absorption bands are much narrower than their separations.

2. *Lattice vibrations.* In addition to shifting electronic excitation energies, the solvent environment also contributes additional degrees of freedom in the form of lattice vibrations which can couple with the solute electronic transition. All of the important spectroscopic consequences of this excitation-lattice vibration (excitation-phonon) coupling can be visualized by regarding the solute molecule plus host system as a very large molecule with the electronic

excitation localized on the solute molecule. The effect of
the host is to provide a large number of additional vibrational degrees of freedom weakly coupled to the electronic transition. To good approximation these vibrations can also be
discussed in terms of the Franck-Condon principle and be described as a set of harmonic oscillators. For molecular
crystals (ordered solids consisting of a single molecular
species) an elegant formalism has been developed to solve for
the eigenfunctions of the normal vibrational modes of the
crystal (9). This formalism capitalizes on the periodic
nature of the crystal lattice via Bloch's theorem and involves
the spatial Fourier transform of the positions of each molecule. The repeating unit in the lattice, the unit cell, may
contain one or more molecules. The variable conjugate to
position (called wave vector) is a good quantum number which
labels each normal vibrational mode of the crystal (called a
phonon). The phonons are completely delocalized throughout
the crystal.

Vibrations involving the relative translation of unit
cells with respect to each other are called acoustic phonons
and have the property that their frequency is linearly related
to wave vector k at small k (i.e., $\omega = ck$ where c is the speed
of sound in the solid). In van der Waals solids these modes
have frequencies in the range 0-80 cm^{-1}. Thus, even at low
temperatures, some acoustic modes may be thermally populated.
Vibrations involving a rotational displacement of the unit
cells are called optical phonons. If the unit cell contains
more than one molecule, optical phonons may also have the
characteristics of relative motions of molecules within the
unit cell. In general $\omega > 0$ at $k = 0$ for optical phonons. In
van der Waals solids their frequencies lie in an energy band
~10 cm^{-1} wide in the range 50-100 cm^{-1}. Collective modes
based on the intramolecular vibrations of the individual
molecules in the solid are also optical phonons. Their
energies vary over a wide range (100 cm^{-1} to 3000 cm^{-1}) and
they occur in bands ~10 cm^{-1} wide.

Of course, in impure crystals k is no longer a good
quantum number and lattice vibrations may to some extent
become localized. However, Montroll and Potts (10) have shown
that, if the solvent and solute molecules differ significantly, then most of the vibrations remain delocalized. Thus,
while the molecular crystal terminology does not strictly
apply, it provides a good first approximation to the disordered solution. Thus, in this article, as in the current
literature, the pure crystal terminology will be used for both
impure and disordered systems. Phonons will be used as a
synonym for normal vibrational modes, acoustic phonons will
refer to a translation type of low frequency mode, etc.

   3. Vibronic spectra. Finally, we review the general
properties of vibronic spectra within the Condon approximation. For concreteness, let us examine electronic states
associated with two harmonic vibrations. In general, both the
equilibrium positions and the frequencies of both vibrational
modes may be different in the different electronic states.
Analytical expressions for the Franck-Condon factors for such
a situation are available (11), so it is a simple matter to
quantitatively describe the spectrum expected for a given displacement and force constant change. Denoting the vibrational
frequencies, in wave numbers, in the ground and excited elec-
tronic states by $\bar{\nu}_{1g}$, $\bar{\nu}_{2g}$ and $\bar{\nu}_{1e}$, $\bar{\nu}_{2e}$ respectively and neglecting the details of line shape for the moment, we would
expect to obtain absorption spectra similar to those schematically indicated in Figure 5. Two important properties of the
spectra shown in Figure 5 should be noted: 1) For small displacements and frequency changes the vibrationless transition
(0-0) is the most intense feature. 2) As temperature is increased to the point where the ground state levels are
significantly populated, the spectrum becomes very complex.
Transitions may be classified as to whether or not they conserve the number of vibrational quanta in a given mode.
Transitions from vibrationally excited molecules that do not
conserve the number of quanta in each mode give rise to new
bands well removed in frequency from those seen at zero temperature. Transitions from vibrationally excited molecules
which do conserve the number of quanta in each mode give rise
to bands separated from the zero temperature bands by linear
combinations of the differences between the ground and excited
state frequencies. Thus, as the temperature is raised new
transitions appear which, together with the zero temperature
lines take on the appearance of narrow bands of lines.

   4. Line shapes. The discussion of the appearance of
vibronic spectra as a function of temperature is perfectly
general within the Condon approximation and may equally well
be applied to lattice vibrations or phonons. Since the interactions between the guest and host molecules will in general
be different when the guest molecule is electronically excited,
we can anticipate that phonon oscillators can be displaced
and have their frequencies modified in the excited state just
as is the case for the oscillators that correspond to intramolecular vibrations of the guest molecule. Thus, a figure
exactly analogous to Figure 4 can be drawn for the phonon
oscillators. As for the localized modes, transitions can be
classified according to whether or not they conserve the
number of quanta in each mode. Because the localized and delocalized modes are well separated in frequency, and because
the number of delocalized vibrations is very large (of order

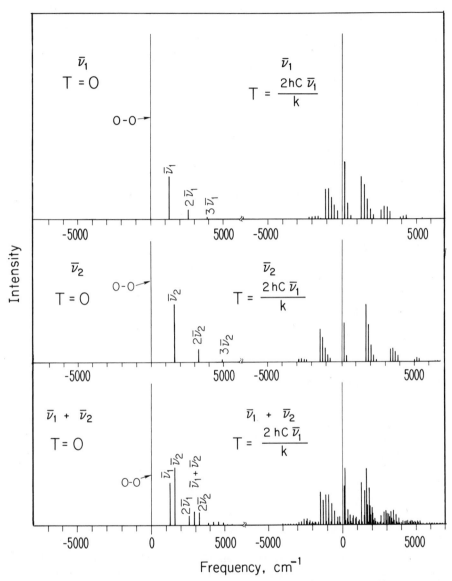

FIGURE 5. *Vibronic structure in condensed phase optical absorption spectra. The left-hand series of spectra show the effects of combining two harmonic modes at zero temperature. The right-hand series of spectra show how both the one- and two-dimensional cases appear at high temperature. The intensities shown are the Franck-Condon factors calculated for harmonic modes with lower state frequencies of 1100 and 1461 cm⁻¹, upper state frequencies of 1300 and 1637 cm⁻¹, displacements of 0.05 Å and reduced masses of 6 atomic mass units.*

3N where N is the number of molecules in the crystal) the
additional lines arising from the delocalized modes just add
a band shape to each of the isolated frozen gas molecule
transitions. That shape in the low temperature limit is just
a Franck-Condon weighted replication of the phonon density of
states. The line shapes for strong and weak excitation-phonon
coupling, that is large and small changes in the phonon oscillators upon excitation, are sketched in Figure 6. A more
detailed analysis than the one presented here (12) gives the
last piece of information needed to understand laser selection
experiments, namely that the ratio of the intensity of the
phonon conserving (zero-phonon) to the sum of the phonon conserving and phonon nonconserving (phonon wing) parts of the
vibronic line shape is given by the Debye-Waller factor

$$F = \exp\{- \sum_i \alpha_i [n_i(T) + 1/2]\}. \tag{II.7}$$

The index i runs over all phonon modes, $n_i(T) = [\exp(\hbar\omega/kT-1]^{-1}$
is the average number of quanta in a given phonon mode, and α_i
is a coupling parameter related to the displacement of phonon
oscillators upon excitation. Since phonon frequencies run
from tens of cm^{-1} down to zero frequency, the Debye-Waller
factor is extremely strongly temperature dependent.

III. SITE SELECTION SPECTROSCOPY

In general, the shape of an absorption band can be described by simply summing up the contributions of all of the
molecules in the sample. For a condensed phase system, all of
the molecules of a given type are expected to have almost
identical spectra in terms of vibronic intervals and relative
intensities but, if molecules exist in a distribution of
environments, these spectra will be shifted with respect to
each other. This is a consequence of the fact that solvent
fields primarily affect the electronic transition energy and
have only a very small effect on the intramolecular vibration
frequencies. The superposition of a large number of similar
but shifted spectra can conveniently be written as a convolution of the basic spectrum referenced to the electronic
transition energy $\bar{\nu}_o$, $s(\bar{\nu}-\bar{\nu}_o)$ with a distribution function
$D(\bar{\nu}_o)$ that gives the probability that a molecule will have
electronic transition energy $\bar{\nu}_o$. Thus the spectrum summed for
all molecules in the ensemble will be $S(\bar{\nu}) = \int D(\bar{\nu}_o)s(\bar{\nu}-\bar{\nu}_o)d\bar{\nu}_o$
(13). Two strategies for obtaining sharp spectra have become
clear. One is to make the solvent shift distribution $D(\bar{\nu}_o)$ as
narrow as possible. This is the approach taken in doped

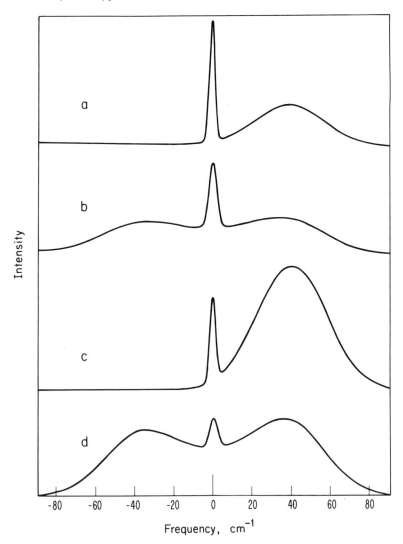

FIGURE 6. *Absorption line shapes for vibronic spectra in solid solution. The sharp features, called the zero-phonon lines, are 0-0 sequence bands for the phonon oscillators. The broader features, called the phonon wings, arise from transitions where the occupation numbers for the phonon oscillators change. At T = 0 [curves a and c] these wings are just the Franck-Condon weighted density of states function. At higher temperatures [curves b and d] or for stronger excitation-phonon coupling [curve c] the intensity of the zero-phonon line relative to the phonon wings decreases. Higher temperatures [curves b and d] also permit anti-Stokes phonon transitions further broadening the line.*

single crystal studies as pioneered by McClure (14) and also
lies behind the successful exploitation of the Shpolskii' (15-
18) effect. The other is to interact with the system in a way
that selects for molecules with a given $\bar{\nu}_O$ and then to obtain
spectra for the resulting subensemble. This is the site
selection approach and can be implemented in a number of different ways. In the sections that follow we shall use the
convolution model to discuss several of the techniques currently in use.

A. General Considerations

The best spectrum that can be obtained by any selection
technique is limited by the intrinsic spectrum of one molecule
plus lattice. Thus, it is advantageous to arrange to have this
intrinsic molecular spectrum as narrow as possible. From the
discussion of line shapes in Section II.B.4, two general rules
immediately follow. First, one would like the excitation-
phonon coupling to be small. This suggests that hosts that
interact only weakly with their guests or extremely rigid hosts
are to be preferred and that guest molecules for which large
conformational changes accompany excitation could pose special
problems. The n-alkane hosts used in the Shpolskii' effect
are a good example of weakly interacting hosts with a relatively narrow $D(\bar{\nu}_O)$. Second, to insure that the zero-phonon part
of the intrinsic line shape is as narrow and as intense as
possible, the temperature should be as close to 0°K as is
economically convenient. For many systems liquid helium temperatures are quite adequate but for others where there is
evidence for considerable molecular distortion upon excitation,
dramatic redistribution of intensity from the phonon wing into
the zero phonon line in the temperature range 1°-2°K has been
observed (19,20). Near zero temperatures will also insure that
the description of the site distribution function as static
(which is necessary for the success of any of these techniques)
remains valid.

To the extent that the Condon approximation is strictly
valid and the local environment affects only the electronic
energy and not the normal-mode vibrational frequencies, the
line shape for all vibronic lines associated with a given
electronic transition will be the same. Thus, in the discussions that follow, we can consider only a single inhomogeneously broadened vibronic line which will be repeated throughout
the spectrum.

Finally, in spite of the label site selection, it must be
realized that these techniques do not in fact select for unique
molecular sites. In the limit they can select for molecules
with a given transition energy. It does not follow that

molecules with identical excitation energies for a given transition occupy identical sites. The selected sites may still exhibit a distribution with regard to variables not strongly coupled to that particular transition. Hence, the selected subensemble may still exhibit inhomogeneous broadening when other electronic transitions are examined.

B. Hole Burning

This is perhaps the simplest and most elegant way to use lasers to interrogate an inhomogeneously broadened system. It will work whenever excited molecules have finite probability of relaxing to a state characterized by a different $\bar{\nu}_0$ or a different spectrum. Possible mechanisms include both photochemistry and local melting. Implementation is easy. One merely irradiates an inhomogeneously broadened absorption band with a laser. The molecules in resonance with the laser eventually get destroyed and the absorption band develops sharp holes that reflect the absorption spectrum of a tiny subensemble of molecules with very similar electronic excitation energies. If the laser is sufficiently monochromatic as to be reasonably represented as a delta function, then in terms of the convolution model, the absorption spectrum gets weighted by $[1 - \delta(\bar{\nu}_L)]$ where $\bar{\nu}_L$ is the laser frequency. Thus the absorption spectrum of the system becomes

$$S(\bar{\nu}) = \int D(\bar{\nu}_0)[1-\delta(\bar{\nu}_L)]s(\bar{\nu}-\bar{\nu}_0)d\bar{\nu}_0 = \int D(\bar{\nu}_0)s(\bar{\nu}-\bar{\nu}_0)d\bar{\nu}_0 -$$

$$- D(\bar{\nu}_L)s(\bar{\nu}-\bar{\nu}_L)$$

$$(III.1)$$

which is just the difference between the original inhomogeneously broadened spectrum and the spectrum of all molecules in sites such that the electronic transition energy has been shifted to $\bar{\nu}_L$. The expectations for such a hole-burning experiment as derived from Equation (III.1) are shown in Figure 7. The qualitative features of Figure 7 correspond exactly with the experimental observations on perylene in ethanol as reported by Kharlamov et al. (21). Voelker et al. (22) have photochemically burned holes 10 MHz wide in free-base porphyrin dilute in n-octane at 1.5°K.

C. Laser Selected Fluorescence

This is one of several techniques based on the subsequent behavior of selectively excited molecules. The idea is to

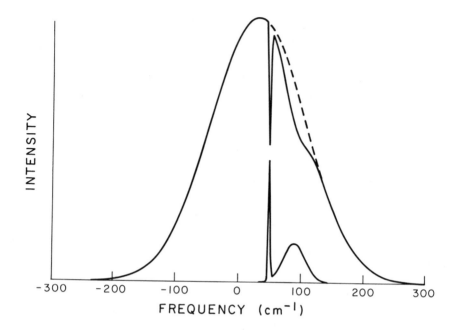

FIGURE 7. *Hole burning. The smooth bell-shaped profile (solid plus broken line) is the absorption line shape that results when the intrinsic molecular line shape shown as curve a in Figure 6 is convolved with the site distribution function $D(\bar{\nu}_O) = exp[-(\bar{\nu}_O/\sigma)^2]$, $\sigma = 100$ cm^{-1}. The upper solid curve shows the absorption expected after intense irradiation 50 cm^{-1} above the most probable $\bar{\nu}_O$ assuming finite yield for photochemical destruction. The difference between the absorption curves before and after irradiation is shown in the lower solid curve. As long as saturation is avoided, the difference curve will have the intrinsic line shape independent of where in the inhomogeneous profile the sample is irradiated.*

use intense monochromatic irradiation as can be provided by lasers to excite a subensemble of molecules with nearly identical solvent shifts. As long as the initial selection is not destroyed by energy transfer or local site melting, the subsequent emission is dramatically sharper than that obtained under more conventional, broad band excitation techniques. Again, these spectra may be conveniently discussed in terms of the convolution model. When a sample with intrinsic absorption spectrum $A(\bar{\nu}-\bar{\nu}_O)$ and site distribution function $D(\bar{\nu}_O)$ is irradiated with a laser whose output spectrum is $L(\bar{\nu}-\bar{\nu}_L)$, the distribution of excited molecules is given by

$$D(\bar{\nu}_O) A(\bar{\nu}-\bar{\nu}_O) L(\bar{\nu}-\bar{\nu}_L). \tag{III.2}$$

If the emission spectrum of a molecule with electronic transition energy $\bar{\nu}_O$ is given by $E(\bar{\nu}'-\bar{\nu}_O)$, then the emission of the laser irradiated sample is given by

$$S(\bar{\nu}') = \iint D(\bar{\nu}_O) A(\bar{\nu}-\bar{\nu}_O) L(\bar{\nu}-\bar{\nu}_L) E(\bar{\nu}'-\bar{\nu}_O) d\bar{\nu}_O d\bar{\nu} . \tag{III.3}$$

If the laser line width is very narrow compared to the intrinsic absorption features, then the laser spectrum can be represented as a delta function at $\bar{\nu}_L$ and the spectrum is given by

$$S(\bar{\nu}') = \int D(\bar{\nu}_O) D(\bar{\nu}'-\bar{\nu}_O) A(\bar{\nu}_L-\bar{\nu}_O) d\bar{\nu}_O. \tag{III.4}$$

Note that, unlike simple hole burning, the output spectrum is not simply related to the intrinsic molecular spectrum but rather represents a convolution of both absorption and emission. For condensed phase line shapes which can be described in terms of narrow zero-phonon and broad phonon-wing components, this leads to both the importance of extreme red-edge excitation and a dramatically stronger temperature dependence than is found for the more simple hole burning situation. Writing the intrinsic spectrum as a sum of zero-phonon (Z) and phonon-wing (W) contributions weighted by the temperature dependent Deybe-Waller factor F we obtain

$$S(\bar{\nu}-\bar{\nu}_O) = (F) Z(\bar{\nu}-\bar{\nu}_O) + (1-F) W(\bar{\nu}-\bar{\nu}_O).$$

The output spectrum is then a sum of four contributions (13)

$$S(\bar{\nu}') = F^2 \int D(\bar{\nu}_O) Z(\bar{\nu}'-\bar{\nu}_O) Z(\bar{\nu}_L-\bar{\nu}_O) d\bar{\nu}_O$$

$$+ F(1-F) \int D(\bar{\nu}_O) Z(\bar{\nu}'-\bar{\nu}_O) W(\bar{\nu}_L-\bar{\nu}_O) d\bar{\nu}_O$$

$$+ F(1-F) \int D(\bar{\nu}_O) W(\bar{\nu}'-\bar{\nu}_O) Z(\bar{\nu}_L-\bar{\nu}_O) d\bar{\nu}_O$$

$$+ (1-F)^2 \int D(\bar{\nu}_O) W(\bar{\nu}'-\bar{\nu}_O) W(\bar{\nu}_L-\bar{\nu}_O) d\bar{\nu}_O. \tag{III.5}$$

These correspond to zero-phonon absorption followed by zero-phonon emission, wing absorption followed by zero-phonon emission, zero-phonon absorption followed by wing emission and wing absorption followed by wing emission. Only the first term contributes to the narrow line component of the emission spectrum. Since for absorption at low temperatures the phonon wing is displaced to the high energy side of the zero-phonon line (only Stokes processes), excitation at the red edge

favors the first term relative to the other three. If the
zero-phonon and phonon-wing components are represented as
Gaussians, the quadratures can easily be done analytically and
the tuning and temperature dependences explored. Figure 8
summarizes the expectations. The verification of this convolution model by measuring the tuning dependence of laser
selected perylene fluorescence has been reported by Abram
et al. (13).

D. *Excitation for Narrow Band Detectors*

This is the complement of the laser selected fluorescence
technique. Instead of recording the emission spectrum for
fixed frequency monochromatic excitation, the emission is
detected over only a very narrow band of frequencies and its

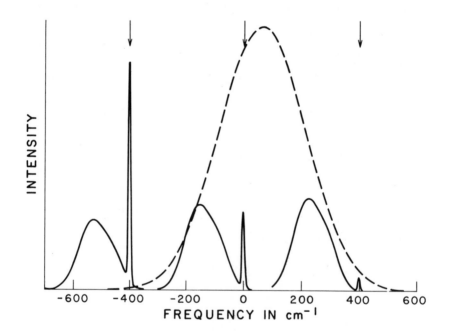

*FIGURE 8. Selective excitation. The dashed curve is the
absorption profile that results when the intrinsic molecular
line shape shown as curve a in Figure 6 is convolved with the
site distribution function $D(\bar{\nu}_o) = exp[-(\bar{\nu}_o/\sigma)^2]$, $\sigma = 200$ cm^{-1}.
The solid curves are the fluorescence line shapes that would
result from monochromatic excitation at the frequencies indicated by the arrows. To better display the red-edge enhancement of the zero-phonon line, the integrated intensities of
the fluorescence lines have been made equal.*

intensity as a function of excitation frequency is determined
by scanning the excitation source. In terms of the convolution model, assuming the detector bandwidth is negligible, the
output spectrum may be written

$$S(\bar{\nu}') = \int D(\bar{\nu}_O) E(\bar{\nu}_L - \bar{\nu}_O) A(\bar{\nu}' - \bar{\nu}_O) d\bar{\nu}_O. \qquad (III.6)$$

Again the output represents a convolution of both the absorption and emission spectra. The consequence of having only
Stokes processes for the phonons is that the output line shape
is the approximate mirror image of that seen in emission. The
line shape can be analyzed in terms of zero-phonon and phonon-
wing components as was done for the laser selected emission
with the resulting prediction that the best spectrum will be
obtained when the detector is set to the extreme blue edge of
the emission spectrum. Figure 8 will do for the excitation
case as well by simply reversing the frequency axis. This
technique can be used to obtain greatly narrowed absorption
spectra (and hence precise information on the excited state
vibrational frequencies), however, because of the extra difficulties involved with scanning lasers rather than monochro-
mators few studies of this type have been made (23).

E. *Double Resonance Techniques*

In the most general type of selection experiment a subset
of the total experimental ensemble is labeled in some way
(usually electronic excitation) and then the properties of
this subset are examined in a subsequent experiment (most
often emission). It is clear that this approach can easily
be generalized by substituting other types of spectroscopy
for the labeling and/or examination part of the experiment.
A study of the interaction of the labeled subensemble with a
second radiation field gives rise to a whole range of double
resonance experiments where monochromatic excitation can
greatly reduce inhomogeneous broadening. Straightforward
possibilities include triplet-triplet absorption spectroscopy
and ODMR spectroscopy. Van Egmond, Kohler and Chan (24)
have discussed the ODMR case and presented results for
quinoxaline. They were able to demonstrate a relationship
between the solvent induced shifts of the 3.6 GHz microwave
transition of triplet quinoxaline in a hydrocarbon glass at
4.2°K and the corresponding shifts of the phosphorescence 0-0
band and interpret their observations in terms of solvent
induced mixing in the quinoxaline triplet manifold. This
experiment also established the possibility of using narrow-
band microwave sources to produce a narrowed modulated phosphorescence spectrum. While the universal application of

double resonance techniques is compromised by the fact that
transition energy selection is not strictly site selection,
the invention and exploitation of novel double resonance
techniques should prove to be an active area in the future.

IV. CONCLUSIONS

Most of the technical difficulties associated with constructing tunable lasers for selectively exciting optical
transitions are now well under control and a variety of
reliable commercial systems are available. With the hardware
problem nearing solution and with at least a qualitative
understanding of the nature of impurity spectra in low temperature solids in hand, it is perhaps reasonable to speculate
on what some of the future developments and applications of
site selection spectroscopy will be.

Most of the past work has centered on fluorescence studies
on organic molecules (25-27). As the number of molecules for
which narrow line emission spectra can be obtained mounts, the
obvious analytical potential of these techniques looks more
and more promising. The extraordinary sensitivity of fluorescence methods has long been appreciated. With laser selection
techniques this sensitivity can be largely retained and at the
same time a high information content spectrum generated. To
the extent that the Condon approximation is valid, this
spectrum, which is built up of the totally symmetric ground
state vibrational frequencies, is sufficiently unique and
independent of local environment as to permit unambiguous
identification of complex molecules in trace amounts. Selective excitation can be expected to make the analysis of rather
complex mixtures of fluorescent substances feasible. To
fully realize this analytical potential, it is imperative that
the sample be maintained at liquid helium or lower temperatures and that both the laser system and the spectrometer be
extremely narrow banded (FWHM 0.1 cm^{-1} or less in order to not
greatly exceed the width of the zero-phonon line). With a
system properly designed to detect weak, sharp emission
features against a broad background, a major advance in trace
analysis should be possible.

The fact that excitation energy depends on the arrangement of neighboring molecules suggests that there may be ways
to control photochemical reactions by exciting only certain
molecular configurations. Such site determined photochemistry
could provide both new synthetic processes and fundamental
understanding. With properly chosen systems it should be
possible to develop a keener appreciation of the role that
relative molecular orientation plays with respect to excited

state reactions. For such studies to be unambiguously interpreted it will be necessary to develop methods for assigning
specific solvent shifts to specific molecular configurations.
Significant progress in this area has been made by combining
optical spectroscopy, EPR spectroscopy and X-ray diffraction
(28). Hopefully, these kinds of data will stimulate a related
effort on the part of theorists to develop efficient methods
for computing excitation energy as a function of configuration.

Finally, the continuation and expansion of current efforts
to collect and interpret high quality site-specific optical
spectra seems desirable. Because of their small rotational
constants, medium sized organic molecules have almost continuous optical spectra in the gas phase under ordinary conditions. Thus the determination of electronic structure for
these systems by gas phase spectroscopy has progressed slowly.
Incorporating these molecules into solutions exchanges rotational band structure for phonon-band structure and, at ambient
temperatures, doesn't significantly simplify life for the
spectroscopist. The advantage that the solid state holds for
these systems is directly related to the ease with which the
temperature of the system can be made extremely low, simultaneously simplifying the spectrum and enhancing the narrow
zero-phonon components. While the rapidly developing nozzle-
beam techniques should extend some of the benefits of low
temperatures to the gas phase (29), there will remain a large
number of molecules which, because of low vapor pressure or
marginal stability, must be studied in solution. This includes a large number of molecules of biological importance
for which a detailed picture of electronic structure is crucial
to a fundamental understanding of how the molecule functions.
For these systems, the application and refinement of site
selection techniques combined with theoretical modeling should
enjoy both more extensive application and success. More
vigorous pursuit of such linked studies could significantly
advance our current understanding of both intra- and inter-
molecular potentials for large polyatomic molecules.

ACKNOWLEDGMENTS

The writing of this review has generated a large debt to
my former students and postdoctoral fellows Izo Abram, Roy
Auerbach, Robert Birge, Jan van Egmond and James M. Stevenson
who did both the theoretical and experimental work that culmi-
nates in the story told here.

REFERENCES

1. Abram, I. I., Ph.D. Thesis, Harvard University, Cambridge, Massachusetts (1975).
2. Born, M., and Oppenheimer, R., *Ann. Physik 84,* 457 (1927).
3. Herzberg, G., and Teller, E., *Z. Physik Chem. (Leipzig) B21,* 410 (1933).
4. Condon, E. U., *Phys. Rev. 28,* 1182 (1926).
5. Franck, J., *Trans. Faraday Soc. 21,* 536 (1925).
6. Duschinskii, F., *Acta Physicochem. U.R.S.S. 7,* 551 (1937).
7. Craig, D. P., *J. Chem. Soc.* 2146 (1950).
8. Amos, A. T., and Burrows, B. L., *Adv. Quantum Chem. 7,* 303 (1973).
9. Text on the solid state including: Craig, D. P. and Walmsley, S. H., "Excitations in Molecular Crystals," Benjamin Inc., New York, (1968); and Ziman, J. M., "Principles of the Theories of Solids," Cambridge University Press, Cambridge, England, (1969).
10. Montroll, E. W., and Potts, R. B., *Phys. Rev. 100,* 525 (1955).
11. Ansbacher, F., *Z. Naturforsch. 14a,* 889 (1959).
12. An excellent general discussion of excitation-phonon coupling is found in: Rebane, K. K., "Impurity Spectra of Solids,"(translated by John S. Shier), Plenum Press, New York, (1970).
13. Abram, I. I., Auerbach, R. A., Birge, R. R., Kohler, B. E., and Stevenson, J. M., *J. Chem. Phys. 63,* 2473 (1975).
14. McClure, D. S., *J. Chem. Phys. 22,* 1668 (1954); *J. Chem. Phys. 24,* 1 (1956).
15. Shpolskii', E. V., Ilina, A. A., and Klimova, L. A., *Dokl. Akad. Nauk SSR 87,* 935 (1952).
16. Shpolskii', E. V., and Klimova, L. A., *Dokl. Akad. Nauk SSR 111,* 1227 (1956).
17. Shpolskii', E. V., *Sov. Phys. Usp. 3,* 372 (1960); *5,* 522 (1962); *6,* 411 (1963).
18. The early Russian work is reviewed in: Nurmukhametov, R. N., *Russ. Chem. Rev. 38,* 180 (1969).
19. Stevenson, J. M., Ph.D. Thesis, Harvard University, Cambridge, Massachusetts, (1975).
20. Anderson, R. J. M., Kohler, B. E., and Stevenson, J. M., to be published.
21. Kharlamov, B. M., Personov, R. I., and Bykovskaya, L. A., *Opt. Comm. 12,* 191 (1974).
22. Voelker, S., Macfarlane, R. M., Genack, A. Z., Trommsdorff, H. P., and van der Waals, J. H., *J. Chem. Phys. 67,* 1759 (1977), and Voelker, S., Macfarlane, R. M., and van der Waals, J. H., *Chem. Phys. Letts. 53,* 8 (1978).

23. Personov, R. I., and Kharlamov, B. M., *Opt. Comm.* 7, 417 (1973).
24. van Egmond, J., Kohler, B. E., and Chan, I. Y., *Chem. Phys. Lett. 34*, 423 (1975).
25. Eberly, J. H., McColgin, W. C., Kawaoka, K., and Marchetti, A.P., *Nature 251*, 215 (1974).
26. Cunningham, K., Morris, J. M., Funfschilling, J., and Williams, D. F., *Chem. Phys. Lett. 32*, 581 (1975).
27. Al'Sehitz, E. I., Personov, R. I., and Kharlamov, B. M., *Chem. Phys. Lett. 40*, 166 (1976) and references cited therein.
28. Merle, A. M., Lamotte, M., Risemberg, S., Haun, C., Gaultier, J., and Grivet, J. P., *Chemical Physics 22*, 207 (1978).
29. Smalley, R. E., Wharton, L., and Levy, D. H., *Chemical and Biochemical Applications of Lasers 2*, 1 (1977).

BIOLOGICAL APPLICATIONS OF RESONANCE RAMAN SPECTROSCOPY
IN THE VISIBLE AND ULTRAVIOLET:
VISUAL PIGMENTS, PURPLE MEMBRANE, AND NUCLEIC ACIDS [1]

Richard Mathies

Department of Chemistry
University of California
Berkeley, California

I. INTRODUCTION

Resonance Raman spectroscopy is a powerful and versatile
technique for examining the vibrational and electronic structure of molecules. Because of its unique sensitivity and
selectivity, resonance Raman is particularly well-suited for
studies of biological macromolecules. The recent, rapid
increase in experimental and theoretical work on the resonance Raman of biological systems provides a clear indication
of the advantages of this technique (1-7). In a resonance
experiment the frequency of the laser that generates the
Raman scattering is selected to be in or near the electronic
absorption band of a chromophoric group. This results in a
resonance enhancement of the scattering from *just those
vibrations that are coupled to the electronic excitation*.
Selective resonance enhancement is particularly important in
studying biological systems because (1) sample concentrations
can be as low as 10^{-6} M, (2) sample volumes as small as a few
microliters are practical, and (3) scattering from the non-
resonant parts of the polypeptide or polynucleotide will not
complicate the vibrational spectrum. Resonance Raman spectroscopy has now been applied to a wide variety of systems

[1] *This work was supported by grants from the National
Institutes of Health, the National Science Foundation,
the American Chemical Society, Research Corporation,
and the Berkeley Biomedical Research Support Program.*

including porphyrins and hemeproteins (8-12), vitamin B_{12} (13,14), chlorophyll (15,16), nucleic acids (17-19), visual pigments (20-23), and the purple membrane from *Halobacterium halobium* (24-27). Extrinsic chromphores have also been profitably used as "Raman probes" for enzymatic active sites (28-29).

In recent years resonance Raman studies on visual pigments have been particularly interesting and productive. The vitamin A (retinal) chromophore provides excellent Raman spectra that are rich in information about the conformation of the polyene chain. Also, the resonance scattering comes from the molecular center directly responsible for visual photochemistry. The photosensitivity of the visual protein provides a unique set of experimental challenges and capabilities that are only now being fully appreciated. Examples of new approaches include the recently developed rapid-flow and time-resolved Raman techniques (27,30,31).

The biological application of resonance Raman in the ultraviolet is also quite new but already a number of studies on nucleic acids have been performed (17-19). These experiments are likely to become even more prevalent as better ultraviolet lasers and ultraviolet Raman apparatus are developed. Ultraviolet Raman has great potential because one can in principle selectively resonance enhance the individual amino acid and nucleic acid components of biopolymers. Also, this UV work has been a fertile ground for new developments in both theory and experiment because a new spectral region is being explored. This article will concentrate on the application of resonance Raman to visual pigments and nucleic acids because these studies provide particularly good examples of what kind of information can be obtained from a resonance Raman experiment, and also what experimental techniques are available.

II. THEORY

The basic concepts of nonresonance Raman as applied to chemical and biological structure have been given in this series and in other texts (1,32,33). Also, the early work on resonance Raman has been described (34). Detailed theoretical descriptions of the resonance effect have been given by Albrecht (35,36), Peticolas (37-39) and Warshel (40). We will qualitatively follow the discussion of resonance scattering given by Spiro (2). The reader is referred to the above citations for the complete derivations.

A. *The Resonance Effect*

The intensity of a Raman transition from state m to n for randomly oriented molecules is given by

$$I_{m,n} = I_o \frac{2^7 \pi^5}{3^2} \left(\frac{\nu_s}{c}\right)^4 \sum_{\rho\sigma} |(\alpha_{\rho\sigma})_{m,n}|^2 \tag{1}$$

where I_o is the incident light intensity, ν_s is the frequency of the scattered light, and $\alpha_{\rho\sigma}$ is the transition polarizability tensor. Here ρ and σ specify the incident and scattered polarizations, respectively. The elements of the polarizability tensor are derived from second order perturbation theory (41),

$$(\alpha_{\rho\sigma})_{m,n} = \frac{1}{h} \sum_e \left[\frac{\langle m|\mu_\rho|e\rangle\langle e|\mu_\sigma|n\rangle}{\nu_e - \nu_m - \nu_o + i\Gamma_e} \right.$$

$$\left. + \frac{\langle m|\mu_\sigma|e\rangle\langle e|\mu_\rho|n\rangle}{\nu_e - \nu_n + \nu_o + i\Gamma_e} \right] \tag{2}$$

where μ_ρ and μ_σ are the dipole moment operators, ν_o is the incident laser frequency, Γ_e is a damping factor, and $|e\rangle$ is the wave function for an intermediate state with energy ν_e. The initial state (m) and the final state (n) have energies ν_m and ν_n, respectively. In the simplest case, n is produced from m by exciting one quantum of a particular vibration. When $\nu_o \ll \nu_e - \nu_m$ both terms in equation (2) are frequency independent and one observes nonresonant Raman scattering where the intensity is proportional to ν_s^4. When ν_o approaches $\nu_e - \nu_m$ for some state e that has a nonzero transition moment, the eth term in the sum becomes dominant and resonance enhancement of the Raman scattering is observed. Since the energy denominator only gets small for the first, resonant term in equation (2), we will neglect the nonresonant second term.

In the analysis of the polarizability tensor the Born-Oppenheimer approximation allows one to write the total wave functions as products of electronic and vibrational parts. For example,

$$\langle m|\mu_\rho|e\rangle = \langle i|M_e^\rho|k\rangle \quad \text{and} \quad \langle e|\mu_\sigma|n\rangle = \langle k|M_e^\sigma|j\rangle \tag{3}$$

where $|i\rangle$, $|k\rangle$ and $|j\rangle$ are the initial, intermediate and final *vibrational* states and M_e is the *electronic* transition moment from the ground to the eth excited electronic state. The co-ordinate dependence of the transition polarizability can be expressed by expanding the electronic transition moment in a Taylor series about the equilibrium geometry as a function of the normal coordinates, Q.

$$M_e = M_e^\circ + (\partial M_e/\partial Q)^\circ Q + \ldots \tag{4}$$

Here we have considered only one of the 3N-6 possible normal coordinates. Introducing the above approximations into equation (2), the leading terms for the Raman polarizability become

$$(\alpha_{\rho\sigma})_{ij} = A + B \tag{5}$$

where

$$A = \frac{M_e^\rho M_e^\sigma}{h} \sum_k \frac{\langle i|k\rangle \langle k|j\rangle}{\nu_{ik} - \nu_\circ + i\Gamma_k} \tag{6}$$

$$B = \frac{M_e^\rho (\partial M_e^\sigma/\partial Q)^\circ}{h} \sum_k \frac{\langle i|k\rangle \langle k|Q|j\rangle}{\nu_{ik} - \nu_\circ + i\Gamma_k}. \tag{7}$$

The analogous part of the B term with ρ and σ reversed has been omitted. Here ν_{ik} is the energy gap between the lowest vibrational level of the ground state and the kth vibrational level of the resonantly enhanced excited electronic state, e.

The A term dominates the resonance scattering for allowed transitions because in this case M_e, the electronic transition moment, is much larger than $\partial M_e/\partial Q$. The vibrational Raman intensities will depend on the vibrational overlap integrals (Franck-Condon factors) in the numerator of equation (6). Only totally symmetric modes will be allowed. A vibrational mode will be strongly enhanced if the distortion of the molecule in the excited electronic state corresponds closely with the vibrational coordinates. Hence, vibrations will be resonantly enhanced if electronic excitation alters the bonding of nuclei whose motions are important in that vibrational mode (3,40).

The B term will be important when the incident radiation is in resonance with a weak transition. Here nontotally symmetric vibrations will be observed that are effective in borrowing intensity from nearby intense transitions. For polyenes and nucleic acids, which have intense transitions, A term enhancement is expected. However, far from resonance the A term will become small and B term effects should be observable.

B. *Preresonance Approximation*

Many systems have been studied with laser frequencies that approach but are not in resonance with an electronic transition. Therefore, it is useful to develop approximate expressions for the A and B terms in the "preresonance" region. Albrecht originally expressed the coordinate dependence of equations (6) and (7) through a vibronic, Herzberg-Teller expansion. In this case, perturbation theory is used to express $\partial M_e/\partial Q$ as a power series expansion in the other electronic states. For example, if only one electronic state s is strongly mixed, then $\partial M_e/\partial Q$ is given by $M_s h_{es}$ $(\nu_s - \nu_e)^{-1}$ where h_{es} is a matrix element of the vibronic coupling operator between states e and s. The sum over k can be performed in the preresonance region by replacing ν_{ik} with ν_e, the average transition energy to state e. When this is done the zero order A term will contribute only to Rayleigh scattering, $i = j$. However, if the coordinate dependence of the denominator in equation (6) is explicitly considered, then a Raman active contribution to the A term is derived which depends on h_{ee} (35). The results have been given by Albrecht and Hutley (36). Functionally the A and B terms become

$$A \propto \frac{M_e^2 h_{ee}}{h^2} \langle i | Q | j \rangle \frac{\nu_e^2 + \nu_o^2}{(\nu_e^2 - \nu_o^2)^2} \tag{8}$$

$$B \propto \frac{M_e M_s h_{es}}{h^2} \langle i | Q | j \rangle \frac{\nu_e \nu_s + \nu_o^2}{(\nu_e^2 - \nu_o^2)(\nu_s^2 - \nu_o^2)} . \tag{9}$$

The A term is still resonant when $\nu_o = \nu_e$; however, the intensity is proportional to $h_{ee} = \langle e | (\partial H/\partial Q) | e \rangle = \partial E_e/\partial Q$. This is just the slope of the excited state potential surface along the normal coordinate when the nuclei are in their ground state equilibrium positions. *For harmonic potential surfaces h_{ee} is directly proportional to the relative displacement between the ground and excited state potential energy surfaces.* This illustrates why C-C and not C-H stretching modes will be enhanced for π-electron excitation. No change in the equilibrium geometry of a C-H stretching mode is expected when the excitation is

on the carbon centers.   In the *B* term we now see a new resonance for intermediate state *s* that is vibronically coupled to
state *e*.   This arises from the energy denominator $(\nu_s - \nu_e)^{-1}$
in the expression for $\partial M_e/\partial Q$. The frequency dependence of
these two terms is quite different and can be used to distinguish between these two sources of resonance enhancement.

C. *Semiempirical Approach*

As experiments are performed closer to resonance, the
preresonance expressions must be used carefully because their
validity on resonance is not guaranteed.   In this case the
difficult sum in equations (6) and (7) must be performed.
Particularly for large biological molecules, the more direct
approach of Warshel may be appropriate (40).   He uses a semiempirical π-electron calculation to determine the coordinate
dependence of the polarizability tensor either numerically or
analytically.   Since the normal modes of the ground and excited states are obtained, Franck-Condon factors can also be
evaluated.   For studies on visual pigments, the pattern of
resonance Raman vibrational intensities is much more important
than the enhancement profiles in the preresonance region.   In
order to interpret these spectra, the effect of conformational
and environmental perturbations on the Franck-Condon factors
in equation (6) must be predicted.   Recent results on visual
pigments and heme proteins indicate that the semiempirical
approach will be useful in this regard (3).

III. VISUAL PIGMENTS

A. *Introduction*

Absorption of a photon by the photoreceptor protein
rhodopsin initiates the excitation of the retinal rod cell
(42).   This 38,000 dalton membrane-bound protein contains an
11-*cis* retinal prosthetic group that is covalently bound as a
protonated Schiff base to a lysine residue (see Fig. 1).   All
known vertebrate visual pigments contain an 11-*cis* retinal (or
dehydroretinal) chromophore.   Following photon absorption,
retinal is released from opsin in the all-*trans* configuration.
It is clear that a *cis* → *trans* isomerization occurs during
visual excitation (42).   However, a number of fundamental
questions about this nominally simple process have still not
been answered.   First, what is the photochemical pathway for
isomerization?   The protein, opsin, is able to increase the
photoisomerization quantum yield of retinal from 0.2 in solution (43) to 0.67 (44).   Second, how does opsin control the

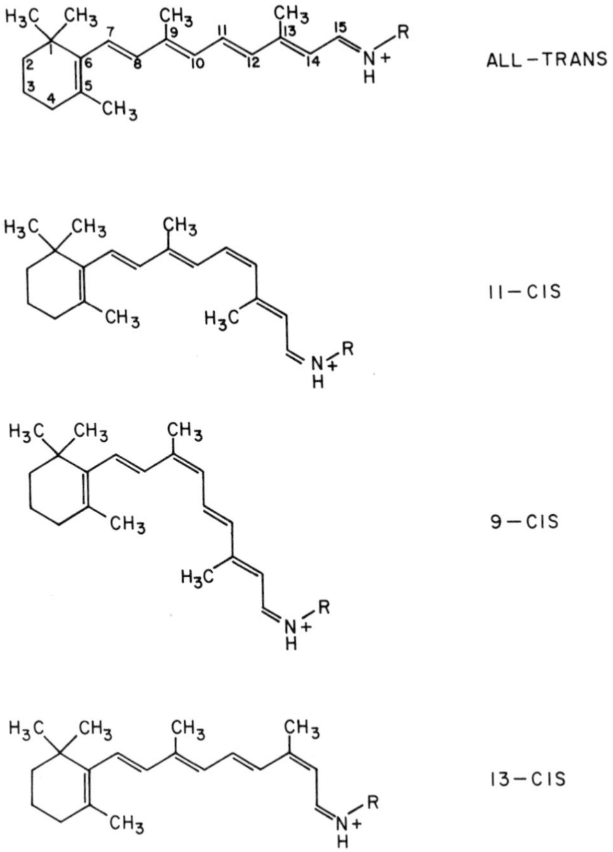

FIGURE 1. *Protonated Schiff base derivatives of all-trans, 11-cis, 9-cis and 13-cis retinal.*

absorption maximum of the complexed retinal? Protonated Schiff base derivatives absorb at 440 nm while rhodopsin absorbs at 500 nm. Vertebrate cone pigments have absorption maxima that range from 432 nm to 562 nm (45). Third, how does the isomerization of retinal initiate the conformational changes in opsin that lead to visual excitation?

In the initial attempts to answer these questions, low temperature absorption spectroscopy was used by Wald and coworkers to characterize the spectral changes that occur following photon absorption (46). They observed a sequence of intermediates and derived the kinetic scheme shown in

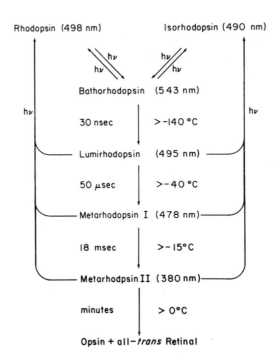

Rhodopsin (498 nm)   Isorhodopsin (490 nm)

Bathorhodopsin (543 nm)

30 nsec   > -140 °C

Lumirhodopsin   (495 nm)

50 μsec   > -40 °C

Metarhodopsin I (478 nm)

18 msec   > -15°C

Metarhodpsin II (380 nm)

minutes   > 0°C

Opsin + all-*trans* Retinal

FIGURE 2. Bleaching sequence for rhodopsin (11-cis retinal + opsin) and isorhodopsin (9-cis retinal + opsin). The absorption maxima, characteristic decay times, and temperatures at which the transitions are blocked have been indicated.

Fig. 2. Following photon absorption, rhodopsin and isorhodopsin (a synthetic 9-*cis* retinal pigment) undergo dark decay through the same sequence of intermediates. Note that light can be used to interconvert these species at almost every stage. A number of detailed reviews of this work are currently available (47-49). It must be noted, however, that much of the available data come from "low resolution" techniques (e.g., absorption spectroscopy) that cannot provide detailed information about the conformation of specific chemical bonds in this macromolecule.

Resonance Raman spectroscopy is a uniquely powerful technique for studying the photochemistry of vision. This is because the resonantly enhanced scattering comes from precisely the chemical group that is undergoing photochemical changes. Furthermore, the observed vibrational spectrum is very sensitive to the details of the retinal conformation (50). This resonance enhancement has been quantitatively

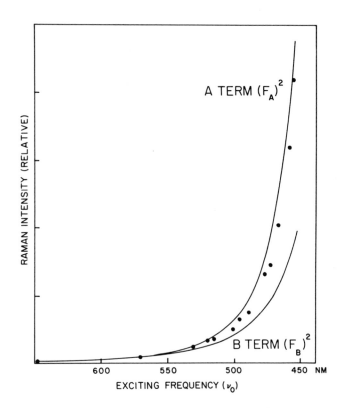

FIGURE 3. Resonance Raman excitation profile for the 1582 cm^{-1} C=C stretching mode of all-trans retinal. The A term was calculated with equation (8) for λ_e = 383 nm. The B term was obtained from equation (9) with λ_e = 383 nm and λ_S = 280 nm (51).

measured for retinal by Doukas et al. (51). In Fig. 3 the Raman scattering intensity of the C=C ethylenic stretching vibration of all-*trans* retinal has been plotted vs. excitation frequency. As the exciting frequency approaches the absorption maximum at 380 nm, a dramatic increase in the scattering intensity is observed (~100$^\times$). This excitation profile shows that the scattering is due to an A term enhancement at 380 nm. It is reasonable to expect that a similar enhancement will occur for visual pigments.

One of the first Raman experiments on visual pigments was performed by Rimai *et al.* in 1970 (20). They cooled an intact bovine retina to -70°C and were able to observe the spectra shown in Fig. 4. Because of baseline drift, several scans were repeated under identical conditions. Scattering was consistently observed in a region (1550 cm^{-1}) that was assigned to the C=C ethylenic stretch of a protonated retinal Schiff base. It was recognized that under these experimental conditions the Raman scattering would be due to a mixture of intermediates including lumirhodopsin (see Fig. 2). From this beginning a great deal of progress has been made in obtaining resonance Raman spectra of visual pigments. The photolability of these species makes it difficult to obtain high-quality spectra of samples with known composition. However, these problems are now well understood and some of the techniques that have recently been developed will now be discussed.

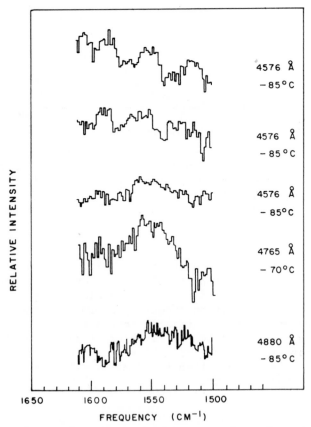

FIGURE 4. *Early resonance Raman spectra of an intact bovine retina (20).*

B. Photolysis Considerations in Experimental Design

In all physical measurements it is important to insure
that the state or molecule being measured is not perturbed.
This is particularly true in resonance Raman spectroscopy.
Even with resonance enhancement, the probability that a visual
pigment molecule will absorb a photon is 10^7 times greater
than the probability that it will scatter a Raman photon.
Also, to observe resonance spectra high light intensities
(100 watt/cm^2) must be used. To maximize the resonance
enhancement, the laser frequency must lie in or near the
absorption band of the retinal chromophore. Exciting on the
red edge of the absorption band can reduce the resulting photolysis to some extent, but even with 600 nm excitation (see
Fig. 2) the photolysis is severe. For example, the first
order rate constant for photoisomerization is given by

$$k = I(\lambda)\sigma_a(\lambda)\phi \tag{10}$$

where $I(\lambda)$ is the light flux (photon/cm^{2}sec), $\sigma_a(\lambda)$ is the
absorption cross-section (cm^2/molecule) and ϕ is the quantum
yield for photoisomerization. For a typical experiment on
rhodopsin with a cw laser power of 25 mW at 600 nm (40 μ
diameter), the rate constant for photoisomerization is 7.2
$\times 10^3$ sec^{-1}. Here $\sigma_a = 3.824 \times 10^{-21} \varepsilon$, $\varepsilon = 470$ cm^{-1}M^{-1} and
$\phi = 0.67$. The lifetime for the light-induced decay of
rhodopsin, τ_i, is just k^{-1} or 140 μsec. Since most Raman
spectrometers require from 15 minutes to several hours to
obtain a spectrum, this light sensitivity is a severe
complication.

1. *Low-Temperature Methods.* The photolysis problem was
first successfully dealt with by Oseroff and Callender (52).
They cooled the rhodopsin sample to 77°K so that only the
first intermediate, bathorhodopsin, could be formed (see Fig.
2). In this way they were able to prepare a stable sample
which contained a photostationary steady state mixture containing only rhodopsin, isorhodopsin and bathorhodopsin. Two
factors must be considered in analyzing the Raman spectra
from such a mixture. First, as shown by Yoshizawa and Wald
(46), the relative proportions are strongly dependent on the
irradiation wavelength. Irradiation with blue light (488 nm)
where rhodopsin and isorhodopsin absorb strongly will produce
a large concentration of bathorhodopsin (λ_{max} = 543 nm).
Irradition with yellow light, where batho absorbs but the
others do not, will deplete the amount of batho in the steady
state. Second, resonance enhancement must be considered. The
laser should have a frequency as close as possible to the

absorption maximum of the molecule to be studied in the mixture. Unfortunately, this will strongly decrease the concentration of the molecule of interest. This difficulty was
overcome by using a second,intense pump laser beam whose frequency is chosen to maximize the concentration of the desired
component. In this way the spectra of isorhodopsin and bathorhodopsin have been obtained. This approach is complicated by
the presence of three components in the mixture. It is difficult to get "pure" spectra of rhodopsin and bathorhodopsin by
this method. In addition, the low-temperature method cannot
be easily extended to later intermediates because the mixture
is more complex and differential pumping and probing will be
less selective.

   2. Rapid-Flow Methods. Obtaining the resonance Raman
spectrum of a single photolabile molecule was made possible
by the development of rapid-flow sampling techniques (30,31).
Here the sample is flowed rapidly through the focused laser
beam so that the fraction of isomerized molecules in the
illuminated volume is very small. For example, if the light
driven isomerization lifetime (τ_i) is 140 μsec, then the
transit time for a molecule through the beam, τ_t, must be
$\sim$ 10 μsec to keep the fraction of isomerized rhodopsin in the
beam less than 5%. This can be accomplished by forming a jet-
stream with the rhodopsin solution that has a velocity of
$\sim$ 400 cm/sec. If the laser beam is focused to a 40 μ spot on
the stream, then the transit time will be $\sim$ 10 μsec. The
geometry of this system is given in Fig. 5. More generally,

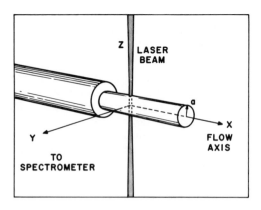

*FIGURE 5. Geometry of a rapid-flow sampling system for
Raman spectroscopy (31).*

if we assume that the stream has a uniform velocity v (cm/sec) and the laser beam a square cross-section of side ℓ (cm), then the transit time, τ_t, is ℓ/v. The fraction of rhodopsin that is isomerized while traversing the beam is given by

$$F = 1 - \exp(-\tau_t/\tau_i) \tag{11}$$

If the amount of isomerization is small ($F \ll 1$) then the exponential can be expanded to give

$$F \cong \tau_t/\tau_i = (I\sigma_a \phi \ell)/v \tag{12}$$

Experimentally, it is more convenient to express F in terms of the laser power P (photon $\sec^{-1}$) which is just $I\ell^2$. Also, the absorption cross-section σ_a (cm^2/molecule) is related to the molar decadic extinction coefficient ε (cm^{-1}M^{-1}) by $\sigma_a = 2.303 \times 10^3 \, \varepsilon/N_o$ where N_o is Avogadro's number. The final expression for F, the *photoalteration parameter*, confirms that F can be made suitably small (< 0.1) by having a sufficiently rapid flow.

$$F = \frac{2.303 \times 10^3 \, P\varepsilon\phi}{N_o \ell v} \tag{13}$$

This derivation has also been carried out by considering the actual gaussian intensity profile of the focused laser beam (31). The results of this calculation are in agreement with the alternative approach of Callender *et al.* (30). In this case the extent to which photolyzed molecules will contribute to the Raman scattering is explicitly considered. First, the steady state spatial concentration distribution of unisomerized molecules in the illuminated part of the stream is computed. Then the contribution of these unisomerized molecules to the total Raman signal is obtained by weighting the composition at each point in the illuminated volume by the light intensity. The result of this calculation is a quantity called the normalized Raman scattering, which represents the fraction of the total scattering (S) that comes from the initial photolabile molecule. S has been plotted vs. the photoalteration parameter in Fig. 6. Now, in the evaluation of F, the effective cross-section of the gaussian laser beam is given by $\sqrt{\pi}\omega$ where ω is the $1/e^2$ radius of the focused beam, and the laser power (P) is given by $\pi\omega^2 I_o/2$ (I_o is the peak light intensity). The final expression for the photoalteration parameter F becomes

$$F = \frac{2.303 \times 10^3 \, P\varepsilon\phi}{N_o \sqrt{\pi} \, \omega v} \quad . \tag{14}$$

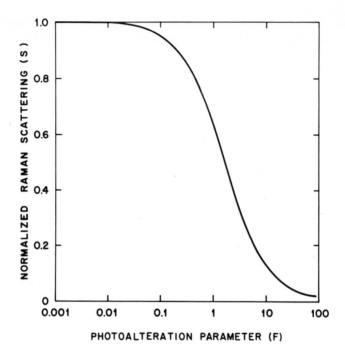

PHOTOALTERATION PARAMETER (F)

*FIGURE 6.   The normalized Raman scattering, S, represents
the fraction of the total scattering that comes from the
initial photolabile molecule in a rapid-flow experiment.   The
photoalteration parameter is a linear measure of the light
exposure experienced by the sample (31).*

When $F \ll 1$, F is simply the fraction of molecules isomerized
in one transit through the beam.   Hence, in this limit

$$S = 1 - F/2 \tag{15}$$

For higher values of photoalteration, F can be interpreted as
the number of absorptions that an equivalent nonphotolabile
molecule would experience while passing through the laser
beam.   As will be discussed later, the functional dependence
of S on F for high values of the photoalteration parameter
will be useful in designing high photoalteration "time-
resolved" Raman experiments.

In order to maintain the requisite flow velocities, flow
rates of ~ 1 ml/sec are usually employed for long periods of
time.   Therefore, the solution must be recirculated to minimize sample volumes.   For visual pigments, recirculation is
not a problem because the bleaching time is fast compared with

sample recirculation times. The bleached molecules disappear
from the experiment since the scattering from all-*trans*
retinal is only weakly enhanced by visible light. The fraction of the molecules in the entire recirculating volume that
have been bleached can be easily calculated. When $F \ll 1$ the
fraction of the entire pool that has been photolyzed is given
by

$$F_{bulk} = \frac{4.606 \times 10^3 \, aP\phi\varepsilon T}{N_o \, V} \tag{16}$$

where a is the radius of the jet-stream, V is the total sample
volume (cm^3) and T is the total irradiation time (23). For
model compounds that isomerize but do not bleach, F_{bulk} is a
measure of the cumulative photoisomerization in the sample and
should be kept $\leqslant 0.05$.

This rapid-flow technique has now been used to obtain
spectra of various visual pigments and several of their photolytic intermediates (30,31,53). The major disadvantage is
that large quantities (~ 100 ml) of solution are needed at
moderate concentrations (10^{-5} M). As will be discussed later,
the use of multichannel detection should reduce these material
requirements by at least a factor of 10 (see Section VI).
Also, the flow method is limited to intermediates that can be
made or trapped in fluid solution for long periods of time.

3. Time-Resolved Methods. It is difficult to obtain
Raman spectra of intermediate structures like lumirhodopsin
because the low temperature photostationary steady state and
the flow methods cannot be directly applied. One possible
solution is to perform time-resolved resonance Raman experiments. These could be modeled after conventional flash
photolysis experiments where absorption spectra are obtained a
specified time after an actinic flash. For example, consider
obtaining the time-resolved Raman spectrum of lumirhodopsin.
First, a high photoalteration laser pulse is used to initiate
the bleaching of rhodopsin. Then, after a few μsec a second
low photoalteration probe pulse would be used to generate the
Raman scattering from lumirhodopsin. The scattering from the
intense bleaching pulse must be gated away and the duration of
the bleach pulse must be short compared to the kinetics studied. The requirement for a low photoalteration probe pulse
puts stringent restrictions on these experiments. The maximum
power permissible in the probe pulse can be calculated using
equation (12) in the rapid-flow section. The molecular transit time (τ_t) is simply replaced by the laser pulse duration

τ. Also, if the laser uniformly illuminates a circular area of radius ω, then the pulse energy E (photons) is given by $\pi\omega^2 I\tau$. With these expressions the photoalteration parameter becomes:

$$F = \tau I \sigma_a \phi = \frac{2.303 \times 10^3 \ E\epsilon\phi}{\pi\omega^2 N_o} . \tag{17}$$

It is reasonable to assume that the resonance enhancement and photoisomerization quantum yield of lumi will be similar to those of rhodopsin. For $\omega = 20$ μ, $\epsilon = 470$ cm^{-1}M^{-1}, and $\phi = 0.67$, a value of $F = 0.1$ is obtained for 1×10^{12} photons/pulse at 600 nm. With 600 nm excitation a typical experiment on rhodopsin will require passing 10^{20} photons through the sample to obtain a reasonable signal-to-noise ratio. Therefore, the bleach-probe sequence must be repeated 10^8 times. Since the lumi → meta I transition is complete in 100 μsec a reasonable repetition rate is 10 kHz. The experiment would still require a very long time, ∼ 3 hours. However, with multichannel detection a 50 fold improvement in efficiency is obtained (see Section VI) which makes this experiment feasible. By simply changing the delay between the two pulses, the kinetics and spectra of these species can be unambiguously obtained. Initial experiments on purple membrane using this method have recently been performed (27). These experiments will be discussed in detail later.

Pulsed lasers are being used a great deal in resonance Raman experiments because (1) they provide time resolution, (2) they are easily frequency doubled, (3) they have the high peak power necessary for nonlinear Raman scattering and (4) pulsed dye lasers have wide tuning ranges. However, the photoalteration problem can be especially severe for these lasers. Properties of some typical commercial pulsed lasers have been listed in Table 1. The photoalteration parameters have been calculated for rhodopsin according to equation (17). The values range from 0.2 to 10^6. The highest values cannot be taken literally since nonlinear processes will take over. However, this calculation illustrates that for resonance experiments with, for example, a nitrogen-pumped dye laser, even very unlikely photochemical events ($\phi = 10^{-3}$) may become a real problem. For example, it has recently been observed that vitamin B$_{12}$ derivatives, which do not have large photochemical quantum yields, undergo photoconversion in resonance Raman experiments (54). For pulsed lasers this problem can be significantly reduced by defocusing the laser beam (see Section V) and/or reducing the laser pulse energy.

TABLE 1. Pulsed Laser Photoalteration Parameters[a]

Laser	τ	Frequency	Pulse Energy		F
Mode-locked Argon Ion	150 psec	80 MHz	12	nJ	0.2
Cavity-dumped Argon Ion	40 nsec	100 kHz	0.5	μJ	10
Nitrogen-pumped Dye Laser	7 nsec	30 Hz	1	mJ	2×10^4
Flash Lamp Pumped Dye Laser	1 μsec	30 Hz	5	mJ	1×10^5
Q-Switched Doubled YAG	10 nsec	10 Hz	100	mJ	2×10^6

[a] Calculated from equation (17) at 500 nm with $\varepsilon = 40,000$ $cm^{-1}M^{-1}$, $\omega = 20$ μ, and $\phi = 0.67$.

C. Visual Pigment Raman Spectra

In Fig. 7 the known resonance Raman spectra of rhodopsin and its photolytic intermediates are compared with the appropriate protonated (PSB) and unprotonated (SB) Schiff base derivatives of retinal. The spectra of rhodopsin, isorhodopsin, metarhodopsin I and metarhodopsin II were obtained using rapid-flow techniques (30,31,53). The spectrum of bathorhodopsin was obtained from a low temperature photostationary steady state mixture by using a 585 nm probe beam to maximize the resonance enhancement of batho (55). The residual contribution of rhodopsin and isorhodopsin to the spectrum of the steady state mixture have been subtracted to produce a "pure" spectrum of the batho intermediate.

These resonance Raman spectra provide information about the vibrational structure of the retinal chromophore. The most intense line is the C=C ethylenic stretch near 1550 cm^{-1}. The frequency of this vibration for different forms of retinal has been correlated with the λ_{max} of the absorption spectrum (53,56). As the electrons become more delocalized, the ground-excited state energy gap decreases and the double and single bond strengths approach one another. This is illustrated by comparing the 1556 cm^{-1} vibration of the 11-*cis* PSB (λ_{max}

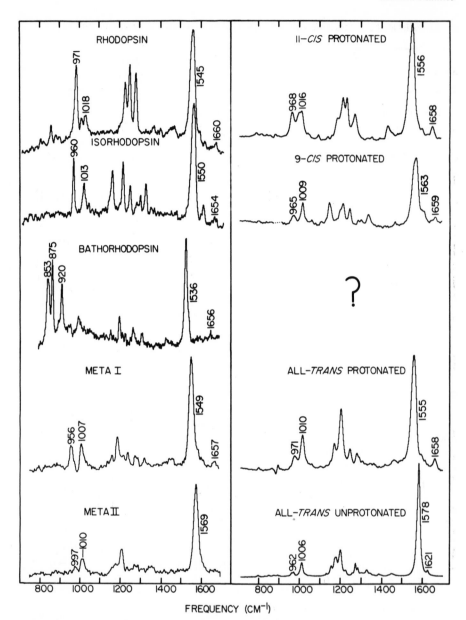

FIGURE 7. *Resonance Raman spectra of rhodopsin and its photolytic intermediates are compared with the appropriate protonated and unprotonated Schiff base derivatives of retinal. The rhodopsin and isorhodopsin spectra are from reference (31), bathorhodopsin is from reference (55), and meta I & II are from reference (53). The protonated Schiff base spectra are from reference (23).*

= 440 nm) with the 1545 cm^{-1} line in rhodopsin (λ_{max} = 500 nm).
The Schiff base C=N stretching vibration is found between
1655 and 1660 cm^{-1} when it is protonated and near 1625 cm^{-1}
when it is unprotonated (52,57). The lines near 1657 cm^{-1}
show that the retinal chromophore is bound as a protonated
Schiff base in rhodopsin, isorhodopsin and meta I (52). In
meta II the C=N vibration is not observed but the absorption
maximum and the pattern of the Raman spectrum are consistent
with an unprotonated Schiff base linkage (53). Presumably the
unprotonated Schiff base vibration is obscured by noise.
Bathorhodopsin also has a protonated C=N linkage. Further
evidence for this will be provided later.

It has been noted by several authors that the fingerprint
region (1100-1300 cm^{-1}) of retinal is very sensitive to the
conformation of the chromophore (23,50). This is illustrated
by comparing the spectra of the 11-*cis*, 9-*cis* and all-*trans*
PSB derivatives of retinal in Fig. 7. These vibrations are
expected to be mixtures of C-C stretching and C-C-H bending
modes of the polyene chain. This conformational sensitivity
has led to the expectation that these spectra would be useful
in analyzing the photochemically induced changes in visual
excitation. Lines in the 1010 cm^{-1} region have been assigned
to C-CH$_3$ stretching modes of the 9-and 13-methyl groups (30,
58,59). Lines near 970 cm^{-1} are expected to be due to mixtures of methyl stretches and C-H bending modes (60).

First, let us compare the spectra of rhodopsin and
isorhodopsin with the 11-*cis* and 9-*cis* model compounds. In
the fingerprint region, the frequencies and intensities are in
generally good agreement. Hence, the ground state conformation of retinal in rhodopsin (or isorhodopsin) is very similar
to the conformation of the 11-*cis* (or 9-*cis*) protonated Schiff
base retinal in solution (23). This implies that *no gross
distortions are introduced in the 11-cis and 9-cis chromophores when they combine with opsin to form a visual pigment.*
Similarly, the spectra of meta I and meta II are found to be
very close to those of the all-*trans* protonated and unprotonated Schiff base derivatives (21,53). Outside the fingerprint region two differences are consistently observed. First,
the ethylenic vibration is ~ 10 cm^{-1} lower in the visual pigment than in the model compound. Second, for the protonated
compounds (excluding bathorhodopsin) the 970 cm^{-1} line is more
enhanced in the visual pigments. Qualitatively, the shift in
the C=C frequency as well as the absorption data show that the
bound retinal chromophore has a more delocalized π-electron
structure. Changes in line *intensities* potentially provide
information about the effect of opsin on the excited electronic state of retinal. The Raman intensities depend
strongly on the displacement of the molecule along that normal

coordinate in the excited state. This suggests that opsin
catalyzes the deformation of the excited state of retinal
along the 970 cm^{-1} vibrational coordinate. Theoretical analysis of this observation may provide detailed information about
the excited state isomerization pathway.

There is presently a great deal of interest in the nature
of the primary photochemical event (e.g., rhodopsin $\longrightarrow$
bathorhodopsin). The fact that rhodopsin contains an 11-*cis*
chromophore that is released in the all-*trans* configuration
following photon absorption shows that a *cis* → *trans* isomerization *must* occur. The Raman spectra show that a complete
isomerization has occurred by the time meta I is formed. A
logical inference is that the conversion to bathorhodopsin
involves a *cis* → *trans* isomerization (42). Also, it has been
argued that since batho is a common intermediate between the
11-*cis* and 9-*cis* chromophores, it must be in an "essentially
all-*trans*" conformation (61,62). While appealing, this argument could represent an oversimplification of the actual
potential *surface* on which isomerization occurs. Therefore,
it is crucial to obtain a direct measure of the conformation
of bathorhodopsin. Unfortunately, the resonance Raman spectrum of bathorhodopsin does not bear a strong resemblance to
any of the PSB model compounds. This was recognized some time
ago by the observation that batho has prominent lines at 856,
877 and 920 cm^{-1} that are not found in the model compound
spectra (52). It appears that some fairly significant changes
in the structure of the retinal chromophore have occurred.
Until recently the usual explanation was that batho contained
a "strained" all-*trans* chromophore.

The applicability of the simple *cis* → *trans* isomerization
model has been questioned by recent experimental results.
First, picosecond absorption measurements have shown that
bathorhodopsin is formed in less than 6 psec at room temperature (63). It is theoretically reasonable to isomerize a
molecule as large as retinal in this time (64). However, if
the active site of opsin sterically hinders the *isomerizing
retinal*, then 6 psec might be insufficient time for a complete
conformational change. Unfortunately, no information is currently available on the structural details of the retinal
binding site. Second, the Raman spectra indicate that batho
has not yet attained a relaxed all-*trans* conformation. Third,
picosecond measurements at helium temperature in protonated
and deuterated media indicate that proton translocation occurs
in the formation of batho (65). Fourth, it has been observed that bleaching deuterated rhodopsin produces all-*trans*
retinal that has been partially deuterated (66). These data
led to the hypothesis that proton translocation occurs in the
formation of bathorhodopsin as depicted in Fig. 8 (65-67).

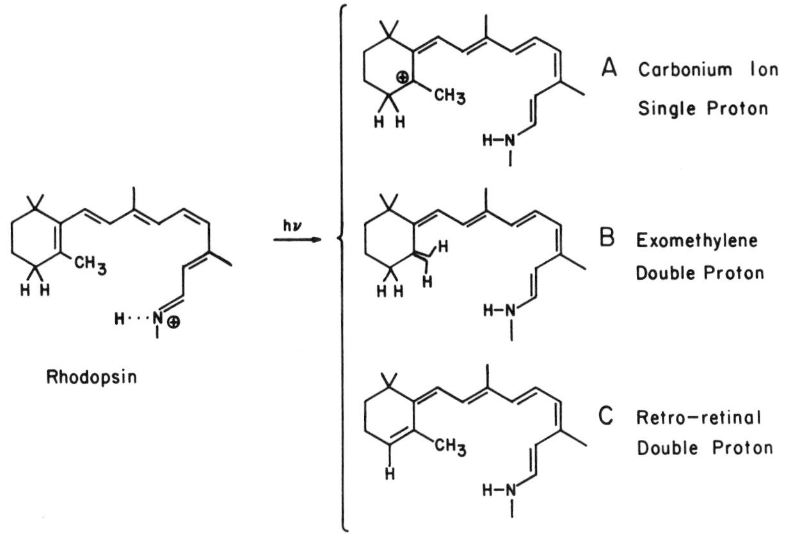

FIGURE 8. Proposed proton translocation models for bathorhodopsin.

A common feature of these models is that the formation of batho is associated with the reduction of the C=NH bond order and the increase of the N-H bond order on the Schiff base nitrogen.

In order to clearly observe the C=N vibration *in batho-rhodopsin* it was necessary to maximize the relative scattering from this intermediate. Low temperature spectra were taken with a 585 nm probe and a 488 nm pump beam in order to maximize both the concentration and the resonance enhancement of bathorhodopsin. Spectra in Fig. 9 were taken in the presence and absence of a 488 nm pump beam in protonated and deuterated media. The Raman spectrum taken with the probe only shows one major pigment band at 1553 cm^{-1}, the C=C stretch of isorhodopsin. When the blue pump beam is turned on, the percentage of bathorhodopsin rises from 0 to 34 percent and the yellow probe produces a spectacular resonance enhancement of the 1535 cm^{-1} C=C stretch of batho. Another resonantly enhanced batho-rhodopsin line appears at 1657 cm^{-1} in spectrum A. This line can be assigned as a C=NH stretch by observing that in deuterated media this line shifts to 1625 cm^{-1}. The C=NH frequency and deuteration shift are exactly those found for rhodopsin and isorhodopsin (52,55). Hence, *no net change in the strength*

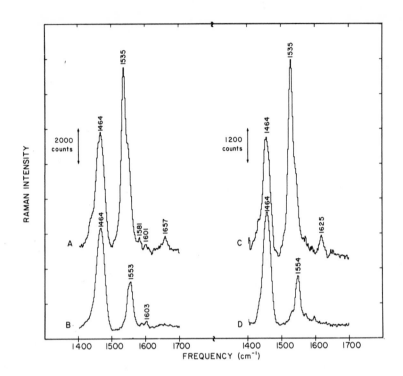

FIGURE 9. (A and B) Resonance Raman spectra of a
photostationary steady state mixture of rhodopsin, isorhodopsin and bathorhodopsin at -160°C. Spectra were taken with
a 585 nm probe beam in the presence (A) and in the absence (B)
of a 488 nm pump beam. Scattering at 1464 cm^{-1} is due to
glycerol. (C and D) Analogous spectra taken in a deuterated
medium (55).

of the C=NH bond or in the degree of protonation of the Schiff
base nitrogen occurs when rhodopsin (or isorhodopsin) is converted to bathorhodopsin. The predictions of the proton
transfer models for the Schiff base bond are clearly
inconsistent with these results.
     It is worthwhile exploring the implications of these
results in more detail. The deuteration and psec kinetics
must be explained. Also, it must be recognized that the
Schiff base Raman data do not explicitly exclude proton transfer. These data merely require that by the time batho is
formed, the C=NH and N-H bonds must be restored to their
original strengths. There are several mechanisms involving
proton transfer that are consistent with these data. First,
if the Schiff base proton is involved in the translocation, it

could move toward the Schiff base nitrogen following vertical
excitation of rhodopsin. Subsequent ground state relaxation
of bathorhodopsin would then involve a reverse proton translocation (22,55). Second, the proton could be translocated
elsewhere on the protein (68,69). Third, the transferred proton could be elsewhere on the retinal. For example, suppose
that a hydrogen on the 5-methyl group is transferred to opsin
when bathorhodopsin is formed (see Fig. 8). Maintaining the
exomethylene structure and the full C=NH bond order at the
same time would be inconsistent with the valence bond structures of retinal in Fig. 8. However, it is possible that the
electronic structure of batho is sufficiently delocalized to
accomodate this (67,70). Note that in all the above mechanisms, concerted conformational changes that lead toward an
isomerization about the 11-12 bond are almost certainly
occurring.

All the details of the arguments for and against the
proton translocation and isomerization models cannot be presented here (71). We have addressed those aspects of the
models that can be criticized via Raman spectra. It is clear
that resonance Raman spectroscopy is capable of providing high
resolution data that introduce severe constraints on the
models that can be proposed. The current Raman data provide
no direct evidence for a proton translocation. However, this
Raman data alone cannot be used to conclusively disprove proton transfer or to demonstrate isomerization because no lines
of bathorhodopsin other than the C=N can be easily interpreted. In particular, it is important to experimentally
characterize a vibrational mode that reports on the bonding in
the ionone ring and/or polyene chain portion of the chromophore in bathorhodopsin. The "low wavenumber" lines (853,
875 and 920 cm^{-1} should be interesting in this regard because
they have been attributed to a strained all-*trans* retinal (2)
or an exomethylene structure (66,67). Further Raman work on
visual pigment analogs and on intermediates such as lumirhodopsin should help to resolve this controversy.

IV. PURPLE MEMBRANE PROTEIN

A. *Introduction*

The purple membrane from *Halobacterium halobium* contains a
26,000 dalton protein which is complexed with a protonated
Schiff base retinal chromophore (72,73). Under anaerobic conditions it uses the energy of light absorbed by the chromophore to actively transport protons across the cell membrane

(74). The chromophore has been shown to undergo light-dependent photoconversions that are similar to those in rhodopsin (75). For this reason, this protein is often called bacteriorhodopsin. There is much interest in this process because it is thought that the proton gradient is used to drive ATP synthesis. The purple protein exists in two forms, light-adapted (BR_{570}) and dark-adapted (BR_{560}), as depicted in Fig. 10. When BR_{570} absorbs a photon, it cycles through a sequence of intermediates in 10 msec and functions as a proton pump. In the dark, BR_{570} spontaneously relaxes to BR_{560}, which can be reconverted to BR_{570} by irradiation. Chemical extraction experiments have shown that BR_{570} contains an all-*trans* chromophore while BR_{560} contains an equimolar mixture of all-*trans* and 13-*cis* retinal (76). Extraction experiments on a long-lived intermediate, M_{412}, produce variable results which indicate that M_{412} is either 13-*cis* or a mixture of 13-*cis* and all-*trans* (76). Resonance Raman spectroscopy has been used to show that retinal in BR_{570} is linked to the protein by a protonated Schiff base linkage that is deprotonated in M_{412} (73). Raman measurements on the purple membrane are of interest because the conformation and state of protonation of the chromophore are thought to be important in the mechanism of proton pumping.

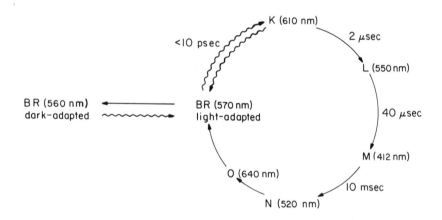

FIGURE 10. Schematic representation of the light-adapted photocycle of bacteriorhodopsin.

B. *Purple Membrane Raman Spectra*

Fig. 11 presents rapid-flow resonance Raman spectra of BR_{570}, BR_{560} and the all-*trans* and 13-*cis* protonated Schiff base derivatives of retinal. The spectra for BR_{570} and BR_{560} are in good agreement with flow data from several other laboratories (26,77,78). Significant differences are observed between the spectra of BR_{570} and BR_{560}. The dark-adapted spectrum has a broader ethylenic line and overlapping lines in the 1200 cm^{-1} region that are indicative of a mixture of retinal conformations. This is expected on the basis of the chemical extraction data (76). BR_{560} is a mixture which cannot be directly compared with the model compound spectra. If we *assume* that the all-*trans* component in BR_{560} has the same Raman spectrum as BR_{570}, then the all-*trans* component can be subtracted out to reveal the scattering from just the "13-*cis* fraction". Spectrum C in Fig. 11 results when BR_{570} is subtracted from BR_{560} until the prominent light-adapted features (e.g., 1526 cm^{-1} line) are reduced to a smooth baseline. This spectrum of the "13-*cis* fraction" is virtually identical to that derived independently by Terner *et al.* (78).

The assignment of the conformations of BR_{570} and BR_{560} *from Raman data alone* is difficult because the 13-*cis* and all-*trans* isomers are qualitatively more similar than are the corresponding isomers for rhodopsin. This has led to a series of conflicting and/or equivocal conclusions in the literature (26,27,78). This is predominantly due to the individual comparison methods because there is now good agreement on all the data presented in Fig. 11. Because Raman spectra of all-*trans* and 13-*cis* are so similar, it is impossible to compare the absolute intensities or frequencies of, for example, the "all-*trans* pigment" with the all-*trans* PSB. A more valid method is to compare the differences between the "all-*trans*" and "13-*cis*" pigments with the differences between the all-*trans* and 13-*cis* model compounds. For example, compare BR_{570} with spectrum C. Both compounds have two intense lines between 1150 and 1200 cm^{-1}. However, the lower frequency line (1182 cm^{-1}) is closer to 1200 cm^{-1} in C than it is in BR_{570}. Likewise this line is closer to 1200 cm^{-1} in the 13-*cis* PSB than in the all-*trans* PSB. A similar analysis of the frequencies above 1200 cm^{-1} show that C and the 13-*cis* PSB have no prominent isolated lines until the 1280 cm^{-1} region. BR_{570} and all-*trans* both have lines near 1250 and 1275. Features between 1300 and 1400 cm^{-1} are weak and cannot be reliably reproduced by all workers. If this comparison method is employed, then BR_{570} is closest to all-*trans* and the difference component in BR_{560} is most like 13-*cis*. The fact that this approach agrees with the chemical extraction results provides a reasonable justificiation of the method. Here

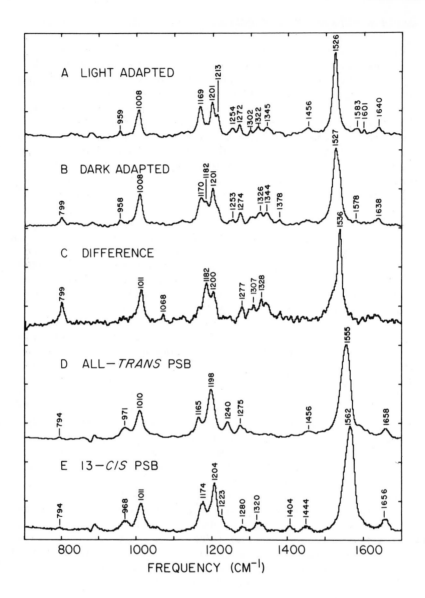

FIGURE 11. *Rapid-flow Raman spectra of purple membrane are compared with the all-trans and 13-cis protonated Schiff base (PSB) derivatives of retinal (A) BR_{570} taken with 0.6 mW at 590 nm. (B) BR_{560} taken with 1.0 mW at 590 nm. (C) Difference spectrum between BR_{560} and BR_{570} scaled to represent the "13-cis component" in BR_{560}. (D and E) All-trans and 13-cis protonated Schiff base spectra (23).*

the *pattern of line shifts* is emphasized rather than their absolute positions. Note that in the visual pigment comparisons, absolute frequency disagreements of as much as 14 cm^{-1} (e.g., isorhodopsin 1153 cm^{-1} line) were not questioned because the *pattern* of lines agreed so well. We are simply suggesting a systematic method for recognizing this pattern. It is clear that an isomerization occurs between BR$_{570}$ and BR$_{560}$.

While the above attributions are reasonable, the fact that the fingerprint line intensities are in poorer agreement for purple membrane than for rhodopsin indicates that more interaction between the chromophore and protein is taking place. This is supported by the fact that the ethylenic and the Schiff base lines in BR$_{570}$ are as much as 35 and 15 cm^{-1}, respectively, away from the corresponding frequencies in the model compounds. Lewis has proposed that this is due to the interaction of another group with the Schiff base nitrogen (79). Such a perturbation could alter the C=N frequency, the π-electron delocalization, and the excited state Franck-Condon factors. Until the purple membrane spectra are compared with the "correct" model compounds (defined as molecules with similar group frequencies), it is premature to decide whether BR$_{570}$ contains a relaxed or nonrelaxed all-*trans* chromophore. This same caveat applies to further interpretation of the 13-*cis* component in BR$_{560}$. It will be important to pursue these studies because chemical extraction cannot provide detailed information about the retinal-protein interactions that facilitate proton pumping.

The rapid-flow spectrum of M$_{412}$ has also been obtained by trapping this intermediate at high pH (26). Fig. 12 compares M$_{412}$ with the all-*trans* and 13-*cis* unprotonated Schiff bases. Here the closer agreement between the C=C and C=N frequencies imply that the solution spectra will provide a better model for the M$_{412}$ intermediate. Also, the fingerprint regions in the model compounds appear to be much more distinctive. However, the 1100 to 1300 cm^{-1} region of M$_{412}$ looks like all-*trans* while the 1300 to 1400 region is remarkably like 13-*cis*. Recall that the chemical extraction results were also indecisive. This may simply represent the fact that high pH-isolated M$_{412}$ has a nonnative chromophore-protein conformation. Alternative methods for viewing M$_{412}$ produce quite different spectra (80). The Raman data presented in Fig. 12 do not prove or disprove the idea that an isomerization about the 13-14 double bond occurs in the production of M$_{412}$. Further work on these intermediates will be necessary because it is important to establish the nature of the retinal conformational changes that occur in the light-adapted photocycle.

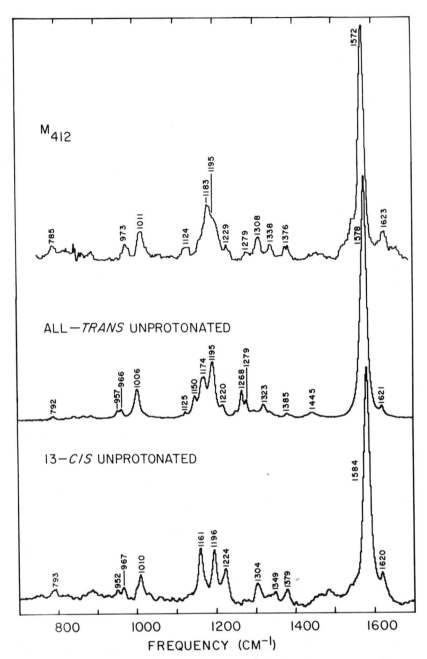

FIGURE 12. *Rapid-flow Raman spectrum of the* M_{412} *inter-mediate of bacteriorhodopsin (26) compared with the all-trans and 13-cis unprotonated Schiff base derivatives of retinal.*

C. Time-Resolved Raman Spectra of Purple Membrane

Recently, a number of authors have demonstrated that "time-resolved" Raman spectroscopy can be used to obtain spectra of the kinetic intermediates of purple membrane. Using 5 nsec pulses from a nitrogen-pumped dye laser, Campion *et al.* (81) were able to observe the C=C stretching frequency of the K intermediate. Under these high light-intensity conditions, a photostationary steady state between K and BR_{570} is produced. In other experiments Marcus and Lewis (25), and Campion *et al.* (82) used μsec illumination to obtain features of the M_{412} and L intermediates. This was done by flowing BR_{570} past a focused laser beam at different flow velocities (25) or by irradiating a stationary sample with variable duration pulses (82). Their results are similar and the data of Campion *et al.* have been presented in Fig. 13. Here, the duration of the laser pulse from a chopped argon-ion laser has been indicated for irradiation frequencies that will strongly (476.5 nm) and weakly (514.5 nm) enhance the scattering from M_{412}. They observe that a line at 1620 cm^{-1} increases in intensity *before* the 1570 cm^{-1} line of M_{412} appears. Since 1620 cm^{-1} is coincident with the unprotonated Schiff base stretching frequency, it was felt that deprotonation of the chromophore occurs before the L $\rightarrow M_{412}$ transition. This would imply that either another unprotonated intermediate occurs between L and M_{412}, or that L is unprotonated. This conclusion depends on the correct assignment of the 1620 cm^{-1} band. Polyenes and carotenoids with no Schiff base bond can have C=C stretching modes in the 1620-1630 region. Therefore, it would be useful to confirm this assignment by using appropriate isotopic substitution.

These single pulse "time-resolved" experiments can be used to identify the spectral features of the intermediates in these systems. However, the experimental design dictates that the photoalteration parameter of each pulse must be increased as the time duration is lengthened. Hence, the kinetics of Raman lines will in many cases be dictated by *light-dependent* rates, not unimolecular decay rates (see Fig. 6). Also, since the irradiation continues for the entire time period, the spectra may be further complicated by photoconversion to other species. This is particularly important since evidence now exists that L and M_{412} are in fact photosensitive (83). These difficulties mean that the single pulse experiments will be most useful when one can insure that either light-dependent or light-independent rates are dominating the observed relaxation processes.

As described earlier, the kinetics of a photosensitive system can be unambiguously examined with a double pulse Raman method. These experiments have been performed by Terner *et al.*

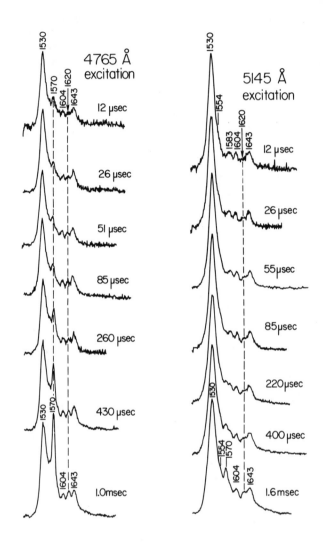

FIGURE 13. Single pulse time-resolved resonance Raman spectra of the L → M$_{412}$ transition in the purple membrane. The times indicate the duration of the laser irradiation produced by chopping a cw argon-ion laser (82).

(27) on the msec time scale to study the decay of the M$_{412}$ intermediate in purple membrane. In Fig. 14 a high photoalteration (0.6 mJ) pulse is used to initiate the photocycle of BR$_{570}$. A weaker probe pulse is used to monitor the composition after a specified time delay. Here the decay of the 1567 cm^{-1} C=C stretching mode of M$_{412}$ is clearly seen. With

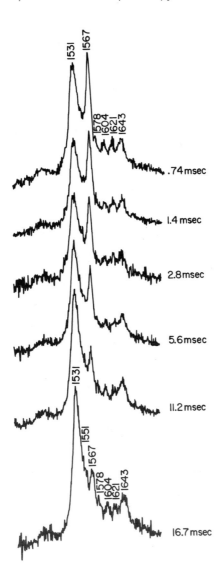

FIGURE 14. Double pulse time-resolved Raman spectra of
the decay of M_{412} in the purple membrane. The time delay
between an actinic laser pulse and a weaker probe pulse at
476.5 nm has been indicated (27).

this experimental design, absolute rates can be determined and
detailed information about the chemical bonds in an intermediate can be *simultaneously* obtained. Because of the msec delay
time between pulses, the repetition rate for the pump-probe
sequence was very low (30 Hz). For shorter delay times these
experiments become easier if the repetition rate can be
increased accordingly. There seems to be no reason why these
double pulse experiments cannot be used to study relaxation
processes on a very rapid (even psec) time scale.

V.   ULTRAVIOLET RESONANCE RAMAN

With the development of commercial ultraviolet lasers,
resonance Raman experiments in the UV have become possible.
A number of experiments have recently been performed on chemical (84-86) and biochemical systems (17-19, 87) in the 350-250
nm region. While small in number, these experiments indicate
the potential of ultraviolet excited resonance Raman. Possible biochemical applications include selectively enhancing
the scattering from individual residues in proteins or specific bases in nucleic acid polymers.

Blazej and Peticolas (17) have recently obtained the UV
excitation profile of adenosine 5'-monophosphate (AMP) from
500 nm (20 kK) to 270 nm (38 kK)[2]. Their results have been
reproduced in Fig. 15. Here the ν_s^4 dependence has been
removed so that only the frequency dependence of the polarizability tensor is displayed. The dramatic observation is that
*the scattering intensity increases by 10^5 when excitation is
tuned on resonance.* The intensity profiles for the 1338 and
1484 cm^{-1} ring models of AMP were fit to an A term expression.
This analysis showed that the enhancement of the 1484 cm^{-1}
mode was due to a transition at 276 nm while the 1338 cm^{-1}
band was enhanced by a transition in the 259-269 nm region.
Since lasers are not available to the blue of the AMP absorption, it was not possible to make experimental observations on
both sides of the absorption band.

A similar study by Johnson on a modified base, 4-thiouri-
dine, provides further indication of the potential of UV Raman
(88). 4-thiouridine has a red shifted absorption spectrum
that facilitates excitation throughout the first and approaching the second electronic transition. Excitation profiles for
the 1242 cm^{-1} and 1620 cm^{-1} vibrations have been fit to

<hr>

[2] *1 kK (kiloKaiser) is equal to 1,000 cm^{-1}.*

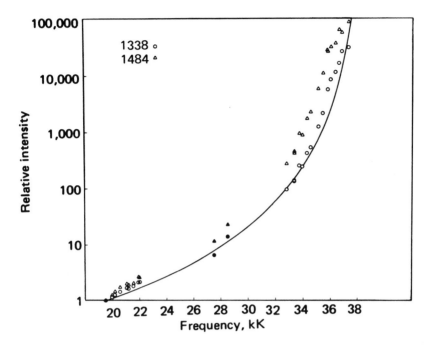

FIGURE 15. *Relative Raman intensity of the 1338 and 1484 cm^{-1} bands of AMP as a function of excitation frequency. The solid line was calculated using a preresonance A term analogous to equation (8) with an electronic transition at 260 nm (17).*

resonance A term expressions and are plotted in Fig. 16. Note that the 1620 cm^{-1} vibration shows resonances at 30 kK and 40 kK while the 1242 cm^{-1} vibration is resonant at only 30 kK. Evidently, *different vibrational modes can be differentially enhanced depending on the nature of the electronic transition that is resonantly excited.* Here, preliminary analysis indicates that the 1620 cm^{-1} band is a C=O stretch that is enhanced by not only the 330 nm $\pi\to\pi^*$ transition, but also by a localized $\pi\to\pi^*$ or n$\to\pi^*$ transition near 245 nm. The 1242 cm^{-1} pyrimidine ring mode is selectively enhanced by the 330 nm $\pi\to\pi^*$ transition. The strongest enhancement for 4-thiouridine occurs in the 30 kK region where the scattering from AMP (Fig. 15) is only slightly enhanced. Therefore, it should be possible to selectively enhance the scattering from 4-thiouridine in a polynucleotide containing a large number of other bases. Such a study has recently been performed on tRNA.

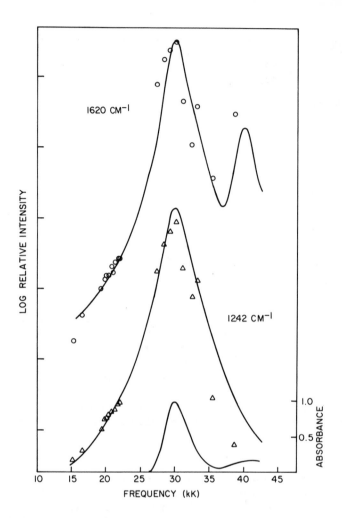

FIGURE 16. Relative Raman intensity of the 1242 and 1620 cm^{-1} vibrations of 4-thiouridine as a function of excitation frequency (88). The 1242 cm^{-1} profile was fit to an A term expression at 330 nm. The 1620 cm^{-1} profile was fit to a sum of A term expressions at 330 nm and 245 nm. The 1620 cm^{-1} profile has been displaced upward for clarity.

In Fig. 17 resonance Raman spectra of formylmethionine tRNA and glycine tRNA from *Escherichia coli* obtained by Nishimura *et al.* are presented (19). They used 363.8 nm excitation from an argon-ion laser to selectively enhance the scattering from a single 4-thiouridine in the presence of ~ 75

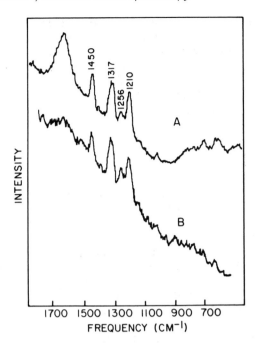

FIGURE 17. *Resonance Raman spectra of 4-thiouridine in
(A) fmet tRNA and (B) gly tRNA from E. coli obtained with
363.8 nm excitation (19).*

other bases. As anticipated from the previous excitation profiles, only scattering from 4-thiouridine is observed. In
this case, the spectra indicate that the 4-thiouridine was
photochemically crosslinked to an adjacent cytidine by the
363.8 nm irradiation. Nishimura *et al.* demonstrated that this
photochemistry could be controlled by flowing the sample.
These results suggest that UV Raman will be a useful technique
for studying the electronic and vibrational structure of
specific bases in polynucleotides.
 UV experiments are difficult because these lasers have
low power and they are difficult to operate. Furthermore,
efficient, achromatic optical systems for gathering the
scattered UV light are not available. Recent advances in
eximer lasers (89), doubled pulsed dye lasers, and doubled
cw lasers will vastly increase experimental capabilities.
However, because most of the UV sources are pulsed, the problems of photolysis, as discussed earlier, must be considered
in detail. Also, molecules with long lived excited state decay times such as 4-thiouridine ($\sim$ 100 nsec) will be subject
to ground state depletion (saturation) during the laser pulse.

We have minimized these problems by using fairly long pulses
(1 μsec) from a flash-lamp pumped dye laser (CMX-4). Nevertheless it was necessary to further reduce the light intensity
by focusing with a cylindrical lens (backscattering geometry)
to eliminate these effects (88). Lasers with shorter pulses
such as the doubled N_2-pumped dye laser will be much more
susceptible to this problem (see Table 1).

VI. MULTICHANNEL DETECTORS IN RAMAN SPECTROSCOPY

A major difficulty associated with spontaneous Raman
scattering, even in resonance, is the time required to obtain a
spectrum. With low irradiation intensities dictated by the
sample (e.g., visual pigments) or by the laser (e.g., UV excitation), it takes hours to scan spectra. Particularly for
biological samples, this is complicated by the presence of
background scattering and fluorescence that is many times the
intensity of the Raman bands. These difficulties are amelior-
ated by the use of multichannel detectors. Because the usual
photon counting detection is a single channel technique with a
low spectral bandwidth (~ 5 cm^{-1}), the overall efficiency for
gathering photons is very low ($\sim 0.5\%$ for a 1000 cm^{-1} scan).
Multichannel detectors have now been developed that can view
1000 cm^{-1} of a Raman spectrum with nearly the same sensitivity
as a photomultiplier. This is accomplished by coupling an
image intensifier to a vidicon camera or diode array. A number of systems have recently been described (27,82,90-96). In
our own work (55,96,97) we have employed a dry ice cooled
"silicon intensified target" (SIT). The SIT does not have the
gain of some systems (1 count = 2 photons), but the ease of
installation and operation make this a very useful detector.
Especially when cooled, the SIT has enhanced detection capability over a photomultiplier because the target responds to
energy rather than power, permitting long term data integration on the detector. For example, we have obtained spectra
with peak count rates as low as 0.1 count/sec/channel. With
extended integration a signal-to-noise ratio (S/N) of 10 can
be obtained with count rates as low as 0.01 counts/sec/channel.
Since all the channels acquire optical signal simultaneously,
mechanical scanning techniques are no longer needed. The majority of the increase in S/N occurs because the detector
has approximately 100 times the efficiency of the conventional
photon counting spectrometer. An additional increase in S/N
will be observed in many cases because the multichannel system
is unaffected by fluctuations in the sample scattering, the
sample background, and the laser intensity.

The design of the spectrometer is particularly important in a multichannel system. For biological samples, sufficient resolution is provided by displaying 200 Å ($\sim$ 800 cm^{-1}) of scattering in 500 channels. Also, the spectrometer must reduce the stray incident laser light by $\sim$ 10^{10}. This has usually been accomplished by using a single spectrograph with colored glass and/or spatial filters to reduce stray light (27,82,90-95). We have found that an optimal solution is obtained with the *double* monochromator shown in Fig. 18. Here a low-dispersion monochromator, which acts as a prefilter, is coupled to a higher dispersion monochromator with the grating oriented and blazed for subtractive dispersion. The second monochromator reduces the stray light but does not increase the net dispersion of the transmitted light. This device has $\sim$ 10^{-10} stray light, the same throughput as a conventional double monochromator, and it can be easily coupled to a dry ice cooled SIT.

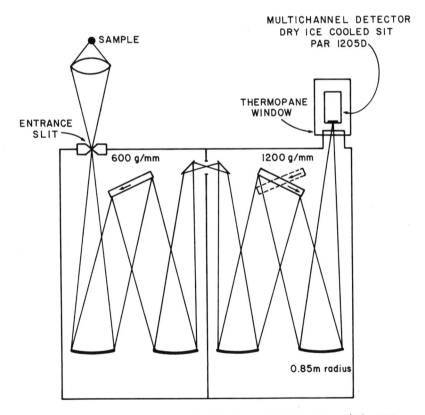

FIGURE 18. Multichannel Raman spectrometer with SIT detection and a mismatched subtractive dispersion double monochromator (96).

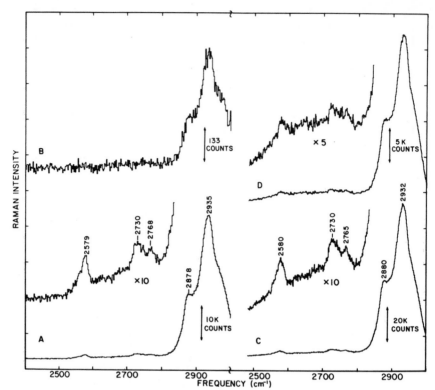

*FIGURE 19. Comparison of SIT and photon counting
detection systems for the high-frequency Raman vibrations of
bovine lens proteins. (A) Photon counting spectrum taken with
100 mW,514.5 nm for 600 sec. (B) Photon counting spectrum
taken with 2 mW,514.5 nm for 600 sec. (C) SIT spectrum taken
with 200 mW,514.5 nm for 25.2 sec. (D) SIT spectrum with
2 mW,514.5 nm for 983 sec. (96).*

An example of the performance of a multichannel system is
shown in Fig. 19. Here the SIT system is compared with a conventional photon counting system under identical conditions.
The sample is an excised bovine lens which was studied at high
(200 mW) and low (2 mW) laser power. With 200 mW the SIT
spectrum (C) is identical to the photon counting spectrum except that the multichannel spectrum was acquired in 30 seconds
(25 times faster). With 2 mW of laser power the lines at
2580, 2730 and 2765 cm^{-1} cannot be observed with the conventional system. The multichannel system is more sensitive at
both high and low light levels. The Raman lines are due to
the bovine lens proteins (α-, β-and γ-crystallin). The oxidation of sulfhydryl to disulfide bonds can be monitored through

the intensity of the S-H (2580 cm^{-1}) and S-S vibrations (508 cm^{-1}). This oxidation has been correlated with the development of certain types of cataracts (98). This study demonstrates that *in vivo* studies of cataract formation can be performed at laser powers low enough to avoid retinal damage.

Multichannel detectors will facilitate the observation of time-resolved Raman spectra. The time-resolved results on purple membrane (Figs. 13 & 14) were obtained with a SIT detector (27,81,82). The spectrum of p-terphenyl anion radical has been obtained with a 600 nsec, 10 mJ pulse (92). The resonance Raman spectrum of cytochrome c has been obtained with a single 7 nsec, 75 mJ pulse at 532 nm from a doubled YAG (94). Bridoux and Delhaye have described a pulsed mode-locked system that they used to obtain the spectrum of neat cyclohexane in 25 psec (90). The ultimate time resolution of these experiments appears to be limited only by the laser pulse duration available.

Multichannel detectors also make it possible to use Raman as a probe in perturbation experiments (temperature-jump, pressure-jump) and in differential experiments (vibrational optical activity). For example, Sturm *et al.* (95) have used 3 J pulses at 694.3 nm to observe the relaxation of a double-helical polynucleotide (polyadenylic acid - polyuridylic acid) following the application of a temperature jump (see Fig. 20). The observed changes in intensity of the 785, 814 and 1100 cm^{-1} bands are indicative of changes in the double-helical A form of the polymer induced by the T-jump (1). Multichannel detection will also be useful in the observation of vibrational optical activity (99). These experiments measure the difference in scattering intensity observed when a chiral sample is irradiated with right and left circularly polarized light. These differences are small (0.1%) but may provide information about the absolute configuration of atoms involved in a particular normal mode. It has recently been demonstrated that a multichannel system can substantially (10-fold) improve the S/N in these experiments.

The above experiments demonstrate that time-resolved, perturbation, and difference Raman measurements can now be performed in a reasonable amount of time with good S/N. Replacing low-resolution spectroscopic probes such as absorption and emission with Raman spectra should provide additional information about many chemical and biochemical systems.

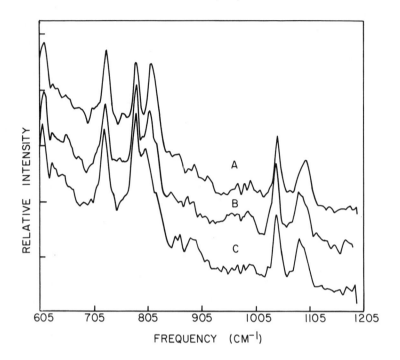

FIGURE 20.   *Temperature-jump Raman spectra of poly A -*
poly U.   (A) poly A - poly U at 56°C.   (B) Poly A -
poly U   2 msec after a 3° T-jump, and (C) 100 msec after
T-jump (95).

VII. CONCLUSION

I hope that this discussion of the present capabilities of
resonance Raman spectroscopy will help others to apply this
technique to new systems.   Resonance Raman is currently an
underdeveloped but already very powerful spectroscopic tool.
Looking back to the first visual pigment Raman spectrum
obtained in 1970 (Fig. 4) illustrates the magnitude of the
progress in recent years.   Some significant experimental
problems still remain.   For example, fluorescent systems are
very difficult to study.   However, nonlinear techniques such
as resonance coherent anti-Stokes Raman scattering (CARS),
which are insensitive to fluorescence, have been successfully
used to study diphenyl polyenes, β-naphthol, cytochrome c,
vitamin B_{12} and flavin (100-103).   Also, temporal discrimina-
tion of fluorescence using pulsed laser excitation has

recently been demonstrated (104). The development of new detectors and laser sources will undoubtedly continue to expand the capabilities of resonance Raman spectroscopy. The biological applications of resonance Raman should be even more exciting in the coming years.

ACKNOWLEDGMENTS

I wish to thank R. Callender, M.A. El-Sayed, A. Lewis, W. L. Peticolas, L. Rimai, L. Stryer, M. Tsuboi and N.-T. Yu for permission to use figures and to cite results from published and/or unpublished work. The unprotonated Schiff base retinal data were obtained at Yale University with T. Freedman and L. Stryer. Raman data obtained at Berkeley was the work of M. Braiman, B. Curry, G. Eyring, and B. Johnson. W. Stoeckenius and R. Lozier generously provided samples of *Halobacterium halobium* and assisted in the initial phases of this work.

REFERENCES

1. Spiro, T.G., in "Chemical and Biochemical Applications of Lasers", Vol. *I* (C.B. Moore, ed.), p. 29. Academic Press, New York, (1974).
2. Spiro, T.G. and Stein, P., *Ann. Rev. Phys. Chem. 28,* 501 (1977).
3. Warshel, A., *Ann. Rev. Biophys. Bioeng. 6,* 273 (1977).
4. Callender, R. and Honig, B., *Ann. Rev. Biophys. Bioeng. 6,* 33 (1977).
5. Spiro, T.G. and Gaber, B.P., *Ann. Rev. Biochem. 46,* 553 (1977).
6. Spiro, T.G. and Loehr, T.M., in "Advances in Infrared and Raman Spectroscopy", Vol. *1* (R.J.H. Clark and R.E. Hester, eds.), p. 98. Heyden, London, (1975).
7. Spiro, T.G., *Acc. Chem. Res. 7,* 339 (1974).
8. Spiro, T.G. and Strekas, T.C., *J. Amer. Chem. Soc. 96,* 338 (1974).
9. Sunder, S., Mendelsohn, R. and Bernstein, H.J., *J. Chem. Phys. 63,* 573 (1975).
10. Shelnutt, J.A., O'Shea, D.C., Yu, N.-T., Cheung, L.D. and Felton, R.H., *J. Chem. Phys. 64,* 1156 (1976).
11. Warshel, A., *Chem. Phys. Lett. 43,* 273 (1976).
12. Asher, S. and Sauer, K., *J. Chem. Phys. 64,* 4115 (1976).
13. Mayer, E., Gardiner, D.J. and Hester, R.E., *Biochim. Biophys. Acta 297,* 568 (1973).
14. Wozniak, W.T. and Spiro, T.G., *J. Amer. Chem. Soc. 95,* 3402 (1973).

15. Lutz, M., *J. Raman Spectrosc. 2*, 497 (1974).
16. Lutz, M., *Biochim. Biophys. Acta 460*, 408 (1977).
17. Blazej, D.C. and Peticolas, W.L., *Proc. Natl. Acad. Sci. USA 74*, 2639 (1977).
18. Chinsky, L., Turpin, P.Y. and Duquesne, M., *Biopolymers 17*, 1347 (1978).
19. Nishimura, Y., Hirakawa, A.Y., Tsuboi, M. and Nishimura, S., *Nature 260*, 173 (1976).
20. Rimai, L., Kilponen, R.G. and Gill, D., *Biochem. Biophys. Res. Commun. 41*, 492 (1970).
21. Sulkes, M., Lewis, A., Lemley, A.T. and Cookingham, R., *Proc. Natl. Acad. Sci. USA 73*, 4266 (1976).
22. Aton, B., Callender, R.H. and Honig, B., *Nature 273*, 784 (1978).
23. Mathies, R., Freedman, T.B. and Stryer, L., *J. Mol. Biol. 109*, 367 (1977).
24. Mendelsohn, R., *Nature 243*, 22 (1973).
25. Marcus, M.A. and Lewis, A., *Science 195*, 1328 (1977).
26. Aton, B., Doukas, A.G., Callender, R.H., Becher, B. and Ebrey, T.G., *Biochemistry 16*, 2995 (1977).
27. Terner, J., Campion, A. and El-Sayed, M.A., *Proc. Natl. Acad. Sci. USA 74*, 5212 (1977).
28. Carey, P.R. and Schneider, H., *J. Mol. Biol. 102*, 679 (1976).
29. Petersen, R.L., Li, T.-Y., McFarland, J.T. and Watters, K.L., *Biochemistry 16*, 726 (1977).
30. Callender, R.H., Doukas, A., Crouch, R. and Nakanishi, K., *Biochemistry 15*, 1621 (1976).
31. Mathies, R., Oseroff, A.R. and Stryer, L., *Proc. Natl. Acad. Sci. USA 73*, 1 (1976).
32. Tobin, M.C., "Laser Raman Spectroscopy", Wiley, New York, (1971).
33. Dollish, F.R., Fateley, W.G. and Bentley, F.F., "Characteristic Raman Frequencies of Organic Compounds", Wiley, New York, (1974).
34. "Raman Spectroscopy", Vols. *1* & *2* (H.A. Szymanski, ed.) Plenum Press, New York.
35. Tang, J. and Albrecht, A.C., in "Raman Spectroscopy", Vol. 2 (H.A. Szymanski, ed.) p. 33. Plenum Press, New York, (1970).
36. Albrecht, A.C. and Hutley, M.C., *J. Chem. Phys. 55*, 4438 (1971).
37. Peticolas, W.L., Nafie, L., Stein, P. and Fanconi, B., *J. Chem. Phys. 52*, 1576 (1970).
38. Johnson, B.B. and Peticolas, W.L., *Ann. Rev. Phys. Chem. 27*, 465 (1976).
39. Johnson, B.B., Nafie, L.A. and Peticolas, W.L., *Chem. Phys. 19*, 303 (1977).

40. Warshel, A. and Dauber, P., *J. Chem. Phys. 66*, 5477 (1977).
41. See,for example, Barnett, G.P. and Albrecht, A.C.,in "Raman Spectroscopy", Vol. *2* (H.A. Szymanski, ed.) p. 207. Plenum Press, New York, (1970).
42. Wald, G., *Nature 219*, 800 (1968).
43. Kropf, A. and Hubbard, R., *Photochem. Photobiol. 12*, 249 (1970).
44. Dartnall, H., in "Handbook of Sensory Physiology", Vol. *VII*, p. 122 (1972).
45. Dartnall, H.J.A. and Lythgoe, J.N., *Vision Res. 5*, 81 (1965).
46. Yoshizawa, T., and Wald, G., *Nature 197*, 1279 (1963).
47. Parson, W.W.,in "Chemical and Biochemical Applications of Lasers", Vol. I (C.B. Moore, ed.) p. 339. Academic Press, New York, (1974).
48. Honig, B. and Ebrey, T.G., *Ann. Rev. Biophys. Bioeng. 3*, 151 (1974).
49. Ebrey, T.G. and Honig, B., *Quart. Rev. Biophys. 8*, 129 (1975).
50. Rimai, L., Gill, D. and Parsons, J.L., *J. Amer. Chem. Soc. 93*, 1353 (1971).
51. Doukas, A.G., Aton, B., Callender, R.H. and Honig, B., *Chem. Phys. Lett. 56*, 248 (1978).
52. Oseroff, A.R. and Callender, R.H., *Biochemistry 13*, 4243 (1974).
53. Doukas, A.G., Aton, B., Callender, R.H. and Ebrey, T.G., *Biochemistry 17*, 2430 (1978).
54. Salama, S. and Spiro, T.G., *J. Raman Spectrosc. 6*, 57 (1977).
55. Eyring, G. and Mathies, R., "Resonance Raman Studies of Bathorhodopsin: Evidence for a Protonated Schiff Base Linkage", submitted to *Proc. Natl. Acad. Sci. USA*.
56. Rimai, L., Heyde, M.E. and Gill, D., *J. Amer. Chem. Soc. 95*, 4493 (1973).
57. Heyde, M.E., Gill, D., Kilponen, R.G. and Rimai, L., *J. Amer. Chem. Soc. 93*, 6776 (1971).
58. Gill, D., Heyde, M.E. and Rimai, L., *J. Amer. Chem. Soc. 93*, 6288 (1971).
59. Cookingham, R., and Lewis, A., *J. Mol. Biol. 119*, 569 (1978).
60. Warshel, A. and Karplus, M., *J. Amer. Chem. Soc. 96*, 5677 (1974).
61. Rosenfeld, T., Honig, B., Ottolenghi, M., Hurley, J. and Ebrey, T.G., *Pure Appl. Chem. 49*, 341 (1977).
62. Hurley, J.B., Ebrey, T.G., Honig, B. and Ottolenghi, M., *Nature 270*, 540 (1977).
63. Busch, G.E., Applebury, M.L., Lamola, A.A. and Rentzepis, P.M., *Proc. Natl. Acad. Sci. USA 69*, 2802 (1972).

64. Warshel, A., *Nature 260,* 679 (1976).
65. Peters, K., Applebury, M.L. and Rentzepis, P.M., *Proc. Natl. Acad. Sci. USA 74,* 3119 (1977).
66. Fransen, M.R., Luyten, W.C.M.M., van Thuijl, J., Lugtenburg, J., Jansen, P.A.A., van Breugel, P.J.G.M. and Daemen, F.J.M., *Nature 260,* 726 (1976).
67. van der Meer, K., Mulder, J.J.C. and Lugtenburg, J., *Photochem. Photobiol. 24,* 363 (1976).
68. Lewis, A., *Proc. Natl. Acad. Sci. USA 75,* 549 (1978).
69. Warshel, A., *Proc. Natl. Acad. Sci. USA 75,* 2558 (1978).
70. Warshel, A., personal communication.
71. Kropf, A., *Nature 264,* 92 (1976). Green, B.H., Monger, T.G., Alfano, R.R., Aton, B. and Callender, R.H., *Nature 269,* 179 (1977).
72. Henderson, R., *Ann. Rev. Biophys. Bioeng. 6,* 87 (1977).
73. Lewis, A., Spoonhower, J., Bogomolni, R.A., Lozier, R.H. and Stoeckenius, W., *Proc. Natl. Acad. Sci. USA 71,* 4462 (1974).
74. Oesterhelt, D. and Stoeckenius, W., *Proc. Natl. Acad. Sci. USA 70,* 2853 (1973).
75. Lozier, R., Bogomolni, R. and Stoeckenius, W., *Biophys. J. 15,* 955 (1975).
76. Pettei, M.J., Yudd, A.P., Nakanishi, K., Henselman, R. and Stoeckenius, W., *Biochemistry 16,* 1955 (1977).
77. Aton, B., Doukas, A.G., Callender, R.H., Becher, B. and Ebrey, T.G., "Resonance Raman Studies of the Dark-Adapted Form of the Purple Membrane Protein", submitted to *Biochim. Biophys. Acta.*
78. Terner, J., Hsieh, C.-L. and El-Sayed, M.A., "Time-Resolved Resonance Raman Characterization of the bL_{550} Intermediate and the Dark Adapted bR_{560}^{DA} Form of Bacteriorhodopsin", submitted to *Biophys. J.*
79. Lewis, A., Marcus, M.A., Ehrenberg, B. and Crespi, H., "Experimental Evidence for Secondary Protein-Chromophore Interactions at the Schiff Base Linkage in Bacteriorhodopsin: A Molecular Mechanism for Proton Pumping", to be published in *Proc. Natl. Acad. Sci. USA.*
80. Marcus, M.A. and Lewis, A., "Resonance Raman Spectroscopy of the Retinylidene Chromophore in Bacteriorhodopsin (BR_{570}), BR_{560}, M_{412} and Other Intermediates: Structural Conclusions Based on Kinetics, Analogs, Models and Isotopically Labeled Membranes", to be published in *Biochemistry.*
81. Campion, A., Terner, J. and El-Sayed, M.A., *Nature 265,* 659 (1977).
82. Campion, A., El-Sayed, M.A., and Terner, J., *Biophys. J. 20,* 369 (1977).
83. Hurley, J.B., Becher, B. and Ebrey, T.G., *Nature 272,* 87 (1978).

84. Hong, H.-K. and Jacobsen, C.W., *Chem. Phys. Lett. 47,*
     457 (1977).
85. Hong, H.-K. and Jacobsen, C.W., *J. Chem. Phys. 68,*
     1170 (1978).
86. Ziegler, L. and Albrecht, A.C., *J. Chem. Phys. 67,*
     2753 (1977).
87. Pézolet, M., Yu, T.-J. and Peticolas, W.L., *J. Raman
     Spectrosc. 3,* 55 (1975).
88. Johnson, B.B. and Mathies, R., to be published.
89. Ewing, J.J., in "Chemical and Biochemical Applications of
     Lasers", Vol. *II* (C.B. Moore, ed.) p. 241. Academic
     Press, New York, (1977).
90. Bridoux, M. and Delhaye, M., in "Advances in Infrared and
     Raman Spectroscopy", Vol. *2* (R.J.H. Clark and R.E.
     Hester, eds.) p. 140. Heyden, London, (1976).
91. Delhaye, M. and Dhamelincourt, P., *J. Raman Spectrosc. 3,*
     33 (1975).
92. Pagsberg, P., Wilbrandt, R., Hansen, K.B. and Weisberg,
     K.V., *Chem. Phys. Lett. 39,* 538 (1976).
93. Woodruff, W.H. and Atkinson, G.H., *Anal. Chem. 48,* 186
     (1976).
94. Woodruff, W.H. and Farquharson, S., *Anal. Chem. 50,* 1389
     (1978).
95. Sturm, J., Savoie, R., Edelson, M. and Peticolas, W.L.,
     Ind. J. Pure Appl. Phys. 16, 327 (1978).
96. Mathies, R. and Yu, N.-T., "Raman Spectroscopy with
     Intensified Vidicon Detectors: A Study of Intact Bovine
     Lens Proteins", to be published in *J. Raman Spectrosc.*
97. Princeton Applied Research 1205D and 1205A.
98. East, E.J., Chang, R.C.C., Yu, N.-T. and Kuck, J.F.R. Jr.,
     J. Biol. Chem. 253, 1436 (1978).
99. Boucher, H., Brocki, T.R., Moskovits, M. and Bosnich, B.,
     J. Amer. Chem. Soc. 99, 6870 (1977).
100. Hudson, B., Hetherington, W., Cramer, S., Chabay, I. and
     Klauminzer, G.K., *Proc. Natl. Acad. Sci. USA 73,* 3798
     (1976).
101. Carreira, L.A., Goss, L.P. and Malloy, T.G., *J. Chem.
     Phys. 68,* 280 (1978).
102. Nestor, J., Spiro, T.G. and Klauminzer, G., *Proc. Natl.
     Acad. Sci. USA 73,* 3329 (1976).
103. Dutta, P.K., Nestor, J.R. and Spiro, T.G., *Proc. Natl.
     Acad. Sci. USA 74,* 4146 (1977).
104. Harris, J.M., Chrisman, R.W., Lytle, F.E. and Tobias,
     R.S., *Anal. Chem. 48,* 1937 (1976).

LASER EXCITATION OF RAMAN SCATTERING FROM
ADSORBED MOLECULES ON ELECTRODE SURFACES

Richard P. Van Duyne

Northwestern University
Evanston, Illinois

I.   INTRODUCTION

In recent years there has been a continuing rate of
growth in the amount of research effort devoted to fundamental
studies of surface and interfacial phenomena (see, for example
(1-5)). In particular there has been intense interest in the
development of detailed atomic and molecular level descriptions of surfaces which include answers to such questions as:

1. What is the chemical identity of the adsorbed
atomic or molecular species?
2. What is the geometric or structural arrangement of
these species with respect to the substrate surface atoms?
3. What is the charge distribution and the energy
level structure of the valence electrons in both adsorbate
and substrate (i.e., the nature of the chemisorption bonding)?
4. What are the vibrational, rotational, and translational motions of all species on the surface?

Several complementary factors can be identified as the cause
of this renaissance in surface science. First, it has been
recognized that surface phenomena are responsible for the
existence or control the performance of many important
technologies and/or devices. Second, a vast array of high
resolution, surface sensitive spectroscopic techniques
capable of providing detailed analytical and structural data
on surfaces have been developed. Third, two-dimensional
interfaces such as the solid-gas (vacuum) or the solid-liquid
(electrolyte) interface constitute a separate phase of condensed matter with unique properties that are only now
starting to be understood.

The technologies most critically dependent on the fundamental understanding of the properties of surfaces are probably those of energy and information. Many forms of energy production, storage, and conversion are based on heterogeneous catalytic processes operating at the surface of a solid in contact with either a gas or liquid phase environment. The improvement of the efficiency of existing energy systems and the development of new energy systems will depend heavily on the fundamental information obtained about the surface structure and/or surface composition vs. system performance relationship (6). Corrosion is an energy wasting process that cost the U.S. economy approximately $70 billion in 1975 (7). Since corrosion is fundamentally a solid-liquid or solid-gas interfacial process, rational approaches to its prevention and the development of new corrosion resistant materials will also depend on the data base provided by fundamental surface studies (8). Information processing and communications technology are now completely based on thin-film microelectronic devices. The fabrication and performance of these solid state devices depends crucially on the ability to control chemical and physical processes occurring at both solid-solid and solid-gas interfaces (9). This is primarily due to the extremely large surface-to-volume ratios inherent to these devices. Fundamental surface studies will contribute not only to improved device fabrication techniques and quality control, but also to the analysis of failure mechanisms.

The techniques used for carrying out surface spectroscopy either on technologically significant surfaces or on scrupulously clean model surfaces involve the interaction of a probe beam of electrons, photons, or ions with the target surface. One then measures the resulting output signal (viz., absorption, emission, or scattering of these particles) which carries away the information content of the probe beam/surface interaction. A spectroscopic technique is considered to be intrinsically surface sensitive if either the input beam or the output beam is limited in its penetration depth to the first few atomic layers of the target surface (10). Low energy (less than 1000 eV.) electrons and ions have penetration depths on the order of 10-20 $\overset{o}{A}$; whereas, photons have much deeper (viz., 100-10,000 $\overset{o}{A}$ depending on photon energy and the nature of the substrate) penetration depths. Consequently the primary techniques for the structural and analytical characterization of surfaces involve energy, intensity, and angular distribution measurements on electron and/or ion beams; and, therefore, require ultrahigh vacuum (UHV) environments for their operation. The most widely used surface spectroscopies include: Auger electron spectroscopy (AES, electrons-in and electrons-out); ultraviolet and x-ray photoelectron spectroscopy (UPS and XPS, photons-in and electrons-out); and secondary ion mass spectrometry (SIMS, ions-in and ions-out). These UHV techniques provide excellent qualitative and quantitative elemental analysis, valence electronic structure determination, depth profiling information, and, in the case of the scanning Auger microprobe (SAM) technique, spatial resolution. The main limitation of these surface spectroscopies lies in the area of molecular surface characterization. Currently it is very difficult to detect, identify, and quantitate individual molecular species adsorbed on a surface (e.g., various hydrocarbons adsorbed on a supported metal catalyst) except in the case of very simple molecules adsorbed on clean, ordered substrates (e.g., CO adsorbed on Pt(111)). An additional limitation involved in the application of UHV surface spectroscopies is the inherent one of being restricted from the direct examination of the solid-liquid (electrolyte) interface such as would be required in the study of corrosion, electrochemical fuel cells, liquid junction photovoltaic solar cells, and the fundamentals of electrochemical reaction mechanisms.

Motivated to overcome these limitations in conventional UHV electron and ion surface spectroscopies, some surface scientists, in particular electrochemists and solid state physicists interested in solid-liquid (electrolyte) interfaces, have sought to develop other types of surface spectroscopy. Two approaches have emerged from this effort: surface UV-VIS electronic absorption spectroscopy and surface vibrational spectroscopy.

Several manifestations of UV-VIS electronic absorption spectroscopy including: specular reflectance (external and internal as well as electroreflectance) spectroscopy, ellipsometry, and transmission spectroscopy (on transparent metal film substrates) have been evaluated experimentally and theoretically for suitability in surface studies. These developments in surface electronic absorption spectroscopy have been reviewed several times by McIntyre (11,12), Hansen (13), Muller (14), and Kuwana (15,16). The main drawback to these techniques is, of course, the fact that they exhibit low intrinsic surface sensitivity due to the large penetration depths of optical frequency photons. In addition the molecular specificity of surface electronic absorption spectroscopy is quite low. The limited surface sensitivity for these optical techniques can be greatly improved by restricting observation to chemisorbed molecules known to reside in monolayer or submonolayer amounts on the surface (17); or, in the more restricted case of solid-electrolyte interfacial studies, using modulation of the interfacial potential (i.e., electroreflectance spectroscopy) to impart surface sensitivity (18). The main strength of optical spectroscopic probes for surface studies is that they are quite universal in nature and can, therefore, be used to study the solid-gas (vacuum) interface as well as the solid-liquid (electrolyte) and solid-solid interfaces. This feature is particularly advantageous in the study of surface electro- chemical processes and high pressure catalytic processes where the UHV surface spectroscopies are either difficult or imposs- ible to apply except in an *ex situ* fashion.

The specificity for determining the chemical identity of adsorbed molecular species can be dramatically improved by employing surface vibrational spectroscopy instead of surface electronic absorption spectroscopy or the conventional UHV surface spectroscopies. The study of the vibrational motions of surface atoms and chemisorbed molecules is a burgeoning new field of surface spectroscopy consisting of the following techniques: infrared transmission spectroscopy (ITS) (19,20), infrared reflection spectroscopy (IRS) (12,21,22), high resolution electron energy loss spectroscopy (ELS) (23-25), inelastic electron tunnelling spectroscopy (IETS) (26,27), and surface Raman spectroscopy (SRS) (28-30). Surface vibra- tional spectroscopy is likely to have a major impact on the business of molecular surface characterization since it has the ability to provide directly interpretable, detailed information on: (1) the structure and orientation of ad- sorbed molecules; (2) the strength and nature of surface chemisorption bonding; (3) the structure of surface active sites and catalytic reaction intermediates; and (4) the dynamics of surface chemical reactions. Unlike ITS, which is restricted to the study of chemisorbed molecules on highly dispersed substrates at the solid-gas interface; IRS, which is restricted to the solid-gas interface but has sufficient sensitivity for the study of monolayer adsorbates on well- defined, smooth surfaces; ELS, which has excellent surface sensitivity but poor vibrational energy resolution and is restricted to the solid-vacuum interface; and IETS, which also has excellent surface sensitivity but must operate under the cryogenic conditions of the Josephson Junction; surface Raman spectroscopy enjoys extreme flexibility in the type of surface situations which may be investigated. In principle SRS can combine the high vibrational energy resolu- tion of the infrared techniques with the universal inter- facial applicability and sampling flexibility of the optical absorption spectroscopies.

The purpose of this chapter is to review the recent advances that have been made in the application of Raman spectroscopy to the molecular characterization of surfaces. The particular emphasis that will be adopted here is that of laser excitation of Raman scattering from molecules adsorbed on reasonably well defined metal surfaces as opposed to the older and more common type of SRS involving adsorbed molecules on highly dispersed metal oxide particle surfaces. This latter field has recently been reviewed elsewhere (31, 32). Surface Raman spectroscopy is the least well developed of the various surface vibrational spectroscopies. This is probably due to the fact that, although SRS enjoys sampling flexibility and high vibrational energy resolution, it lacks sensitivity when only the conventional spontaneous Raman scattering process is operative. Research is currently underway in several laboratories to overcome this sensitivity problem by increasing the Raman scattering efficiency by several orders of magnitude through the application of the resonance Raman (RR) and coherent anti-stokes Raman (CARS) scattering processes. Special interest in surface Raman spectroscopy has recently developed as a result of the reports from two research groups (33,34) which suggest the possibility that when a molecule is appropriately adsorbed on or near a metal surface, its Raman scattering efficiency can be increased by factors of 10^3 to 10^6. Such enormous enhancements of the normal Raman (NR) scattering process completely overcome the traditional limitation of SRS -- its low sensitivity. The experimental context in which surface enhanced Raman scattering (SERS) was first recognized was that of an electrochemical cell. Consequently this chapter will largely be devoted to the discussion of laser excited Raman scattering from molecules adsorbed on metal electrode surfaces. Section II will briefly review some of the basic principles of Raman scattering and discuss in some detail the sensitivity aspects of Raman scattering as applied to problems in surface science. Section III reviews the first successful experimental observation of SRS for molecules adsorbed on electrode surfaces, the verification of these observations, and the relevant variables that control SRS in an electrochemical environment. Section IV presents a survey of the current experimental results concerning the SERS phenomenon and then summarizes the hypotheses (theories?) which have been put forth to explain the large Raman cross sections for adsorbed molecules on metal surfaces. The reader is asked to bear in mind that recognition of the SERS phenomenon only occurred two years ago, new research groups are starting work on the problem with increasing frequency, and much new data needs

to be acquired before a comprehensive picture of SERS is at
hand. Nevertheless the bare outlines of SERS can now be seen.
Finally, Section V considers the prospects for the future of
SERS and for related forms of laser excited surface Raman
spectroscopy. The reader more broadly interested in chemisorbed molecules at metal-gas (vacuum) interfaces as well as
at metal-liquid (electrolyte) interfaces will be heartened to
know that some of the theoretical and experimental results
pertaining to the SERS effect suggest that it will not be
restricted only to an electrochemical environment.

II.   SENSITIVITY REQUIREMENTS FOR THE IMPLEMENTATION OF
         SURFACE RAMAN SPECTROSCOPY

Surface Raman spectroscopy holds considerable promise
for providing the molecular structure information (viz.,
adsorption induced changes in geometry, orientation of the
adsorbate with respect to the substrate, direct observation
of surface chemisorption bonding, etc.) that is complimentary
to the atomic and electronic structure information about monolayer and sub-monolayer coverage adsorbates on clean, well-
defined metal surfaces provided by the UHV particle spectroscopies. Before SRS can be successfully implemented and widely
used in this context, two serious sensitivity limitations
must be overcome. First, the surface number density of adsorbed molecules, N_{surf}, is only on the order of 10^{14} to 10^{15}
molecules cm^{-2}. Secondly, the Raman scattering cross sections
associated with the NRS process, $d\sigma/d\Omega$, are intrinsically very
small (viz., ca. 10^{-30} cm^2 sr^{-1} molecule^{-1}). This ineffi-
ciency of NR scattering arises fundamentally from the fact
that the laser irradiates the target molecule in a transparent
region of its electromagnetic absorption spectrum. Thus, the
laser is only weakly interacting with the target molecule.
This combination of low number density of Raman scatterers
and the small scattering cross section is so severe a
sensitivity limitation that, to date, no high signal to noise
ratio, S/N, surface Raman spectra have been obtained for
monolayer molecular adsorbates on metal surfaces based on the
normal spontaneous Raman effect.

In comparison surface IRS has recently been successfully
implemented for monolayer and submonolayer coverages of CO
on a variety of clean polycrystalline and single crystal
metal surfaces (21,22,35). Although IRS must contend with
the same low value of N_{surf} as SRS, it has the substantial
advantage that infrared absorption cross sections are
typically 10-12 orders of magnitude greater (e.g., 2 x 10^{-18}
cm^2 molecule^{-1} for CO adsorbed on Pt) than the corresponding
NRS cross section (e.g., 3.3 x 10^{-30} cm^2 sr^{-1} molecule^{-1} for

gas phase CO irradiated with 4880 Å laser light (36)).
Nevertheless the current state of the art in IRS instrumentation restricts the applicability of this technique to extremely strong ir absorbers such as CO and furthermore restricts
one to the observation of relatively high frequency vibrational bands, $\bar{\nu} > 1000$ cm^{-1}. Thus part of the motivation for
implementing SRS is to develop a photon based, surface vibrational spectroscopy that will have a broad enough observa-
tional window to encompass both the high frequency and low
frequency domains. This latter region is particularly important from the viewpoint of directly observing chemisorption
bonding.

A.   *Quantitative Intensity and S/N Analysis for a CW SRS Experiment*

A quantitative discussion of Raman scattering from molecules adsorbed on low-area surfaces is necessary in order to
evaluate possible instrumentation strategies based on modern
laser and photodetector technology for overcoming the intrinsic sensitivity barriers to SRS. The objective in this
section is to calculate the anticipated intensity of the
Raman scattering signal from a monolayer of adsorbed molecules
on an atomically smooth (i.e., idealized) metal surface. This
intensity calculation will be based on the cross sections
representative of the normal Raman scattering process. For
purposes of this calculation a standard instrumental configuration for Raman spectroscopy will be assumed (viz., CW
argon ion laser, double monochromator, large aperture backscattering collection optics, high quantum efficiency photomultiplier tube such as the RCA C31034A, and photon counting
detection electronics). Numerical evaluation of the scattering intensity expression will be carried out for the
specific case of CO adsorbed on the (111) surface of a
platinum single crystal. This example was chosen to permit
direct comparison with the IRS experiment and because this
system is a favorite of surface physicists who have studied
it with the complete arsenal of UHV surface spectroscopies.
The intensity of a Raman line corresponding to a transition between an initial state, i, (usually the 0th vibrational
level of the ground electronic state -- $|g,0>$) and a final
state, f, (the 1st vibrational level of the ground electronic
state in the case of a Stokes Raman fundamental -- $|g,1>$) is
given by (37-39):

$$I_{if}(\omega_S) = \frac{2^3 \pi}{3^2 c^4} \omega_S^4 I_L(\omega_L) \sum_{\rho,\sigma} |(\alpha_{\rho\sigma})_{if}|^2 \qquad (1)$$

where $\omega_S = 2\pi c \bar{\nu}_S$ (in cm^{-1}) is the frequency of the Stokes
Raman scattered light, $I_L(\omega_L)$ is the flux of the incident
laser source in units of photons sec^{-1} cm^{-2}, ρ and σ are the
polarizations of the Raman and incident light respectively,
and $\alpha_{\rho\sigma}$ is an element of the Raman scattering tensor which
represents the probability amplitude for a transition from
state i to state f. Furthermore it should be noted that
$I_{if}(\omega_S)$ represents the total light intensity (in photons
sec^{-1}) per molecule scattered into 4π steradians and that
$\omega_S = \omega_L - \omega_R$ where $\omega_R = 2\pi c \bar{\nu}_R$ (in cm^{-1}) is the frequency of
the Raman active, vibrational, normal mode excited by the in-
elastic scattering process. Since differential Raman
scattering cross sections rather than individual Raman
scattering tensor elements are the intensity variables usually
measured experimentally (36, 40-42), a more convenient form
of Eq. (1) can be obtained by recognizing that (38):

$$\Omega \frac{d\sigma(\omega_S)}{d\Omega} = \frac{2^3 \pi}{3^2 c^4} \omega_S^4 \sum_{\rho,\sigma} |(\alpha_{\rho\sigma})_{if}|^2 \qquad (2)$$

so that Eq. (1) simply becomes:

$$I_{if}(\omega_S) = \Omega \frac{d\sigma(\omega_S)}{d\Omega} I_L(\omega_L) \qquad (3)$$

Converting $I_L(\omega_L)$ to units of laser power and $I_{if}(\omega_S)$ to units
of photoelectron counts sec^{-1} $molecule^{-1}$ (i.e., Hz $molecule^{-1}$)
and assuming that the quantum efficiency of the photomultiplier tube is Q, the transmission of the double monochromator at ω_S is T_m, and that the transmission of the
collection optics at ω_S is T_o, we get:

$$I_{if}(\omega_S) = \Omega \frac{d\sigma(\omega_S)}{d\Omega} P_L(\omega_L) \varepsilon(\omega_L)^{-1} Q T_m T_o \qquad (4)$$

where $\varepsilon(\omega_L) = hc\bar{\nu}_L$ is the energy of an incident laser photon
in joules.

In order to compute $I_{if}(\omega_S)$ for a monolayer of adsorbed
molecules, one must incorporate certain features of the
atomic arrangement of the substrate and the adsorption geometry into Eq. (4) so that N_{surf} can be specified. If N_{sub}
is the number density of atoms exposed at a particular metal
substrate surface (e.g., single crystal surface plane

with specified Miller indices) and n is the ratio of surface atoms to molecules involved in the bonding to a specific adsorption site, then for an adsorption system with only one type of adsorption site:

$$N_{surf} = \frac{N_{sub}}{n} \qquad (5)$$

Converting $I_{if}(\omega_s)$ to the total expected scattering intensity in Hz from the number of molecules located on the specified surface within the focal area, A, of the incident laser beam gives:

$$[I_{if}(\omega_s)]_{surf} = N_{surf} \; A\Omega \; \frac{d\sigma(\omega_s)}{d\Omega} \; P_L(\omega_L)\epsilon(\omega_L)^{-1}QT_mT_o \qquad (6)$$

It should be understood that Eq. (6) represents the Raman intensity integrated over ω_s for a specified Raman active vibration. In addition we have not treated here certain detailed molecular orientation features peculiar to the surface environment or accounted for the optical reflectivity of the substrate which can modify the total detected intensity by factors of 2 or 3 (43).

For the specific case of CO adsorbed on the (111) face of platinum only in "on-top" positions (reference 21 cites recent IRS, ELS, and LEED data which indicates that CO actually adsorbs on two different adsorption sites but this complication is ignored for purposes of the present calculation), one gets a maximum value for N_{surf} = 1.5 x 10^{15} CO molecules cm^{-2} where N_{sub} = $N_{Pt(111)}$ = 1.5 x 10^{15} atoms cm^{-2} and n = 1 Pt atom per CO adsorption site. The gas phase Raman cross section for the C≡O stretch Stokes fundamental ($\bar{\nu}_R$ = 2145 cm^{-1}) at $\bar{\nu}_L$ = 20491 cm^{-1} (4880 Å) is 3.3 x 10^{-30} cm^2 sr^{-1} molecule^{-1}. Assuming Ω = 1 steradian, a line focus for the incident laser beam with a focal area of 5mm x 100μ = 5 x 10^{-3} cm^2 (this is a typical situation when backscattering collection optics are used (44)), a CW laser power of 1.0 watt at the sample, a photomultiplier tube with Q = 0.30 at all detected ω_s, collection optics with T_o = 0.90, and a monochromator transmission at $\bar{\nu}_s$ = 18346 cm^{-1} of T_m = 0.01, and a photon counting system that records all photoelectrons produced by the photomultiplier, we would expect a total detected signal integrated over the CO vibrational band of $[I_{if}(\omega_s)]_{CO,surf} = 32.8$ Hz.

This integrated surface Raman signal can be converted to a peak count rate, I_{max}, by considering that the lineshape of the SRS band is Lorentzian:

$$[I_{if}(\bar{\nu}_s - \bar{\nu}_{if})]_{surf} = I_{max} \frac{\Gamma_R^2}{\Gamma_R^2 + (\bar{\nu}_s - \bar{\nu}_{if})^2} \qquad (7)$$

where frequency is measured in cm^{-1} and Γ_R is the half width at half maximum height in cm^{-1}. The area of this lineshape function which is then equal to the 32.8 Hz integrated Raman signal is:

$$\int_{-\infty}^{+\infty} [I_{if}(\bar{\nu}_s - \bar{\nu}_{if})]_{surf} d\bar{\nu} = \pi \Gamma_R I_{max} \qquad (8)$$

So that for a typical value of $\Gamma_R = 5$ cm^{-1}, $I_{max} = 2.09$ Hz/cm^{-1}. From this value we can calculate the peak S/N for this CW SRS experiment since (45):

$$S/N = \frac{I_{max}\Delta t^{1/2}}{[I_{max} + I_{dark}]^{1/2}} \qquad (9)$$

where Δt is the photon counting dwell time per cm^{-1} of monochromator scan and I_{dark} is the dark count rate of the photomultiplier tube. Assuming $\Delta t = 1.0$ sec and $I_{dark} = 10$ Hz (this is a typical dark count for the RCA C31034A photomultiplier tube operated at $-20°C$), we get a S/N for the CO on Pt(111) experiment of only 0.60. The standard definition of detection limit is usually considered to be a S/N of 2.0 or greater so that a Δt photon counting interval of 1 sec is apparently insufficient to adequately define the Raman signal from adsorbed CO. A Δt of at least 16 sec is required for the peak S/N to exceed the detection limit.

The net conclusion from this S/N analysis is that detection of a monolayer of adsorbed CO on a low-area metal surface using normal spontaneous Raman scattering is at best a tenuous proposition. Extremely long data acquisition times will be required to record a single Raman band and this in turn requires that the surface sample be thermally and photochemically stable under the intense laser radiation for this time period. Furthermore this analysis assumes that the only noise source is photomultiplier dark counts and that no fluorescence or Rayleigh scattering is involved. In most practical situations this will not be the case. Under these conditions it is concluded that SRS based on the NRS process is of marginal practicality. Clearly what is needed is some means of enhancing I_{max} by several orders of magnitude.

B.   *Raman Enhancement Techniques for Overcoming the Sensitivity Limitations of SRS*

Based on Eqs. (6) and (9) one can propose several possible strategies designed to increase I_{max} in a SRS experiment. These strategies fall into the following categories:

1. Optimizing the performance of the Raman spectrometer
2. Increasing the surface number density of Raman scatterers
3. Increasing the Raman scattering cross section

Optimizing a Raman spectrometer for the SRS (or any other high sensitivity) experiment would involve: (1) maximizing Q, T_o, Ω, and T_m; (2) minimizing the photomultiplier tube dark count; (3) using as high a laser power density as possible; and (4) counting for long data acquisition periods and/or signal averaging on repetitive monochromator scans. Possibilities (1) and (2) are not very promising since they are either technologically very difficult or would result in only small increases in I_{max}. Possibilities (3) and (4) are experimentally tractable but at best would improve the S/N by only a factor of 10 and would require the sample to withstand enormous photon doses. This would probably result in restricting the applicability of the SRS technique to very few samples (viz., refractory materials). Increasing N_{surf} by studying only multilayer adsorbates or monolayers on high surface area support materials is the traditional method of improving the SRS signal intensity (31,32). For monolayer adsorbates on metals one could consider various roughening procedures such as mechanical abrasion, electrochemical anodization, or ion sputtering to increase the effective surface area. It is doubtful that this approach could result in more than a factor of 100 increase in I_{max}. Consequently the most likely approach to the dramatic improvement of I_{max} is to try and increase the effective Raman scattering cross section of the adsorbate molecule. The known methods for increasing $d\sigma/d\Omega$ are: (1) choosing an adsorbate molecule with the largest possible NRS cross section; (2) employing near uv, uv, or vacuum uv laser excita-tation frequencies to take advantage of the fact that the cross section is directly proportional to the fourth power of the Stokes frequency (Eq. (2) $\omega_s = \omega_L - \omega_R$); and (3) employing a laser excitation frequency that is coincident with an electronic transition in the adsorbate/metal system to take advantage of the resonance Raman effect.

CO does not have a particularly large value of $d\sigma/d\Omega$ so it might be possible to significantly increase the surface Raman scattering intensity simply by judicious choice of the adsorbate system to be studied. Table 1 shows the value of $[I_{if}(\bar{\nu}_s)]_{surf}$ calculated from Eq. (6) for various diatomic and polyatomic adsorbates with larger cross sections than CO. Although this is not an exhaustive listing, it does serve to illustrate a common trend. Large Raman cross sections tend to be found in molecules which would occupy relatively large surface areas in the adsorbed state. Thus there are offsetting factors which limit the intensity improvement to be obtained by choice of adsorption system.

TABLE 1

Integrated SRS Intensities for Various Diatomic and Polyatomic Normal Raman Scatterers Adsorbed on Pt(111)

Molecule	$\bar{\nu}_R$ cm^{-1}	$d\sigma/d\Omega \times 10^{30}$ $cm^2\ sr^{-1}\ molecule^{-1}$ a	$N_{surf} \times 10^{14}$ $molecule\ cm^{-2}$ b	$[I_{if}(\bar{\nu}_s)]_{surf}$ Hz c
H_2	4161	7.94	15	78.9
CO	2145	3.31	15	32.8
O_2	1556	4.30	15	42.7
$C_6H_5NO_2$	1345	103	2.1	143.2
C_6H_6	992	32.5	2.1	45.2
C_5H_5N	991	18.5	2.1	25.7

a. $\bar{\nu}_L = 20492\ cm^{-1} = 4880\ \text{Å}$.

b. Adsorption geometry assumed to give maximum value of N_{surf} for monolayer coverage.

c. All other parameters in Eq. (6) have the same values used in the calculation for CO.

The use of uv laser systems to increase the NR cross section through the ω_s^4 factor is potentially promising only if one can use high average power lasers with very deep uv outputs. Excellent candidates are: (1) the fourth harmonic of the Nd:YAG laser at $\bar{\nu}_L = 37594$ cm^{-1}; (2) the KrF excimer laser at $\bar{\nu}_L = 40160$ cm^{-1}; and (3) the ArF excimer laser at $\bar{\nu}_L = 51813$ cm^{-1}. Assuming that constant laser flux could be delivered to the sample and that all other parameter values in Eq. (6) remain constant, $[I_{if}(\bar{\nu}_s)]_{surf}$ for the systems listed in Table 1 would be increased by factors of 11.3, 14.7, and 40.9 respectively. If only constant laser power (i.e., 1 watt) can be delivered to the sample then these factors are reduced by the $\varepsilon(\bar{\nu}_s)^{-1}$ factor to 6.2, 7.5, and 16.2 respectively making this approach much less appealing.

So far all of the Raman enhancement techniques discussed have at best resulted in projected factors of 10-100 increase in the integrated surface Raman signal intensity. These small improvements, while useful, are not sufficient to impart general applicability to SRS. We now turn our attention to the last of the known methods for enhancing $d\sigma/d\Omega$ - the resonance enhanced Raman effect. Several excellent reviews on this subject are available which discuss theory (46, 47); experimental aspects (48, 49); and applications to inorganic chemistry (50,51), radical ion chemistry (52), and biochemistry (39,51,53-55). The primary distinction between the NR and RR effects as well as the origin of the cross section enhancement in RRS can easily be seen by examining the well known expression for the Raman scattering tensor element, $(\alpha_{\rho\sigma})_{if}$, derived from second order, time-dependent perturbation theory (37, 38, 46, 47):

$$(\alpha_{\rho\sigma})_{if} = \sum_{j}{}' \frac{<i|\rho|j><j|\sigma|f>}{(\varepsilon_j-\varepsilon_i)-\varepsilon(\omega_L)-i\Gamma_j}$$

$$+ \frac{<i|\sigma|j><j|\rho|f>}{(\varepsilon_j-\varepsilon_f)+\varepsilon(\omega_L)-i\Gamma_j}$$

(10)

where $|j>$ represents an intermediate state in the Raman scattering process (usually a particular vibronic state of an electronically excited state of the sample); $\varepsilon_i,\varepsilon_j,\varepsilon_f$ are the energies of the initial, intermediate, and final states respectively; Γ_j is the Lorentzian linewidth associated with the finite lifetime of $|j>$; the conditional sum over j excludes terms where $|j> = |i>, |f>$; and all other symbols re-

tain their previous definitions. The NRS regime is defined
by the condition:

$$\varepsilon(\omega_L) \quad << \quad (\varepsilon_j - \varepsilon_i) \qquad (11)$$

Thus for NRS no one term in the sum over intermediate states
(see Eq. (10)) dominates and furthermore all the denominators
are large so that $(\alpha_{\rho\sigma})_{if}$ is small. Consequently $d\sigma/d\Omega$ takes
on its characteristically small value of ca. 10^{-30} cm^2 sr^{-1}
molecule^{-1}. When $\varepsilon(\omega_L)$ approaches the energy difference
between the initial state and a particular intermediate state
of the sample, either by judicious choice of the sample molecule so that its electronic absorption spectrum overlaps a
fixed frequency laser or by using a tunable laser system so
that one can tune into a selected molecular resonance, one
enters the resonance Raman scattering regime which is defined
by the condition:

$$\varepsilon(\omega_L) \quad \approx \quad (\varepsilon_j - \varepsilon_i) \qquad (12)$$

In this circumstance the denominator of the first term in
Eq. (10) becomes small (i.e., limited only by the value of
the intermediate state linewidth, Γ_j) and consequently the
first term(s) corresponding to the $|j>$ state(s) selected by
the incident laser dominates the scattering tensor expression.
In RRS the value of $d\sigma/d\Omega$ corresponding to the resonantly
enhanced value of $(\alpha_{\rho\sigma})_{if}$ can be 10^2 to 10^6 times greater than
the corresponding NRS case. The magnitude of this enhance-
ment factor depends strongly on the degree of resonance
between the laser light and the vibronic intermediate levels
of the target molecules. In addition it should be noted that
the enhancement factor also depends on the value of the
numerator in Eq. (10), which is effectively the extinction
coefficient of the target molecule measured at ω_L. Thus in
general one expects that the larger the extinction coefficient
of the sample, the more exactly condition (12) is satisfied,
and the smaller is Γ_j, the larger will be the value of the
resonance Raman scattering cross section, $(d\sigma/d\Omega)_{RR}$. It
should be pointed out that there are at least three distinct
types of electronic transitions in adsorbate/metal scattering
systems that could be used to provide resonance enhancement
for an SRS experiment. First, there are electronic transitions which are localized on the adsorbate molecule. These
are the most direct and obvious type of transitions to try to
excite although there is the distinct possibility that one
will not be able to observe RR enhancement in vibrations
associated with the surface chemisorption bonding since these

are not necessarily associated with an electronic transition localized on the adsorbate. Second, there are electronic transitions that are shared between the adsorbate and the substrate and are, in effect, created by the process of adsorption. An example of such a transition might be a charge transfer transition where an electron originally on the metal substrate might be excited by the incident laser light into an orbital whose description involved substantial adsorbate character. Such transitions would be expected to strongly resonance enhance vibrations associated with surface chemisorption bonding. Third, there are electronic transitions localized within the metal substrate (viz., interband transitions, surface plasmon transitions and bulk plasmon transitions). If the nature of the adsorption interaction is such as to mix the molecular electronic states of the adsorbate with the electronic states of the metal substrate to form essentially a continuum of intermediate levels for the RR scattering process, then laser excitation of a predominantly metal transition could result in RR scattering from the adsorbate. This mechanism for RR scattering was first proposed by Philpott (56). Only RR scattering from the first type of electronic transition (viz., adsorbate or molecule localized) has been demonstrated experimentally to date.

In order to provide a quantitative appreciation for the extent to which RR enhancement can overcome the sensitivity limitations inherent to SRS, Table 2 shows the SRS signal intensities for a few adsorbate/Pt(111) systems calculated using Eq. (6). I_2 irradiated with a 1.0 watt CW dye laser at $\bar{\nu}_L = 18135$ cm^{-1}, which is within the $B^3\Pi_{0u}^+ \leftarrow X^1\Sigma_g^+$ transition, shows such a large enhancement factor that the peak S/N = 1100 as calculated from Eqs. (8) and (9) assuming $\Gamma_R = 5$cm^{-1} $\Delta t = 1.0$ sec, and $I_{dark} = 10$ Hz. Clearly it can be concluded that at least for some molecules with strongly allowed electronic transitions in the visible and ultraviolet regions that RRS does provide enough sensitivity improvement to reliably implement SRS. One word of caution should be added, however, and that is that we assume that when employing RRS to enhance the SRS signal, the sample can withstand the photon fluxes involved. This is a more stringent assumption than in the NRS case since we are now irradiating a strong electronic transition and the excitation energy is being degraded to heat via nonradiative processes.

TABLE 2

Integrated SRS Intensities for Various Resonance Raman Scatterers Adsorbed on Pt(111) [a]

Molecule	$\bar{\nu}_R$ cm^{-1}	$\bar{\nu}_L$ cm^{-1}	$(d\sigma/d\Omega)_{RR}$ $cm^2\ sr^{-1}\ molecule^{-1}$ [b]	$[I_{if}(\bar{\nu}_s)]_{RR,surf}$ Hz [c]
NO_2	1320	21990	5.6×10^{-27}	5.2×10^4
O_3	1103	34892	1.1×10^{-26}	6.4×10^4
I_2	213	20492	4.4×10^{-28}	4.4×10^3
I_2	213	18315	1.7×10^{-24}	1.9×10^7

a. $N_{surf} = 1.5 \times 10^{15}$ *molecules* cm^{-2}.

b. *RR cross sections are taken from reference (57).*

c. *All other parameters in Eq.(6) have the same values used in the calculation for CO.*

III. THE SURFACE ENHANCED RAMAN EFFECT

A.   *The First Successful Observation of SRS from Molecular Adsorbates on Metals*

As indicated in the previous section, the most attractive strategy for implementing SRS based on the currently known principles of Raman spectroscopy was to employ the RR effect. While various laboratories, including our own, were pursuing this approach, a totally unexpected development occurred. In a series of pioneering papers (29, 58-61), Fleischmann, *et al.* succeeded in observing quite high quality Raman spectra for several molecular adsorbates on metal surfaces. The unique feature of these experiments, as compared to previous attempts to observe SRS for adsorbates on metals (28), was that the metal surface was immersed in an aqueous electrolyte solution. Furthermore, the metal surface acted as the

working electrode in a three-electrode electrochemical cell
(62). The working electrode potential was controlled with
respect to a reference electrode (e.g., the saturated calomel
electrode (SCE) which is +0.246 V vs. the normal hydrogen
electrode (NHE)) with an operational amplifier circuit known
as a potentiostat (i.e., a device for maintaining a controlled
potential) (63). The adsorbate/electrode systems studied
by this group included: Hg_2Cl_2/Hg(Pt), Hg_2Br_2/Hg(Pt), and
HgO/Hg(Pt) (58); pyridine/Ag (29, 59, 61); and pyridine/Cu
(60). Hg(Pt) refers to amalgamated platinum electrode and
Ag and Cu were polycrystalline wire or foil electrodes.
Typically 50-100 mW of 5145 Å and 4880 Å argon ion laser power
was backscattered from these samples resulting in SRS signals
with S/N $\geq$ 10.

In the cases of Hg_2Cl_2, Hg_2Br_2, and HgO on Hg(Pt),
useful Raman spectra were obtained only for multilayer (5-10
layers) deposits even though mercurous and mercuric salts are
extremely strong NR scatterers. This work then cannot qualify
as a successful SRS experiment.

Since the sensitivity of NRS was not apparently great
enough to give high quality spectra for monolayers of mercury
salts on the smooth Hg(Pt) electrode, Fleischmann, *et. al.*
reasoned that monolayer sensitivity could be achieved if the
surface area of the electrode were increased by ca. a factor
of 10 so that the number of Raman scatterers intercepted by
the focal area of the laser beam was correspondingly increased.
Silver was selected as the next electrode material to try
because: (1) it is a typical "soft" metal which is expected
to strongly adsorb "soft" (i.e., highly polarizable and therefore strong Raman scattering) adsorbates (64, 65), and (2) the
surface area of silver can readily be increased by an electrochemical roughening procedure. This roughening procedure,
electrochemical anodization (oxidation), involved the application of a cyclic triangular voltage waveform to sweep the
potential of the Ag working electrode from -0.30 V vs. SCE
to + 0.20 V vs. SCE at a rate of 0.5 V sec^{-1}. At potentials
more positive than ca. +0.08 V vs. SCE, Ag metal is oxidized
to Ag^+ which in the presence of Cl^- (viz., 0.1 M KCl/H_2O was
the supporting electrolyte solvent system used in these
experiments) forms a AgCl film on the electrode. Cycling back
to potentials more negative than ca. -0.22 V vs. SCE results
in the reduction of AgCl to Ag metal. After 450 such
anodization cycles the Ag electrode had been roughened, its
surface area was estimated to have increased by at least a
factor of 10, and the color of the electrode had dramatically
changed from that of a reflective,metallic color to a pale
cream (perhaps semiconducting) color. Hereafter this roughening procedure will be known as the triangular waveform

anodization method. Pyridine (pyr) was selected as the test
adsorbate because (1) it is a strong Raman scatterer (see
Table 1); (2) its adsorption at the solid (i.e., SiO_2 and
Al_2O_3)/gas interface has been extensively studied by infrared and normal Raman spectroscopies (31, 32); and (3) it was
known from previous electrochemical studies (66) to adsorb
from aqueous electrolyte solutions on to Ag electrodes with
a saturation coverage of ca. 2×10^{14} pyr cm^{-2} at a bulk
pyridine concentration of $3.5 \times 10^{-5}M$.

The Raman spectra reported by Fleischmann and co-workers
(29) for the 0.05M pyr/Ag/0.1 M KCl/H_2O system are shown
in comparison with the NR spectra of neat pyridine and 0.05 M
aqueous pyridine in Fig. 1. There are two particularly
noteworthy features of the adsorbed pyridine spectra. First,
the intensities are quite high (ca. 10^3 Hz/cm^{-1}) considering
that only ca. 100 mW of 5145 Å laser power was incident on
the electrode. Second, the intensities and perhaps the
frequencies of the Raman bands are dependent on the applied
electrode potential. At this point let us concentrate on
the intensity of these bands while deferring the discussion
of the significance of the potential dependence. Apparently
in their original paper Fleischmann *et al.* did not find the
intensities particularly remarkable since they attribute their
success in observing these bands primarily to increasing the
electrode surface area by the triangular wave anodization, and,
perhaps secondarily, to the possibility that they were in fact
observing multilayers of adsorbed pyridine since the bulk
concentration of pyridine in their experiments was 1430 times
greater than the bulk concentration required for monolayer
coverage in the experiments of Conway and Barradas (66).
Eqs. (6) and (8) can be used to semiquantitatively compare
Fleischmann's results with our *a priori* expectations of
surface Raman intensities. This comparison is considered to
be semiquantitative only because many of the values of the
instrumental parameters required by Eq. (6) were not specified
in reference 29. We assume then that Ω, Q, T_m, and T_o are
closely approximated by the values used to calculate the
SRS intensity for CO on Pt(111). Eq. (6) should be slightly
modified to take into account a surface roughness factor, R,
and the possibility that, m, monolayers of pyridine are
actually being sampled by the deeply penetrating 5145 Å laser
beam. This modification is accomplished by replacing N_{surf} A
in Eq. (6) with mN_{surf} times RA_{geo}. where A_{geo}. is the
geometric focal area of the laser beam. Using an incident
laser power of 20 watts cm^{-2} (i.e., 100 mW) and R = 10,
one calculates that $[I_{if}(\bar{\nu}_s)]_{surf}$ = 22.0 Hz. This corresponds to I_{max} = 1.40 Hz/cm^{-1} if Γ_R = 5 cm^{-1} as before.
Clearly the experimental intensities are on the order of

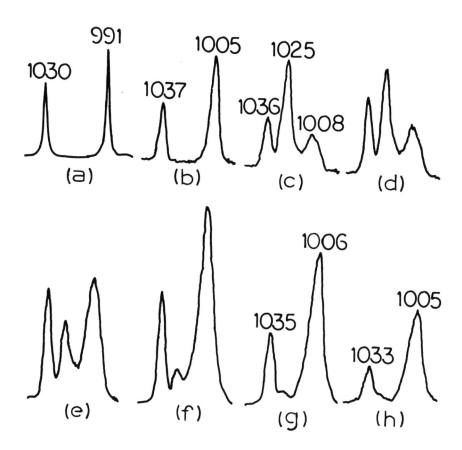

FIGURE 1. *Raman spectra of pyridine in solution and adsorbed at the silver-0.10 M KCl/H$_2$O interface. (a) liquid pyridine; (b) 0.05 M aqueous pyridine; (c) Ag electrode potentiostatted at 0.00 V vs. SCE; (d) -0.20 V vs SCE; (e) -0.40 V vs. SCE; (f) -0.60 V vs SCE; (g) -0.80 V vs. SCE; (h) -1.00 V vs. SCE (29). Frequencies given in cm^{-1}.*

10^3 times greater than the calculated intensities based on assumption that mR = 10.

The analogous 0.05 M pyr/Cu/0.1 M KCl/H$_2$O system showed very much weaker intensities compared to the Ag system. It was concluded (60) that the intensity of the observed lines at 1005 cm^{-1} and 1037 cm^{-1} could be accounted for entirely by scattering from the 0.05 M bulk aqueous pyridine. A new Raman feature at 1015 cm^{-1} that is not present in the bulk

spectrum was, however, observed to grow in as the Cu
electrode was potentiostatted at potentials between +0.1 V vs.
SCE and +0.3 vs SCE. This band corresponds to the 1025 cm^{-1}
band observed in the Ag system and may represent a signal
from a pyridine monolayer or multilayer on the Cu electrode.

These early experiments of Fleischmann and co-workers,
especially those for the pyr/Ag system, were quite thought
provoking. Viewed pessimistically the intensity disparity
between calculation and experiment could be explained by
errors in estimating the product mR (i.e., these are not
actually SRS experiments). On the other hand, when viewed
optimistically the intensity disparity can only be explained
by: (1) a "hidden" resonance Raman enhancement or (2) some new
type of Raman enhancement mechanism. The RRS explanation for
the intensity disparity is described as "hidden" because
pyridine itself absorbs only at wavelengths shorter than ca.
280 nm (67) so that it is not immediately apparent why one
should expect a resonance enhancement mechanism to be operative at laser excitation wavelengths as long as 514.5 nm.
To be general we have labeled any new type of Raman enhancement mechanism, which occurs only for molecules adsorbed
on surfaces, as a surface enhanced Raman (SER) effect.

Although our original research goal had been to conduct
resonance enhanced SRS experiments, we decided in the fall
of 1975 to carry out some exploratory studies similar to
those of Fleischmann and Hendra while our surface resonance
Raman spectrometer was being completed. The goal of these
exploratory studies was to determine whether the pessimistic
or the optimistic assessment of the pyr/Ag SRS experiment was
correct. The experimental program that was planned included
both a verification of the Fleischmann-Hendra experiment
and, if that were successful, a systematic evaluation of the
variables that controlled the observed intensity. Section
III.B will discuss the verification experiments and Section
III.C will outline some of the fundamental electrochemical
variables and surface phenomena that could be responsible for
the SRS intensity.

B. *Verification of the Fleischmann-Hendra Experiment*

1. *Apparatus.* The verification measurements on the
0.05 M pyr/Ag (polycrystalline)/0.1 M KCl/H$_2$O system were
carried out in a manner similar to but not identical with
that described by Fleischmann, *et al.* (29, 58-61). The
electrochemical cell used for our SRS studies on polycrystalline wire electrodes is shown in Fig. 2. The electrochemical properties of this cell do not differ in any signifi-

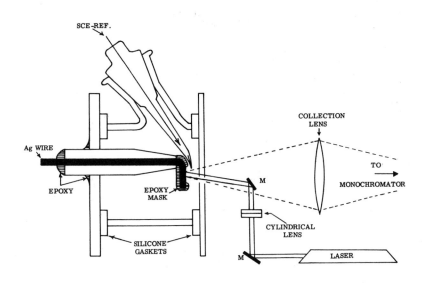

*FIGURE 2. Surface Raman electrochemical cell for use
with polycrystalline wire electrodes and optics associated
with the backscattering geometry.*

cant way from the cell pictured in ref. (29). The Ag working
electrode is masked to an area of ca. 0.05 cm^2 and is placed
as close as possible (viz., 1-2 mm) to the front window of
the cell without distorting the current distribution due to
"thin-layer" effects (68). The distance from the electrode
surface to the window is not a critical parameter unless
the intervening solution contains a Raman scatterer with
sufficiently large cross section to produce an interfering
signal. The third electrode of this three-electrode cell
(i.e., the auxiliary electrode) is simply a platinum wire
coil which surrounds the working and reference electrodes
(but does not obstruct the collection lens view of the
working electrode surface) and enters the cell perpendicular
to the plane of the drawing. The entire cell is clamped
together with an external metal frame (not shown in Fig. 2)
using silicone rubber gaskets. The main advantage of this
"sandwich" type of cell construction is that it can be

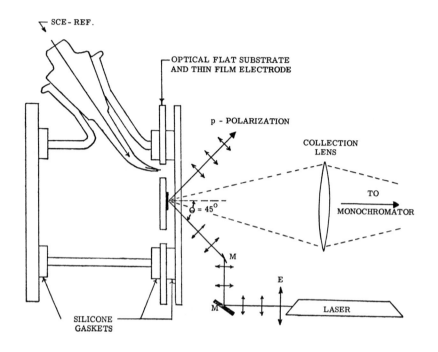

FIGURE 3. Surface Raman electrochemical cell for use with vapor deposited thin film electrodes and optics associated with the backscattering geometry.

readily modified to accept working electrodes with different geometries. Such a modification to adapt the cell to a vapor deposited thin film electrode is shown in Fig. 3.

All SRS experiments carried out in our laboratory are done in backscattering geometry with the laser focused to a line image using a cylindrical lens. Line focus is recommended so that the monochromator slits can be filled and so that the laser power density is reduced compared to a spot focus at the electrode surface. The lower laser power intensity should reduce the possibility of thermal damage to the sample. Typical experiments are done with p-polarized light and an angle of incidence, θ, of approximately 45°. Preliminary experiments have shown that the SRS experiments are not terribly sensitive to the laser polarization or the angle of incidence when one uses rough polycrystalline materials as electrodes. We anticipate greater sensitivity to these optical variables on smooth, single crystal or epitaxial thin film electrodes.

A more detailed schematic view of the electrochemical equipment, laser system, monochromator, detection electronics, and computer system used in our SRS experiments is shown in Fig. 4. The potentiostat is of conventional design (63)

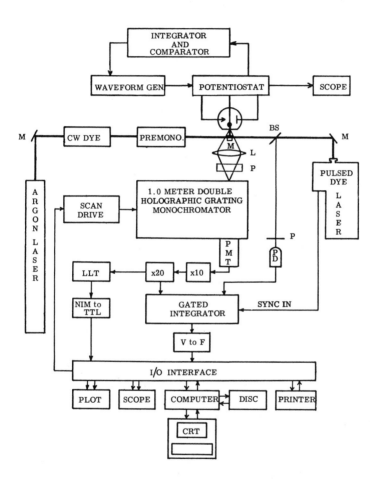

Figure 4. Schematic diagram of the surface Raman spectroelectrochemistry system.

and although this one was "homebuilt" there are similar devices available commercially. The integrator/comparator system permits one to pass a predetermined amount of charge through the electrochemical cell under controlled potential conditions. This is a desirable feature since charge, not current, is the electrochemical quantity directly related by Faraday's law to the number density of electroactive species undergoing the particular electrode process selected by the value of the applied potential. CW argon ion and dye lasers as well as flashlamp-pumped, pulsed dye lasers are used for the SRS experiments. This combination of lasers permits one to select any wavelength of exciting radiation in the range 440.0 nm to ca. 750.0 nm. In addition the argon ion laser can be operated at two discrete uv wavelengths (viz., 351.1 nm and 363.8 nm) and a limited amount of tunable uv power is available by intracavity frequency doubling of the flashlamp pumped dye laser (viz., 1-10 mW from 280.0 nm to ca. 340.0 nm). When the dark current is much less than the signal, the S/N of pulsed dye laser excited SRS is essentially identical to that of CW dye laser excited spectra provided that the same average power is delivered to the sample. Pulsed operation does allow the elimination of dark current through gated detection, and thus S/N improvement for weak signals. All optics are SlUV grade quartz to permit uv operation of the system. In addition a coated, chromatic aberration corrected camera lens is used as the main collection lens to optimize performance in the visible. The double monochromator and computer data acquisition system are standard devices.

   2.   *Results.*   Figure 5 shows a repeat of the Fleischmann-Hendra experiment done in our laboratory. The existence of SRS for the pyr/Ag system is corroborated. All conditions used to obtain the data in Fig. 5 were the same as those in the original experiment except that $\bar{\nu}_s$ was scanned over a much broader range (viz., 100 cm^{-1} to 1700 cm^{-1} in our work as compared to 1000 cm^{-1} to 1045 cm^{-1} in their work) and a different anodization procedure was used to pretreat the electrode surface. In our experiments the Ag electrode surface was prepared by mechanical polishing with a 600 mesh Al_2O_3/H_2O slurry, air drying after a distilled water rinse, and immersion in the electrochemical cell followed by the application of a single square voltage waveform (a double potential step), from an initial potential of -0.60 V vs. SCE to +0.20 V vs. SCE. The actual value of the initial potential is not critical as long as it is negative of the potential at which AgCl starts to form. The time duration of the voltage pulse is controlled by the integrator/comparator. When a charge corresponding to 25.0 millicoulomb cm^{-2} (other values of charge can be preset although an empirical study showed that the most intense SRS were obtained after passage of this amount of charge) of AgCl has passed through the

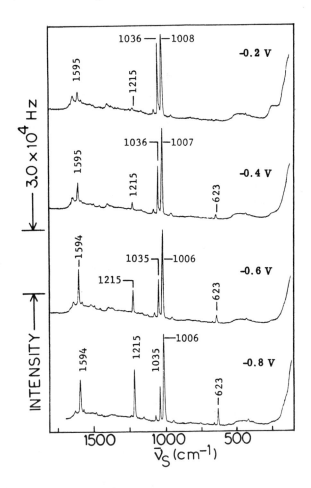

FIGURE 5. Raman spectra of 0.05 M pyr/Ag/0.1M KCl/H₂O
as a function of applied electrode potential (vs. SCE).
Laser power = 80 mW at 514.5 nm. Electrode surface prepared
by a single double potential step anodization. Spectra were
recorded with 2 cm⁻¹ bandpass and a 1.0 sec counting
interval (33).

cell, the voltage output of the integrator trips the preset
voltage of the comparator and the potential applied to the
cell is reset to its initial value. Typically 99.9% of the
charge passed in the AgCl formation step is recovered in the
reduction step indicating that the anodization process is
reversible under these conditions.

The main feature to note in comparing the experimental results shown in Figures 1 and 5 is that although much less surface roughening was done in the single double potential step anodization as compared to the 450 triangular wave anodization cycles, the surface Raman intensities in Fig. 5 are 30-100 times greater than those in Fig. 1 d-g. This apparent inverse correlation between surface roughness and surface Raman intensity argues strongly against Fleischmann's original high surface area explanation for the observation of such strong Raman signals. The unusual behavior of this Raman scattering system is further emphasized by comparing the experimental intensity of the 1008 cm^{-1} totally symmetric breathing mode in Fig. 5 with the calculated intensity from Eqs. (6) and (8). With the incident laser power density of 16 watts cm^{-2} used in our experiment, mR = 1, and all other parameters the same as the previous calculations, Eq. (6) predicts that $[I_{if}(\bar{\nu}_s)]_{surf}$ = 1.76 Hz and Eq. (8) gives I_{max} = 0.112 Hz for Γ_R=5 cm^{-1}. Thus, the experimentally found intensity of the 1008 cm^{-1} line of adsorbed pyridine which is 4.1 x 10^4 Hz is about *3.5 x 10^5 times greater* than the calculated intensity. Similarly if one compares the surface Raman intensity for adsorbed pyridine with the normal Raman intensity of pyridine in solution, one finds that the surface signals are possibly as much as 10^6 times more intense (33). These enormous intensity enhancements are far too large to be simply explained away on the basis that the value of mR is orders of magnitude greater than 1-10.

The discovery of these extremely intense surface Raman signals and the fact that the key factor in their production was the single anodization pulse as compared to the surface roughening procedure of Fleischmann was first reported in May 1976 (69) and subsequently published (33,70). Since that time several other laboratories have also verified these observations. Albrecht and Creighton (34) independently reported intensity enhancements on the order of 10^5 for pyr/Ag. Subsequently Yeager, *et al.* (71) reported similar observations and Pettinger (72) recently has gone on to show that similarly intense Raman spectra can also be obtained from single crystal Ag (100) and Ag(111) surfaces.

The verification of the existence of these anomalously intense or surface enhanced Raman spectra from pyridine adsorbed on Ag electrodes by several independent laboratories now raises some key questions.

1. What experimental variables control the intensity of the SERS effect and how can one optimize S/N in such an

experiment?

2. Are these signals true surface Raman spectra as defined by the criterion that a maximum of one monolayer of Raman scatterers is involved?

3. What is the molecular information content of SERS?

4. Is SERS general for all molecules, metals, and surface environments or is it limited to certain molecules on certain metals immersed in the electrochemical environment?

5. What is the origin or mechanism of the intensity enhancement?

6. What new kinds of experiments or applications of SERS can be suggested based on the predictive capability of a theory of SERS?

Section IV will survey the experiments performed in several laboratories that are directed toward answering these questions. However since the design of many of these experiments, particularly those directed toward optimizing the SRS S/N and proving that these spectra are true SRS, is dependent on a knowledge of certain properties of electro-chemical interfaces and adsorption at such interfaces, that discussion will be preceded by a brief review of the relevant features of interfacial electrochemistry.

C. *The Structure of and Adsorption at Electrochemical Interfaces*

The solid(electrode)-liquid(electrolyte) interface is an intrinsic feature of any electrochemical experiment. The observation of the strongly enhanced Raman scattering from molecules adsorbed on an electrode in an electrochemical cell is, to say the least, unusual. Therefore, it is quite likely that the special properties of the electrode interfacial region strongly influence the Raman observables and may, in fact, be responsible for the existence of SERS. Consequently the design and interpretation of experiments to be used in probing for the SERS mechanism should at least in part be based on a knowledge of the properties of the electrode-electrolyte interface. The structure of electrochemical interfaces, their electrical properties, and the nature of adsorption at these interfaces has been extensively studied by electrochemists. Several excellent reviews of this material are available (73-79) and should be consulted for more detailed information than can be presented here.

1. *Basic Structural Features and Electrical Properties
of the Electrode-Electrolyte Interface.* In the SERS experiments discussed above, a Ag/0.1 M KCl/H$_2$O interface is involved. To begin, imagine that the Ag electrode, after
mechanical polishing and the anodization pretreatment, is
potentiostatted at a value of the applied potential, $E = \phi_m$,
such that the surface of the metal takes on a net positive
charge (i.e., $q^M = +$). In such a case the positive electrode
attracts the anions of the electrolyte because of long-
range electrostatic and short-range van der Waal's forces.
In the simplest situation the anion and its surrounding solvation sheath come into contact with the electrode at a distance of closest approach corresponding to the radius of
the anion plus the diameter of the solvent. Such ions are
said to be nonspecifically adsorbed. In other cases the
anions may actually bond to the electrode with the orbitals
of the adsorbate interacting directly with metal orbitals
protruding out into solution. Ions interacting in this way
are said to be specifically adsorbed. Solvent molecules
play a prominent role in the detailed description of the
electrode-electrolyte interface. If the interaction between
the positively charged electrode and the electrolyte anion
is sufficiently strong, solvent molecules will be displaced
from the interface where they may have been weakly adsorbed
and partially oriented due to permanent dipole interactions
with the electrode. In addition, specifically adsorbed anions
will probably lose part or all of their solvation sheaths as
they become bonded to the electrode. Anions that are specifically adsorbed therefore can approach the electrode more
closely than nonspecifically adsorbed anions with their
solvation sheaths. As a result a layer of essentially non-
solvated anions will accumulate at a distance x_1 from the
electrode surface forming what is known as the inner
Helmholtz layer or plane (IHP) and a layer of solvated anions
will accumulate at a distance x_2 forming the outer Helmholtz
plane (OHP). The combination of the metal surface charge
and the IHP and OHP is called the compact double layer.
Moving farther away from the electrode surface, one enters
a three-dimensional region, the diffuse double layer, which
extends from the OHP out into the bulk of solution. In the
diffuse layer there is an inhomogeneous distribution of electrolyte ions with a net non-zero charge. In the case where
$q^M = +$, the concentration of anions in the diffuse layer
is in excess with respect to the bulk and the concentration
of cations is in defect. The diffuse layer ion distribution
as a function of distance from the electrode surface is
determined by the competition between electrostatic ordering
and thermal agitation causing disordering. This entire

collection of charged ionic layers is loosely called the
electrical double layer and is pictured schematically in
Figure 6. Reversing the sign of q^M reverses the ordering of
anions and cations within the various double layers. At
some intermediate value of E known as the potential of zero
charge E_z, the charge of the electrode surface is exactly
balanced by the charge in the compact layer and represents
a convenient reference state for discussing the potential
dependent properties of the double layer.

Several electrical properties of the double layer
structure are relevant to the discussion of SERS. In particular we will focus on the interfacial capacitance and the
potential, ϕ, vs. distance, x, profile.

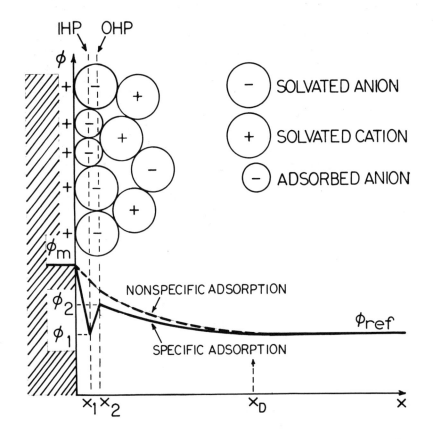

*FIGURE 6. Schematic model of the electrode-electrolyte
interface showing the potential vs. distance profiles for a
positively charged electrode in the case of nonspecific anion
adsorption as well as specific anion adsorption.*

The electrode-electrolyte interface can be regarded as
a molecular scale parallel plate capacitor. The electrode
surface plane and the OHP form the plates of the capacitor
and the ions and solvent contained within the compact layer
form the dielectric material. The magnitude of the double
layer capacitance in farad cm^{-2}, $C_{d.l.}$, can be estimated
from Eq.(13)

$$C_{d.l.} = \frac{K_D \, \varepsilon_o}{x_2}$$ (13)

where K_D is the dielectric constant of the material in the
compact layer, ε_o is the permittivity of free space $= 8.849$
$\times 10^{-14}$ farad cm^{-1} and x_2 is the distance of the OHP from
the electrode surface. For $K_D = 78$ (e.g., the value for
water) and $x_2 = 10\text{Å}$, one calculates that $C_{d.l} = 69$ µfd cm^{-2}
which compares favorably with experimentally measured values
that range from 20 to 50 µfd cm^{-2}. By typical electronic
component standards this is a huge capacitance. The charge,
$Q_{d.l.}$, required to establish a 1.0 volt potential drop across
the compact layer (i.e., $E - \phi_2 = 1.0$ V) is given by:

$$Q_{d.l.} = C_{d.l.} \, (E - \phi_2)$$ (14)

and is of the order of 20-50 µC cm^{-2} $volt^{-1}$ using the experimental values for $C_{d.l.}$.

The spatial distribution of the potential in the diffuse
layer can be calculated from the Gouy-Chapman-Stern (GCS)
theory (73). For a symmetrical electrolyte ($z_+ = z_- = z$) the
potential distribution is given by:

$$\phi(x) = \frac{4RT}{ze} \, tanh^{-1} \{exp \, [p - \kappa x]\}$$ (15)

the Debye reciprocal length κ is:

$$\kappa = 2ze \, [2\pi C_b / \varepsilon RT]^{\frac{1}{2}}$$ (16)

and

$$p = ln \, \{tanh \, [ze \, \phi_2 / 4RT]\} + \kappa x_2$$ (17)

where ε is the solvent dielectric constant, C_b is the bulk
concentration of the electrolyte, ze is the charge on the
electrolyte ions, and ϕ_2 is the potential at the OHP. The

shape of this profile for the case of nonspecific anion adsorption is shown qualitatively in Figure 6. For small values of ϕ_2 or large values of x, Eq. (15) reduces to:

$$\phi(x) \cong \phi_2 \, exp \, (- \kappa x) \qquad (18)$$

Equation (18) is convenient for estimating the thickness of the diffuse layer. For example the thickness, x_D, within which 99.99% of the $\phi_2 - \phi_{ref}$ (ϕ_{ref} is the potential of the reference electrode) potential drop occurs for a 0.1 M KCl solution is ca. 88 Å.

The value of ϕ_2 as a function of E can be determined from another GCS theory result which relates ϕ_2 to q^M:

$$\phi_2 = \frac{2RT}{zF} \, sinh^{-1} \, [\frac{q^M}{(8RT\epsilon C_b)^{\frac{1}{2}}}] \qquad (19)$$

q^M is in turn related to an experimental observable, $C_{d.l.}$ by:

$$q^M(E) = \int_{E_z}^{E} C_{d.l.} \, dE \qquad (20)$$

Thus, from a set of $C_{d.l.}$ measurements made as a function of E, one can numerically integrate to obtain q^M as a function of E and then calculate ϕ_2 vs. E from Eq. (19). The form of the ϕ_2 vs. E curve for a symmetrical electrolyte/Hg interface is shown in Figure 7 as a function of C_b. So far similar measurements have not yet actually been made for the Ag/KCl interface used in the SERS experiments.

There is no adequate theory for calculating the potential distribution in the compact layer. However, once ϕ_2 is known from GCS theory, one can simply assume a form for the ϕ vs. x profile to connect ϕ_m with ϕ_2. A linear form was assumed for the purpose of constructing Figure 6. Since the range of $E-E_z$ used in the SERS experiments is +0.7 V vs. SCE to -0.3 V vs. SCE (E_z for Ag is ca. -0.7 V vs. SCE (80)), ϕ_2 can be estimated from Figure 7 as -0.10 V < ϕ_2 < 0.10 V. Thus we can conclude that the voltage drop, $\phi_m - \phi_2$, across the compact layer is on the order of 0.1 - 0.5 volts across 10 Å, so that the magnitude of the electrode field strength, $d\phi/dx$, (assuming a linear ϕ vs. x profile) is ca. 1-5 x 10^6 volt cm^{-1}.

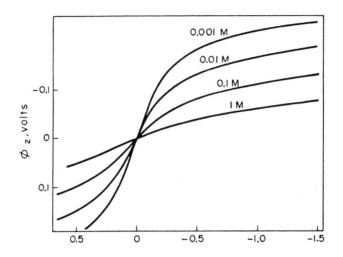

FIGURE 7. Potential drop across the diffuse layer as a function of applied potential. The horizontal potential scale is $(E-E_z)$ in volts. The parameters for the different curves are values of C_b (74).

To summarize, we have reviewed the following properties of an electrochemical interface such as the Ag/KCl interface used in SERS: (1) the interface has a large capacitance requiring ca. 20-50 μC cm^{-2} $volt^{-1}$ to charge it; (2) the potential applied to the metal electrode by the potentiostat, $E = \phi_m$, is dropped across the compact and diffuse double layers which extend out into solution a maximum of ca. 100 Å for 0.1 M KCl electrolyte; (3) most of the applied potential is dropped across the small space of the compact layer which is of the order of 10 Å thick; and (4) an enormous electrode field strength exists in the compact layer of the order of 1-5 x 10^6 volt cm^{-1}.

2. Adsorption at the Electrode-Electrolyte Interface. In the context of the SERS experiments, it is known experimentally that both Cl⁻ (81) and pyridine (66) are adsorbed from aqueous solutions at the Ag electrode. The preceding discussion of the properties of the electrical double layer took into account the specific adsorption of Cl⁻ but did not consider the adsorption of neutral molecules such as pyridine. Neutral molecules are adsorbed on electrode surfaces primarily because of hydrophobic forces (65). In general the less soluble an organic molecule is in the aqueous medium, the

more strongly it is adsorbed. Pyridine also has the possibility for actual covalent bonding to the vacant orbitals
on the Ag surface via the nitrogen lone-pair electrons. This
combination of hydrophobic and covalent bonding driving forces
for adsorption will cause pyridine to enter the compact
double layer and some fraction of those pyridines will be in
direct contact with the Ag electrode surface. From the extensive literature pertaining to neutral organic adsorption
on Hg electrodes, it is known that the double layer parameters E_z, $C_{d.1.}$, and ϕ_2 are all likely to be functions of
the fractional electrode coverage, θ, of the neutral organic
(82, 83). $\theta = \Gamma/\Gamma_s$ where Γ is the surface excess of the
adsorbate in moles cm^{-2} and Γ_s is the value of Γ at saturation coverage. In general E_z shifts to more negative potentials, $C_{d.1.}$ decreases, and ϕ_2 is decreased (i.e., the
functional form of the ϕ_2 vs. $E-E_z$ plot is the same as that
shown in Figure 7 (74) but all the plots are flattened out)
as θ increases. Although the quantitative features of the
electrical double layer are altered by the adsorption of a
neutral organic in a way that has not yet been experimentally
measured for the Ag/pyr/KCl system, the qualitative aspects of
the picture presented in Section III.C.1. above should remain
valid.

Equilibrium adsorption at electrodes is characterized by
an adsorption isotherm in much the same way that adsorption
at the solid-gas interface is characterized. The general
form of this adsorption isotherm is:

$$f(\theta) = \beta a \qquad (21)$$

where $f(\theta) = \theta(1-\theta)^{-1}$ if there are no interactions among the
adsorbed molecules except that the occupancy of an adsorption
site by one molecule denies that site to all other molecules
(i.e., a Langmuir isotherm); a is the activity of the adsorbate in the bulk of solution which is usually approximated
by its bulk concentration, C; and β is the adsorption equilibrium constant which is related to the standard free energy
of adsorption, ΔG^o_{ads}, by:

$$\beta = exp[-\Delta G_{ads}/RT] \qquad (22)$$

There are numerous other forms of $f(\theta)$ that can be used to
describe electrosorption (76). For example, if the free
energy of adsorption is not independent of θ due to adsorbate-
adsorbate interactions, the Frumkin isotherm

$$f(\theta) = (1-\theta)^{-1} exp (A\theta) \qquad (23)$$

may yield a more accurate description of the adsorption
process. A in Eq. (23) is a constant whose sign and value
depend on whether the adsorption process is dominated by repulsive or attractive adsorbate-adsorbate interactions.
Figure 8 shows the form of these particular isotherms in
terms of θ vs. $\log_{10} C$ plots.

The feature that most strongly distinguishes electrosorption from other adsorption phenomena is that ΔG^{O}_{ads} is
a strong function of the externally applied cell potential,
E. Anson (65) has proposed an adsorption classification
scheme based on the morphology of Γ vs. E curves. Within
the framework of this scheme, Cl^- is a class IB adsorbate
since Γ_{Cl^-} has been found from ellipsometry studies to
decrease monotonically as E is made more negative (80). This
data is shown in Figure 9. Although Γ_{pyr} vs. [pyr] measurements have been made (66) the form of the Γ_{pyr} vs. E plot
at constant [pyr] does not appear to have been experimentally
determined. We assume, however, that pyridine is a typical
class II adsorbate and has a Γ_{pyr} vs. E. characteristic
similar to that shown schematically in Figure 10. The fact
that class II adsorbates tend to exhibit their largest adsorption near the potential of zero charge and exhibit
decreasing adsorption in both the negative and positive
directions can be qualitatively understood by recalling the
role of solvent in the overall adsorption equilibrium. In
order for pyridine to adsorb, it must displace water molecules from the electrode surface. At high surface charge
densities, either positive or negative, the interaction
energy of the water dipoles with the interfacial electric

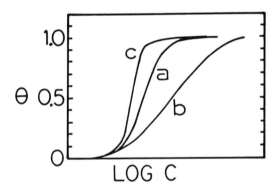

FIGURE 8. Adsorption isotherms: (a) Langmuir isotherm;
(b) Frumkin isotherm with repulsive interactions between adsorbed molecules; (c) Frumkin isotherm with attractive interactions between adsorbed molecules (65).

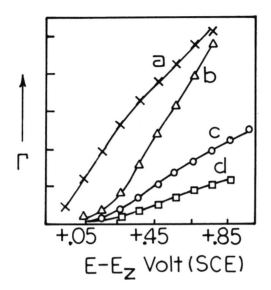

FIGURE 9. Adsorption of anions from 10^{-2} M aqueous solutions on a silver electrode: (a) Br^-; (b) Cl^-; (c) $SO_4^=$; (d) ClO_4^- (81).

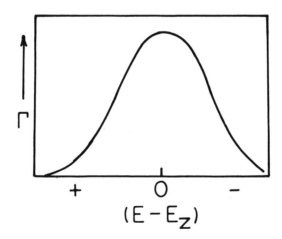

FIGURE 10. Schematic dependence of the adsorption of neutral organic adsorbates on the applied potential referenced to the potential of zero charge (65).

field becomes so large that water is preferentially adsorbed
and therefore pyridine is desorbed.

Another feature of electrosorption that should be
mentioned is coadsorption (84). In a number of electrosorption systems, it has been found experimentally that the
free energy of adsorption for one component is a function of
the coverage of another component. In the SERS system it is
certainly possible, based on other studies of anion induced
adsorption (65), that $\Delta G^O_{ads,pyr}$ is a function of θ_{Cl^-} which
is in turn a function C_{Cl^-}.

To summarize then, we should anticipate that the
standard adsorption free energy of pyridine on a silver
electrode in chloride media could be dependent on θ_{pyr}, θ_{Cl^-},
and E. Furthermore one should remember that the overall
adsorption behavior of the SERS electrolyte system will be
a superposition of the competition between pyridine, water,
chloride ions, and probably even potassium ions for electrode
surface sites.

IV. SURVEY OF EXPERIMENTAL RESULTS AND DISCUSSION

This section reviews a number of recent experimental
and theoretical studies designed to develop a more complete
understanding of the scope and mechanism of the surface enhanced Raman effect. Since the case of pyridine adsorbed on
silver electrodes immersed in an electrochemical environment
is the most thoroughly studied to date, the bulk of the
data reported in this section deals with this model system.
Fortunately recent results show that in addition to the
pyridine/silver system, SERS can be observed with many other
molecules, a few other metals, and in environments other than
the electrochemical one. Otherwise interest in this phenomenon would be rather limited. In addition to the SERS
experiments, recent results concerning the application of
resonance enhanced Raman spectroscopy to the study of adsorption on surfaces will be reviewed. In most cases the
resonance enhancement takes place through an electronic
transition localized on the adsorbate molecule. Some data
will be presented to show that the surface Raman intensities
derived from the SERS effect and the RR effect can be added
together to achieve quite extraordinary surface sensitivity.

A. Surface Enhanced Raman Spectroscopy

1. *Optimization of Signal Intensity in the Pyridine/ Silver Model System.* In SERS experiments there are a large number of surface, chemical, electrochemical, and optical variables that can control the observed intensity of the Raman signals from the molecular adsorbate. The other Raman observables including vibrational frequency, depolarization ratio, linewidth, and line shape are probably functions of these variables also. From the studies that have been made to date (33,34, 69-72, 85,86), the following variables that control these observables have been identified: (1) method of electrode pretreatment-anodization; (2) metal substrate structure; (3) the concentrations and chemical nature of the electrolyte solution species; (4) the applied electrode potential; (5) laser power; (6) laser excitation frequency; (7) angle of incidence of the exciting laser; and (8) incident laser polarization. Mapping out the complete set of these dependencies for the Raman observables so that a unique optimization of the S/N in a SERS experiment can be achieved is clearly a formidable task and has not yet been completed. What has been done is to start with the experimental conditions used to record the spectra shown in Figure 5 and explore the range of each variable systematically. In addition to the goal of optimizing the S/N in SERS, these experiments serve to provide the data base for constructing a theory of the SERS effect. A summary of these optimization experiments follows.

a. *The effect of electrode anodization.* In comparing the results in Figures 1 and 5, which involved 450 triangular wave (TW) anodization cycles and one double potential step (DPS) anodization cycle respectively, it was noted that the DPS anodization procedure produced 30-100 times greater SERS intensity. In addition to these overall intensity differences, one also notices that intensity ratios such as that of the 1025 cm^{-1} band (Fig. 1) relative to the 1036 cm^{-1} and 1006 cm^{-1} bands appear to be sensitive to the anodization pretreatment. Several questions therefore arise from these observations.

1. What anodization conditions maximize the SERS intensity?

2. Is anodization absolutely necessary to observe SERS?

3. What is the function of the anodization pretreatment?

In order to determine the anodization conditions that
maximize the SERS intensity, several interrelated parameters
must be investigated including the shape of the anodization
waveform, the total charge passed in the AgCl formation step,
the anodic potential limit, and the solution conditions. For
polycrystalline Ag wire electrodes, we qualitatively investigated the effects of waveform shape, charge, and potential
limit for fixed solution conditions (viz., 0.05 M pyridine/
0.1 M KCl/H_2O) (33, 70). The SERS intensity was found to be
extremely sensitive to the total charge passed but relatively
insensitive to variations in waveform shape (provided that
the total charge passed was constant) and anodic potential
limit over the range +0.20 V vs. SCE to +0.10 V vs. SCE.
Only this small range of anodic potential was investigated
because at potentials more positive than +0.20 V vs. SCE
extensive, irreversible electrode discoloration (i.e.,
blackening) took place; whereas, at potentials more negative
than +0.10 V vs. SCE the amount of charge that could be
passed was so small that the time required to complete an
anodization with total charge on the order of 25 mC cm^{-2}
became excessively long. For a DPS anodization cycle with an

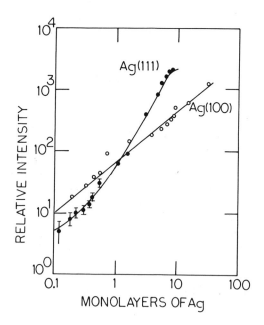

FIGURE 11. *Relative intensity of pyridine (1008 cm^{-1})
on Ag single crystal electrodes of (100) and (111) orientation as a function of the amount of Ag oxidized and reduced
in the anodization pretreatment (85).*

anodic limit of +0.15 V vs. SCE, the SERS intensity was found
to increase with the amount of charge passed up to ca. 30 mC
cm^{-2}, plateau until 50 mC cm^{-2}, and then decrease with further
charge transfer. Since one monolayer on a Ag(111) surface
contains va. 1.5×10^{15} atoms the charge passed corresponding
to a monolayer of Ag metal dissolved is 0.240 mC cm^{-2}. Thus
the 30 mC cm^{-2} charge that produced the maximum SERS intensity
corresponds to the dissolution of ca. 125 monolayers of Ag.
Charge recovery during the AgCl reduction step is always
monitored in our experiments; and,we find that for Ag dissolution charges up to 50 mC cm^{-2}, greater than 99.9% is
recovered so that at most 0.1% of the number of Ag monolayers
dissolved remain on the surface as AgCl after the anodization.
Pettinger (85) has studied the effect of anodization charge
transfer on the SERS intensity of the 1006 cm^{-1} pyridine line
in a more quantitative way for single crystal Ag(100) and
Ag(111) electrodes. His results are qualitatively consistent
with ours on polycrystalline electrodes, and they are shown
in Figure 11.

The question of whether or not anodization is a prerequisite for the observation of SERS has been addressed by
examining Ag electrodes that have only received the mechanical
polishing pretreatment. Figure 12a shows the Raman spectrum
of adsorbed pyridine on an unanodized electrode. This electrode was mechanically polished as described above and
immediately transferred to the electrochemical cell and
potentiostatted at -0.6 V vs. SCE. The intensity of the
1006 cm^{-1} pyridine line is 350 counts per second above background corresponding to an enhancement factor of ca. 10^4 with
respect to the 0.01 M pyridine in the bulk electrolyte solution. That we are seeing adsorbed pyridine rather than a
signal from the bulk is demonstrated in Figure 12b. Changing
the applied potential from -0.6 V vs. SCE which is near E_z
where pyridine is maximally adsorbed to 0.0 V vs. SCE which
produces a positive q^M and maximally adsorbs Cl^- displacing
pyridine causes the SER signal on the unanodized electrode
to disappear below background. The response of alleged SERS
signals to applied electrode potential is a sensitive test of
their surface nature since the influence of the potential
extends to a maximum distance of ca. 100 Å from the electrode
surface as discussed above.

Since it is apparently not absolutely necessary to
anodize the Ag electrode in order to observe SERS, the
anodization must serve the role of enhancing the intensity
in some way. An hypothesis that we have pursued is that
anodization in the electrochemical SERS experiment is the
analog of the Ar^+ sputtering technique used to clean metal
surfaces in UHV surface spectroscopy experiments. To test

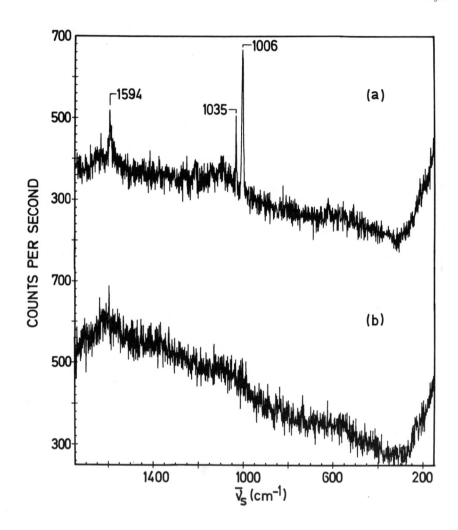

FIGURE 12. Raman spectrum of 0.01 M pyr/Ag/0.1 M KCl/H$_2$O. Laser power = 60 mW at 514.5 nm. Electrode surface was prepared by mechanical polishing only: (a) -0.6 V vs. SCE; (b) 0.0 V vs. SCE.

this hypothesis we have used AES to examine three Ag electrodes subjected to the following pretreatments: (1) mechanical polishing only; (2) mechanical polishing plus a 20 mC cm^{-2} anodization cycle; and (3) mechanical polishing plus a 300 mC cm^{-2} anodization cycle. AES was used to monitor the surface contamination. On the unanodized surface AES showed the presence of S, C, Ca, and O in addition to Ag; and the C/Ag and

O/Ag ratios were 2.3 and 0.3 respectively. After the 20 mC cm^{-2} anodization the C/Ag = 1.5 and the O/Ag = 0.2 showing that the anodized electrode was "cleaner" in the sense that a greater fraction of the surface atoms were Ag. Further anodization leads to reduction of C/Ag and O/Ag to values of 1.1 and 0.1 respectively. Thus we conclude that one of the functions of anodization is to increase the surface Ag concentration by the redeposition of fresh Ag during the AgCl reduction step. Another function of anodization appears to be surface roughening. Scanning electron microscope photographs of these electrodes show a slight increase in surface roughness (perhaps as much as a factor of 2) in comparing the 20 mC cm^{-2} anodized electrode to the unanodized electrode. A large increase (ca. a factor of 20-30) in surface roughness is observed in comparing the 300 mC cm^{-2} anodized electrode to the unanodized electrode.

To summarize, our conclusions about the effect of electrode anodization are. (1) anodization is NOT a prerequisite for observing SERS; (2) the SERS intensity can be

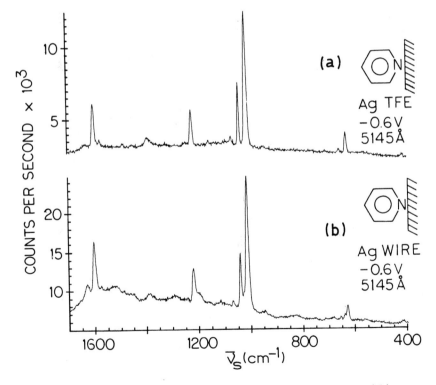

FIGURE 13. *Comparison of SERS for adsorbed pyridine on polycrystalline wire and evaporated thin film electrodes*

increased by factors of ca. 100 in going from an unanodized
electrode to an optimally anodized electrode which requires
ca. 25-30 mC cm^{-2}; and (3) it appears likely that this
increase in SERS intensity is related to some combination
of "cleaning" and roughening.

   b. The effect of Ag substrate structure. SERS for
pyridine have now been observed on polycrystalline Ag wire
or foil (33,34,69-71, 86), polycrystalline Ag thin film electrodes (TFE) formed by high vacuum vapor deposition on glass
substrates (87) as shown in Figure 13, Ag(100) and Ag(111)
single crystals (72,85,88) (see Figure 14), and Ag(111)
epitaxial TFE formed by high vacuum vapor deposition on mica
substrates (72). Only slight changes in the pyridine SERS
are observed in comparing the spectra on polycrystalline
wires vs. the TFE. The intensities on the TFE sample are
ca.a factor of 2 less than on the bulk wire sample probably
due to the smaller surface roughness factor on the TFE.
Pettinger's studies on the silver single crystal electrodes
show that the SERS intensities are comparable to those on
polycrystalline electrodes and that the Raman observables
are quite sensitive to crystal orientation. Furthermore
Pettinger has shown that the polycrystalline SERS can be
interpreted in terms of a superposition of the Raman observables for pyridine adsorption at surface sites with different
orientation.

   *c. The effect of the concentrations and chemical nature
of the electrolyte solution species.* In order to determine
the electrolyte solution conditions that optimized the SERS
intensity, we studied the effect of the bulk concentration
of pyridine and Cl$^-$ on the intensity of the strongest pyridine band -- the 1006 cm^{-1} totally symmetric breathing mode
(33). In addition the effects of: (1) the chemical nature
of the supporting electrolyte anion; (2) the solvent; and
(3) the pH of the aqueous solution on the 1006 cm^{-1} band
intensity were qualitatively investigated.
   Figure 15a shows the intensity of the 1006 cm^{-1} pyridine
band as a function of the bulk concentration of pyridine.
This can be regarded as a Raman detected adsorption isotherm.
Typical saturation behavior is observed for [pyridine] > 50mM.
Thus the solution conditions for this component were very
nearly optimized in the original experiment of Fleischmann
and Hendra (29). A particularly interesting point to
note here is the disparity between the bulk concentrations of
pyridine required to achieve saturation in the Raman detected
isotherm as compared to that required for saturation in the
isotherm measured by Barradas and Conway (66) using UV-VIS

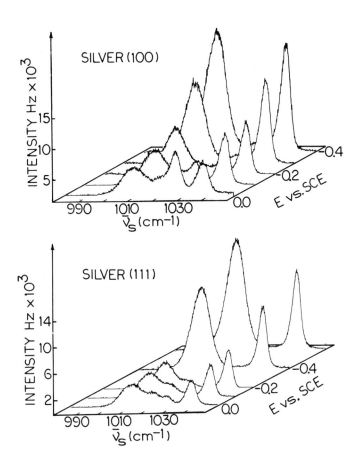

FIGURE 14. SERS for pyridine adsorbed on single
crystal Ag(100) and Ag(111) electrodes as a function of
applied potential (88).

absorption spectrophotometry. Conway's results showed that
saturation could be achieved at [pyridine] = 3.3 x 10^{-5} M or
1430 times lower than the saturation value shown in Figure
15a. Following the presentation of some other relevant data
we will return to this point and try to reconcile these
apparently contradictory data.

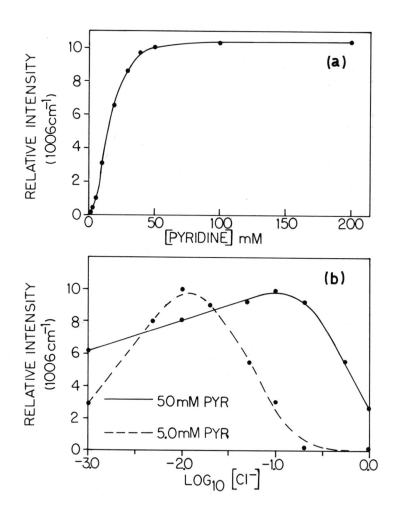

FIGURE 15. *Relative intensity of SERS for pyridine*
(1006 cm^{-1}) as a function of: (a) bulk pyridine concentration
with Cl$^-$ = 0.1 M; and (b) bulk chloride ion concentration
at two different pyridine concentrations. All measurements
were done with 50 mW at 514.5 nm and E = -0.6 V vs. SCE (33).

Figure 15b illustrates the dependence of the 1006 cm^{-1}
line intensity on the bulk [Cl$^-$] at two different values of
bulk [pyridine]. In both cases maximum SERS intensity occurs
when the ratio of the Cl$^-$ to pyridine concentrations is 2:1.
Since bulk concentrations cannot be related to surface con-

centrations without a detailed knowledge of the adsorption isotherms for each component, it is difficult to interpret the 2:1 concentration ratio in terms of a specific surface structure. The qualitative appearance of these curves can, however, be understood by assuming that $\Delta G^{o}_{ads,pyr}$ is a strong function of θ_{Cl^-}. When $[Cl^-] < 2[pyr]$, Cl^- is acting to induce pyridine adsorption and when $[Cl^-] > 2[pyr]$, Cl^- is competing against pyridine for the available adsorption sites. Again the solution conditions used in the original Fleischmann-Hendra experiments were fortuitously close to the optimum.

Substitution of other anions for chloride ion in the supporting electrolyte or the use of a nonaqueous solvent has a substantial effect on the SERS intensity. For a series of solutions where the concentration of the electrolyte anion was twice the bulk pyridine concentration, the 1006 cm^{-1} SERS intensity followed the order:

$$I^- >> Br^- \simeq Cl^- > SCN^- > HPO_4 > SO_4^= > ClO_4^-$$

This ordering probably reflects the nature of the interactions leading to the anion induced adsorption of pyridine. Substitution of other cations for K^+ had essentially no effect on the SERS intensity. When nonaqueous solvents such as CH_3OH, CH_2Cl_2, or CH_3CN are used in place of water, no SERS signal could be observed for the 0.05 M pyridine/Ag/0.10 M tetraethylammonium chloride system. This result is probably a consequence of the greater solubility of pyridine in these solvents as compared to water. Preliminary experiments with pyridine derivatives that are insoluble in water but have solubilities in CH_3CN that are comparable to that of pyridine in water have shown that weak SERS signals can indeed be obtained. Further work to optimize nonaqueous SERS signals is needed.

Variation of the solution pH over the range of 1-13 for the model system 0.05 M pyr/Ag/0.1 M KCl/H_2O leads to substantial changes in the 1006 cm^{-1} SERS intensity. As the pH is decremented from pH = 13 to pH = 5 the SERS intensity is approximately constant. Between pH = 5 and pH = 3 the SERS intensity drops to zero and remains at zero (i.e., below the background) as the pH is further reduced. The interpretation here is that near pH = 5 pyridine is protonated to form the pyridinium cation. Since the potential in these experiments was on the positive side of the potential of zero charge, the pyridinium ion is probably not strongly adsorbed. Experiments in which the intensity vs. pH profile is measured as a function of applied potential are needed to verify this interpretation.

In summary, we can state that: (1) saturation of the
SERS intensity occurs for [pyridine] > 50 mM as required for
a surface phenomenon but such data do not exclude the possibility of finite multilayer adsorption; (2) anion induced
adsorption of pyridine appears to play a role but this does
not mean that pyridine is complexed to Cl^- on the surface,
rather we envision that both pyridine and Cl^- are specifically
adsorbed within the compact double layer and that $\Delta G^o_{ads,pyr}$
is a function of θ_{Cl^-}; (3) SERS signals are not seen in
nonaqueous media simply because pyridine is not adsorbed
from such media; and (4) neutral pyridine is the adsorbed
species in the SERS experiments carried out in unbuffered
solution.

     d. *The effect of applied electrode potential.* The SERS
data obtained by Fleischmann *et al.* (29), Jeanmaire and Van
Duyne (33,70), and Pettinger and Wenning (88) shown in Figs.
1, 5, and 14, respectively, all show that the Raman intensities, number of Raman lines observed, and possibly the
vibrational frequencies of the SERS lines are all functions
of electrode potential. Considering that ΔG^o_{ads} is a strong
function of E and that a 1-5 x 10^6 volt cm^{-1} electric field
exists in the compact double layer these observations are not
too surprising. Jeanmaire and Van Duyne further defined the
effect of E on the SERS intensity by studying the response
of the SERS peak intensities to TW and DPS voltage waveforms.
Figure 16 shows the response of the six most intense,
totally symmetric, intramolecular pyridine vibrations to a
triangular voltage waveform swept from 0.0 V vs. SCE to
-1.0 V vs. SCE at ca. 1.0 volt sec^{-1}. These data were
obtained by signal averaging in a multichannel analyzer
operating in the multichannel scaling mode over ten potential
scans. It can be seen that the variation of E over this
range produces at most a factor of 4-5 change in the SERS
intensity. The general shape of these curves is similar
to the Γ vs. (E-E_z) plot for a class II adsorbate shown in
Figure 10. Maximum SERS intensity corresponding to maximum
Γ_{pyr} does occur near E=E_z. The experimental intensity vs.
E curves are, however, asymmetric. Similarly shaped Γ vs.
E curves have been observed for the adsorption of other class
II adsorbates (e.g., n-butanol (73)) on Hg. A way of
rationalizing this asymmetry is to remember that pyridine
adsorption on Ag is to some extent controlled by the θ_{Cl^-}.
When the electrode is positively charged, the coverage of
chloride is relatively high so that more pyridine is induced
to adsorb than would otherwise be expected if only the
competition between pyridine and water for the surface sites
was involved. When the electrode is negatively charged and

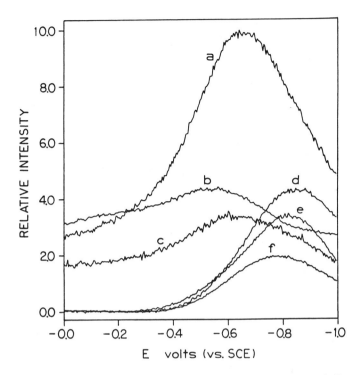

FIGURE 16. Relative intensity of SERS for pyridine as
a function of applied electrode potential for the following
vibrational bands: (a) 1006 cm^{-1}; (b) 1035 cm^{-1}; (c) 3056 cm^{-1}
(d) 1215 cm^{-1}; (e) 1594 cm^{-1} and (f) 623 cm^{-1}. All measurements were done with 50 mW at 514.5 nm and a 20 cm^{-1} spectrometer bandpass to minimize the error in intensity measurements due to a change in vibrational frequency as a function
of potential (33).

the coverage of chloride is low, pyridine adsorption cannot
be induced and in addition, water competes quite successfully
with pyridine for the surface. While this explanation does
account for the asymmetry in the shape of the SERS intensity
vs. E profile, it does not account for the observation that
the SERS intensity maximum is different for different Raman
bands. More work is needed to explain this effect.

Figure 17 shows the SERS intensity vs. time response for
the 1006 cm^{-1} pyridine line to a DPS waveform. The initial
potential was -0.1 V vs. SCE and the final potential corres-
ponded to the potential of maximum SERS intensity (i.e., ca.
-0.6 V vs. SCE). This waveform was repetitively applied at

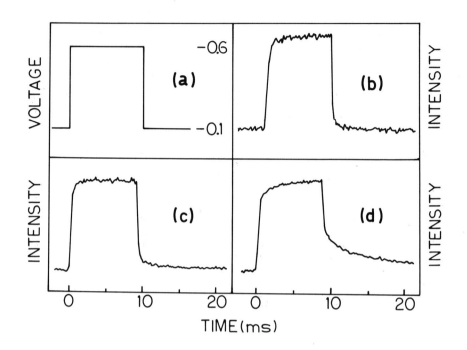

FIGURE 17. SERS intensity (1006 cm⁻¹) vs. time response
for pyridine/Ag/0.1 M KCl using 50 mW of 514.5 nm laser power:
(a) double potential step perturbation waveform, initial
potential =-0.1 V vs. SCE, final potential = -0.6 V vs. SCE,
10 msec. pulse width, 20% duty factor; (b) 0.5 mM pyridine,
10^4 averages; (c) 5.0 mM pyridine, 10^4 averages; (d) 50 mM
pyridine, 10^4 averages   (33).

20 Hz and had a duration of 0.010 sec. at -0.6 V vs. SCE
(i.e., 20% duty factor). Comparison of the shape of the
Raman intensity response with that of the perturbing voltage
waveform shows that at low pyridine concentration (i.e.,
0.5 mM) the Raman intensity tracks the DPS. The rise and fall
of this signal are not perfectly sharp due to the finite
double layer charging time (ca. 1.0 millisec.) for our
potentiostat. At higher bulk concentrations of pyridine
longer rise and fall times are observed. Recently Albrecht,
Evans, and Creighton (89) have interpreted these results as
being indicative of capillary adsorption phenomena taking
place at the rough Ag surface formed when AgCl is reduced in
the anodization step. This may or may not be the case.
Recent results in our laboratory have shown that finite

rise and fall times at these concentrations are observed both
on unanodized electrodes and on very smooth Ag TFE.

To summarize it is found that: (1) the SERS intensity
is indeed modulated by E and that the effect of the applied
potential is probably to alter the surface population of
the various adsorbed species; (2) the shifts in vibrational
frequency with E are extremely small (i.e., 2-3 cm^{-1} and,
in fact, it is very difficult to tell from the data in Figs.
1, 5 and 14 if they are real or artifactual; and (3) the
SERS intensity responds rapidly to step changes in the applied
potential so that the possibility now exists for following
surface kinetic phenomena in Raman active adsorbates.

e. The effect of laser power. After the anodization,
solution, and applied potential conditions that optimized
the SERS intensity had been found, surface enhanced Raman
spectra could be routinely obtained with laser powers less
than 50 mW at 514.5 nm. The majority of the results presented
in this chapter were obtained with such power levels. A
laser power dependence study was undertaken in order to:
(1) determine if SERS is a linear or nonlinear Raman process
and (2) determine the laser power damage threshold for SERS
(90).

To test for linearity in SERS both CW and flashlamp
pumped pulsed lasers were used. Figure 18 shows that the
SERS intensity is a linear function of 514.5 nm laser power

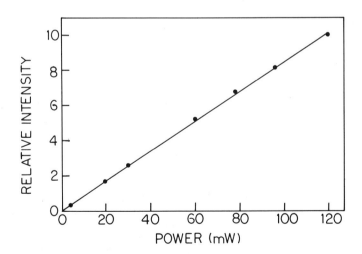

FIGURE 18. *Relative intensity of SERS for pyridine/Ag
(1006 cm^{-1}) as a function of 514.5 nm laser power. E = -0.6
V vs. SCE (90).*

over the range 0-120 mW. Similar results were obtained
using other CW excitation wavelengths. Parallel CW and
pulsed experiments were carried out to determine the sensitivity of SERS to peak vs. average power. Identical SERS
results have been obtained for pyridine on Ag using a CW,argon
ion pumped,dye laser operating at 50 mW (600.0 nm) and a
flashlamp-pumped,pulsed,dye laser operating at 10 mJ per
pulse (10 kW peak power) and 5 pps (600.0 nm). We conclude
that over this limited range of peak and average powers that
SERS is a linear Raman process.

The ultimate sensitivity of any surface spectroscopy is
limited by the surface damage threshold of the target with
respect to the intensity of the source beam. It has been
found that the pyridine/Ag system is quite resistant to surface damage at laser powers up to 100 mW at 514.5 nm for
exposure times on the order of 1-5 hours when the electrode
is potentiostatted at potentials more negative than -0.60 V
vs. SCE. Surface damage can be induced in the pyridine/Ag
SERS system by exposure of the sample to greater than 100 mW
of power for several hours at potentials in the range of -0.2
V vs. SCE or more positive. The manifestation of this surface
damage process is that the 1037 cm^{-1} pyridine line decreases
in relative intensity while the 1006 cm^{-1} line grows in
intensity. The rate of this damage process has not been
studied in detail but we do know that it is faster at more
positive electrode potentials and at laser excitation wavelengths in the 600.0 nm to 680.0 nm range. It has not yet
been determined whether this damage process is of thermal,
photochemical, or photoelectrochemical origin. These observations are simply included here as a guide for future
experiments and to serve as a word of caution for those
experiments which may require long duration laser exposure,
high laser powers, and red excitation wavelengths while
potentiostatting at positive potentials.

f. The effect of laser excitation frequency. The
dependence of Raman intensity on ω_L (i.e., excitation profile)
is diagnostic of the scattering mechanism. For example NRS
is characterized by the ω_L^4 dependence indicated in Eq.(1);
whereas, RRS is characterized by the more complex frequency
dependence described by Eq.(10). In cases where Γ_j in Eq.(10)
is large, one observes a RR excitation profile with a shape
that parallels that of the electronic absorption spectrum of
the target molecule (47,51). When Γ_j is small (91) or when
more than one electronically excited state participates in the
RR enhancement and interference effects occur (51), one can
observe fine structure in the excitation profile that is not
present in the electronic absorption spectrum of the target.

Generally, departures from the ω_L^4 scattering law indicate the existence of pre-resonance or rigorous resonance Raman scattering. Consequently the SERS excitation profile was studied in order to: (1) find the optimum value of ω_L and (2) provide diagnostic information on the SERS mechanism.

At least three groups have now studied the SERS excitation profile for the pyr/Ag system (33,70,85,86,92). In the earliest SERS excitation profile studies (33,70), no strong departure from the ω_L^4 scattering law was reported over the range 457.9 nm < ω_L < 650.0 nm. This set of experiments did not use an internal NRS intensity standard and was only designed to demonstrate that 514.5 nm excitation was not unique for SERS. In addition, these experiments showed that there was no dramatic onset wavelength for the observation of SERS and that there were no order of magnitude changes in intensity with variation in ω_L. Figure 19 presents a

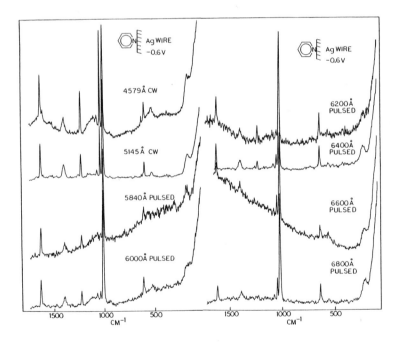

FIGURE 19. *Pyridine Ag SERS taken at various laser excitation wavelengths with CW and pulsed lasers. E = -0.6 V vs. SCE. Noisy backgrounds in the 5840, 6200 and 6600 Å spectra are due only to the effect of prolonged irradiation and are not intrinsically significant.*

collection of SER spectra taken at various ω_L with both CW
and pulsed lasers. Although there is no onset wavelength
or order of magnitude intensity changes in this range, there
is a systematic variation in the relative intensity of the
1037 and 1006 cm^{-1} modes. This relative intensity change is
not observed for either neat or aqueous pyridine in bulk
media. This observation suggests that one type of Raman
scattering mechanism provides the surface enhancement effect
at red excitation wavelengths and this is modulated by a
second mechanism at blue excitation wavelengths. In addition
to the spectra shown in Fig. 19, we attempted to observe SERS
using the 363.8 nm and 351.1 nm argon ion laser uv lines as
well as tunable uv radiation in the 280.0 to 340.0 nm range
from a frequency doubled, flashlamp-pumped, dye laser. In
all cases the results were negative. No uv excited SERS
have been observed to date.

Creighton *et al.* (86) have recently published pyr/Ag
excitation profiles for the 1008, 1026, and 1036 cm^{-1} lines
over the range 457.9 nm < ω_L < 632.8 nm. The 935 cm^{-1} band
of NaClO$_4$ added to the electrolyte and the broad 3450 cm^{-1}
O-H stretch of water served as internal NRS intensity
standards. The intensity of the SERS bands relative to the
NRS intensity standard showed a factor of 20 *increase* as ω_L
decreased. The data were highly scattered but the trend was
unmistakable. These results suggest that SERS does not
follow an ω_L^4 scattering law.

Pettinger, Wenning, and Kolb (85) have independently
verified the SERS excitation profile reported by Creighton.
Their results are shown in Figure 20 a. In addition, these
authors have measured the differential reflectance spectrum
of their single crystal electrodes (Fig. 20 b) and have
found an electronic absorption band that peaks at about 750
nm. The parallelism between this absorption feature, produced only after anodization of the electrode, and the SERS
excitation profiles led these authors to assign the SERS
scattering mechanism as a RR effect. While this conclusion
may be correct, experiments have not yet ruled out the
possibility that this is a RR effect superimposed on an
underlying SERS mechanism.

SERS excitation profile measurements have now been
carried out in our laboratory (92) using an internal standard
system that will be described in section IV.A.2. These
profiles are shown in Figure 21. Although we observe the same
general trend for the 1037 cm^{-1} and 1008 cm^{-1} bands reported
by other workers, the 1215 cm^{-1} appears to follow rigorous
ω_L^4 behavior. Also it should be noted that the deviations
from the ω_L^4 law that we observe for the 1037 and 1008 cm^{-1}
lines are smaller than those reported by Creighton (86) and

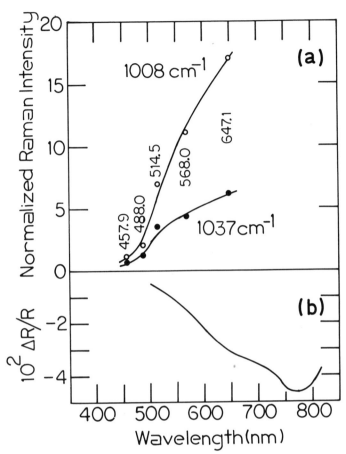

FIGURE 20. (a) Relative Raman intensities for pyr/Ag(111) as a function of excitation wavelength. The intensities are normalized with respect to the ω_L^4 law using the 935 cm^{-1} band of added $LiClO_4$ as a NRS internal standard. (b) Relative reflectance change, $\Delta R/R$, for Ag(111) as a function of wavelength. ΔR = reflectance before anodization - reflectance after anodization. Anodization charge was the equivalent of 8 monolayers of Ag. $E = -0.2$ V vs. SCE (85).

Pettinger (85). In an attempt to fit all of these data into a consistent picture, we have made a preliminary investigation of the extent of the ω_L^4 deviation as a function of applied electrode potential and extent of laser exposure. In general we have found that these deviations are more pronounced the more positive the applied potential and the longer the excitation wavelength. The data in Fig. 21 is taken

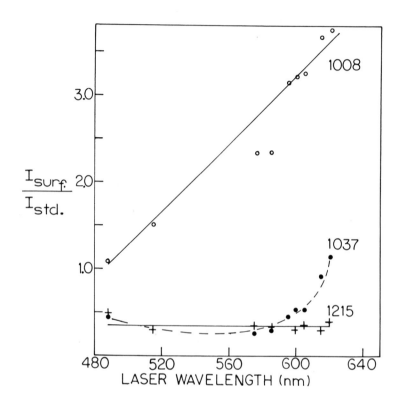

FIGURE 21. *Relative Raman intensities for pyr/Ag(poly-cryst.) as a function of excitation wavelength. The intensities are normalized with respect to the ω_L^4 law using the 966 cm^{-1} band of pyr-d_5 as a NRS intensity standard. E = -0.6 V vs. SCE (92).*

at -0.6 V vs. SCE as compared to ca. -0.2 V vs. SCE for Creighton's and Pettinger's studies. In our work, care was taken to minimize laser power and exposure time (consistent with obtaining high quality S/N spectra) at red excitation wavelengths. In addition we have found a dependence of the ω_L^4 deviations on the extent of electrode anodization. An over-anodized electrode (i.e., $Q_{anod} > 50$ mC cm^{-2}) is more susceptible to laser induced ω_L^4 intensity deviations. In Creighton's work ca. 300 mC cm^{-2} of charge was passed in the anodization step as judged from a repeat of their experiment in our laboratory (i.e., no measurement of Q_{anod} was made in ref.(86)). We conclude that some of the ω_L^4 deviations are artifacts of laser induced surface damage and some are intrinsic to the SERS process. Further experiments are necessary to make a quantitative separation of these effects.

g. *The effect of laser polarization and angle of incidence.* These effects have not yet been systematically studied. The majority of the experiments reported in this chapter have been carried out with p-polarized light and an angle of incidence near 45°.

2. *Is SERS a Surface Spectroscopy?* The optimization experiments presented above have not proved that SERS is a true surface sensitive technique in the sense that only one monolayer of adsorbate is responsible for the observed signal strength. This question arises primarily from two observations that could be interpreted in terms of multi-layer adsorption. First, there is the data of Conway and Barradas (66) showing saturation coverage of pyridine on Ag at a bulk pyridine concentration 1430 times smaller than that required to achieve saturation of the Raman detected isotherm (Fig. 15 a). Second, Albrecht, *et al.* (89) have interpreted the results shown in Fig. 17 in terms of capillary adsorption. This section discusses two experiments designed to: (1) show that all SER scatterers are located at least within 100 Å and probably within 10 Å of the electrode surface; and (2) determine the maximum value of N_{surf} for SERS.

The first experiment is based on the spatial distribution of the potential in the double layer (see Fig. 6). The basic concept is to implement a two-channel, potential dependent form of SERS (93). Kiefer (94) has recently reviewed several other forms of two-channel Raman spectroscopy. In all two-channel Raman techniques, a comparison intensity measurement is made between a sample and a standard at all wavelength increments of the scanning double monochromator. The advantage of this approach, as compared to making two uncorrelated scans of the sample and the standard, is that nonlinearities or irreproducibilities in the monochromator drive mechanism are cancelled out in the two-channel method. In the two-channel SERS experiment, which we call SER potential difference spectroscopy (SERPDS), two SER spectra are simultaneously recorded at two different values of applied potential, E_1 and E_2. After anodization, a repetitive square wave with voltage limits of E_1 and E_2 and a frequency of 1-10 Hz is applied to the electrode. The two channels of

the Raman spectrometer are gated ON and OFF so that one
channel records the E_1 spectrum and the other records the E_2
spectrum. Following the data acquisition scan, the data
is fed to an on-line computer where a difference spectrum
is computed. Only those Raman scatterers located within
the region of the $\phi_m - \phi_{ref}$ potential drop will contribute
to the SERPDS. A schematic diagram of the SERPDS apparatus
is shown in Fig. 22 along with the appropriate gating signals
and voltage excitation waveform. Figure 23 shows the results
of a SERPDS experiment. A nonzero difference spectrum does,
in fact, exist showing that the SER scatterers exist at
least within the diffuse double layer which, for 0.1 M KCl,
has a thickness of < 100 Å. Since most of the applied
electrode potential drop occurs within the compact layer,
this experiment may actually show that the SER scatterers are
within ca. 10 Å of the electrode surface. Additional information is contained in the data shown in Fig. 23. The deri-
vative-like pattern of the SERPDS is indicative of the fact
that in addition to potential dependent intensity changes
both the vibrational frequency and the linewidths are
potential dependent for the 1006 and 1037 cm^{-1} bands. The
frequency shifts of both lines are -3 cm^{-1} while the FWHM

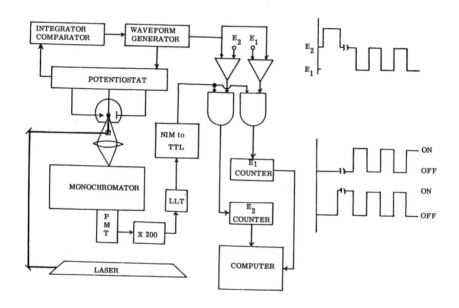

*FIGURE 22. Schematic diagram of the SER potential
difference spectrometer.*

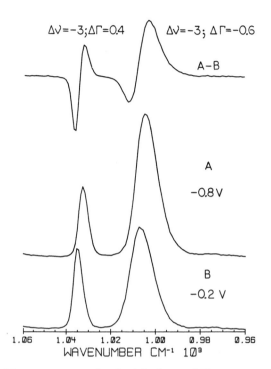

FIGURE 23. SERPDS of adsorbed pyridine. $E_1 = -0.2$ V vs. SCE; $E_2 = -0.8$ V vs. SCE; $\bar{\nu}_s = 960$ cm^{-1} to 1060 cm^{-1}; laser power = 50 mW at 514.5 nm.

(i.e., 2 Γ_R) shifts are -0.60 cm^{-1} and $+ 0.40 \pm .15$ cm^{-1}. The origin of these shifts will be considered in the section dealing with the comparison of SERS and NRS for pyridine.

The second experiment designed to show the surface nature of SERS involves the electrochemical technique of chronocoulometry (95–97) to directly measure N_{surf} for a SERS system (98). Chronocoulometry is restricted to electroactive adsorbates since it depends on the measurement of the charge consumed in the oxidation or reduction of an adsorbate. Consequently chronocoulometry is not directly applicable to the pyr/Ag SERS system since the reduction potential for pyridine (99) is more negative than the potential required for H_2 evolution on Ag in aqueous electrolytes (ca. -1.4 V vs. SCE). To circumvent this difficulty, we have examined the chronocoulometry of 4-acetylpyridine (4-Apyr) since it is reduced in a 2e$^-$ reaction at a potential of ca. -1.0 V vs. SCE (99). The SERS observables for 4-Apyr, particularly the SERS intensity vs. bulk (4-Apyr), are expected to be identical to those

for unsubstituted pyridine. In chronocoulometry one applies
a potential step from E_1 to E_2 and measures the resulting
charge density vs. time transient. E_1 is the potential
at which the adsorption is measured and E_2 is the potential
at which reduction of both the adsorbate and solution phase
electroactive species occurs. For the case of 4-Apyr E_1 =
-0.6 V vs. SCE and E_2 = -1.1 V vs. SCE. The theory of
chronocoulometry shows that the Q vs. t transient is represented by:

$$Q = 2n\ FC^*(Dt/\pi)^{\frac{1}{2}} + nF\ \Gamma_{Ads}(E_1) + Q_{d.l.} \qquad (24)$$

where n is the number of electrons transferred in the redox
process, F is Faraday's constant, C^* is the bulk concentration of the electroactive species and D is its diffusion
coefficient, and $\Gamma_{Ads}(E_1)$ is the surface excess of the adsorbate in moles cm^{-2} at E_1. According to Eq.(24) a plot
of Q vs. $t^{\frac{1}{2}}$ will be linear with an intercept proportional
to the amount of electroactive pyridine adsorbed on the Ag
electrode. The results of a chronocoulometry experiment
on the 4-Apyr/Ag system gave a value of $nF\Gamma_{ads}(E_1)$ = 67 ∓
0.5 $\mu C\ cm^{-2}$ after subtraction of $Q_{d.l.}$ from the intercept
of the Q vs. $t^{\frac{1}{2}}$ plot. $Q_{d.l.}$ was determined from an identical
experiment in which no 4-Apyr was present (i.e., blank
experiment). This adsorbed charge density corresponds to
$\Gamma_{ads}(E_1)$ = 3.24 x 10^{-10} moles cm^{-2} or $N_{surf}(E_1)$ = 1.95 x 10^{14}
molecules cm^{-2} which in turn is equivalent to ca. 1 monolayer
(see Table 1).

    3. Accurate Measurement of the SERS Enhancement Factor.
The enhancement factor, EF, for the SERS experiment has been
estimated to be between 10^4 and 10^6. These estimates were
based on the comparison of SERS intensities to calculated
intensities or to experimental intensities for NRS. An
experiment has now been devised to accurately measure EF(100).
The basic concept is to use a NRS internal standard of comparable intensity to the SERS signal, provide well defined
scattering areas and volumes for the adsorbate and internal
standard respectively, and use the value of N_{surf} determined
in the chronocoulometry experiment. Figure 24 a shows a
schematic diagram of the internal standard holder/Ag electrode
arrangement used to measure EF as well as the data shown in
Fig. 21. The internal standard holder is 1.6 mm i.d. pyrex
tubing that has been masked off with vapor deposited Nichrome.
The Ag electrode is a polycrystalline foil band wrapped
around the tubing. This entire assembly is mounted on the
face of an electrode holder similar to that shown in Fig. 2.

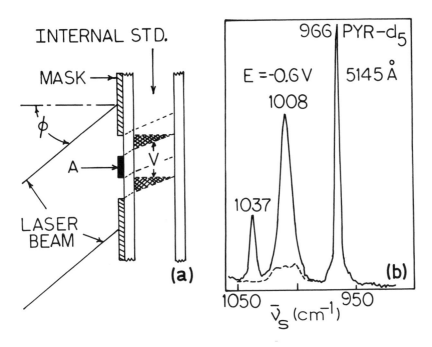

FIGURE 24. (a) Schematic diagram of internal standard holder/Ag foil electrode and (b) Composite pyr-h$_5$/Ag SERS and pyr-d$_5$ NRS. Dotted line indicates the minor overlap of the pyr-d$_5$ spectrum with the pyr-h$_5$ SERS.

The foil band extends back through the electrode holder so that contact can be made. Epoxy seals prevent solution leakage. The laser beam strikes this assembly at an angle of incidence, ϕ. This beam is line focused to ca. 3 mm x 100 μ. The Nichrome mask defines a scattering volume, V, that is not particularly sensitive to ϕ. For $30°$ < ϕ < $50°$, V = 1.7 x 10^{-4} cm^3. The illuminated geometric area of the Ag electrode in this arrangement is 7.6 x 10^{-4} cm2. The internal standard material chosen for this experiment was deuterated pyridine (pyr-d$_5$). A combined pyr-h$_5$/Ag SER spectrum and pyr-d$_5$ NR spectrum is shown in Fig. 24b. The EF is given by:

$$EF = [I_{if}(\bar{v}_s)]_{surf} N_{std} V / [I_{if}(\bar{v}_s)]_{std} N_{surf} A \qquad (25)$$

Since the experimentally determined ratio of the integrated band intensities $[I_{if}(1008 \text{ cm}^{-1})]_{surf}/[I_{if}(966 \text{ cm}^{-1})]_{std} = 1.14$, $N_{std} = 7.03$ x 10^{21} molecules cm^{-3}, $N_{surf} = 1.95$ x 10^{14} molecules cm^{-2} (from the chronocoulometry results), EF is calculated to be 1.3×10^6. Thus we believe that the magnitude of the enhancement in SERS is as high as 10^6 rather than 10^4-10^5 estimated in other studies (34,72,101).

    4. Comparison of the Raman spectra of Pyridine: SERS vs. NRS. A complete, detailed, three-way comparison of all the Raman observables (viz., intensity, totally symmetric frequencies, nontotally symmetric frequencies, depolarization ratios, and lineshapes) has not yet been carried out for neat pyridine, aqueous solutions of pyridine, and pyridine adsorbed on the Ag electrode. Parts of this study have been completed and will be discussed here. Fig. 25 is a comparison of the aqueous pyridine NRS and an optimized SERS. The intensity comparison between such spectra has already been discussed at length. C_5H_5N with 11 atoms has 28 normal modes classified in the following C_{2v} symmetry types: $10A_1 + 3A_2 + 9B_1 + 5B_2$.

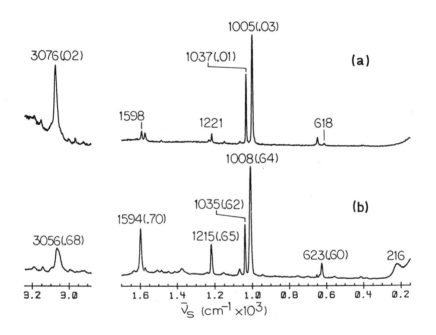

FIGURE 25. *Comparison of the Raman spectra of pyridine: (a) 2.5 M pyridine in water--NRS, laser power = 50 mW at 514.5 nm; (b) 0.05 M pyridine/Ag/0.1 M KCl/H_2O--SERS, laser power = 50 mW at 514.5 nm, E = -0.6 V vs. SCE.*

a. *Totally symmetric vibrations.* Seven strong A_1 modes are seen in the NR spectrum: ν_2 = 3076, ν_3 = 3076 (i.e., ν_2 and ν_3 are a degenerate pair), ν_4 = 1598, ν_6 = 1221, ν_8 = 1037, ν_9 = 1005, and ν_{10} = 618 cm^{-1}. The mode numbers correspond to those used in the normal coordinate analysis (NCA) of pyridine-h_5 made by Sverdlov, et al. (102). The corresponding bands in the SER spectrum are: ν_2 = 3056, ν_3 = 3056, ν_4 = 1594, ν_6 = 1215, ν_8 = 1035, ν_9 = 1008, ν_{10} = 623 cm^{-1}. The frequency shifts between these two spectra are small (ca. 2-20 cm^{-1}) indicating that the molecular and electronic structure of adsorbed pyridine is not dramatically perturbed by the 1-5 x 10^6 V cm^{-1} electric field in the compact double layer. In addition to this set of corresponding modes, a new, moderately strong vibration is seen in the SERS at 216 cm^{-1}. This band has no solution couterpart and is therefore assigned to a Ag-N vibration indicative of the adsorbate/electrode bonding. None of the NCA's of pyridine predict a low frequency vibration in this range (102-104). Metal-nitrogen vibrations have been reported in this range for metal-pyridine coordination complexes (105). The 216 cm^{-1} band should not be confused with the 239 cm^{-1} feature observed in SERS when E is more positive than ca. -0.4 V vs. SCE. Freshly precipitated AgCl shows a band at 239 cm^{-1} with an asymmetric lineshape identical to that of the 239 cm^{-1} in SERS. As E is made more negative, the 239 cm^{-1} band decreases in intensity as residual AgCl reduces on the surface to reveal a 216 cm^{-1} band that grows in in intensity due to the potential dependence of the SERS intensity as shown in Fig. 16. It should be noted that the 216 cm^{-1} band can be observed at very negative potentials ca. -1.1 V vs. SCE. If this is indeed a Ag-N vibration then we conclude that pyridine is N bonded (i.e., an end-on conformation) even at very negative E. This conclusion is quite different from that of Fleischmann (29) where they concluded that at potentials more negative than E_z, the pyridine N is only hydrogen bonded to water. Figure 26 is a times 5 vertical scale expansion of Fig. 25. Many weak vibrational features are now clearly revealed. Two of the remaining three A_1 modes can be seen in the NR spectrum at: ν_5 = 1489 cm^{-1} and ν_7 = 1069 cm^{-1}. The corresponding bands in the SER spectrum are ν_5 = 1504 (or 1482) cm^{-1} and ν_7 = 1065 cm^{-1}. The tenth A_1 mode is likely to be contained within the band envelopes of the broad lines labeled 3076 cm^{-1} (NRS) and 3056 cm^{-1} (SERS).

b. *Nontotally symmetric vibrations.* The complete set of SERS vs. NRS correspondences for the 17 nontotally symmetric vibrations has not yet been completed. In this section it will be demonstrated that: 1) nontotally symmetric modes are in fact seen in SERS; and 2) at least one

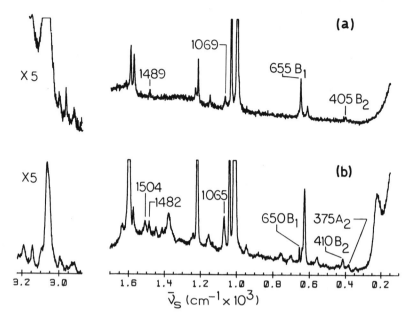

FIGURE 26. *Five-fold vertical scale expansion of*
Fig. 25

band from each symmetry type can be identified. Based on the
NCA of (102), we identify the following bands in the NRS in
Fig. 26: $\upsilon_{22}(B_1)$ = 655 cm^{-1} and $\upsilon_{27}(B_2)$ = 405 cm^{-1}. The
corresponding SER bands are υ_{22} = 650 cm^{-1} and υ_{27} = 410 cm^{-1}.
From the data in Figs. 25 and 26 no clear A_2 bands can be
seen in either the NRS or the SERS. This is not too surprising considering that the A_2 modes are the weakest bands
in the spectrum of neat pyridine. According to Sverdlov
(102) the A_2 modes in neat pyridine are located at: $\upsilon_{11}(A_2)$=
979 cm^{-1}; $\upsilon_{12}(A_2)$ = 883 cm^{-1}; and $\upsilon_{16a}(A_2)$ = 377 cm^{-1}. A
possible SERS A_2 mode for υ_{16a} is 375 cm^{-1} in Fig. 26. A
better example is illustrated in Fig. 27, where one sees a
possible υ_{12} band at 884 cm^{-1} in the neat pyridine NRS and a
possible υ_{12} = 899 cm^{-1} in the SERS.

    c. Depolarization ratios. The depolarization ratios,
ρ, shown in parentheses in Fig. 25 for the A_1 modes are
vastly different between NRS and SERS. In all cases where
the Raman bands were intense enough to measure depolarization ratios, the SERS ρ was 0.6 to 0.7 as compared to the
NRS ρ of 0.01-0.03. A control experiment showed that the
p-polarized laser light was not depolarized merely by

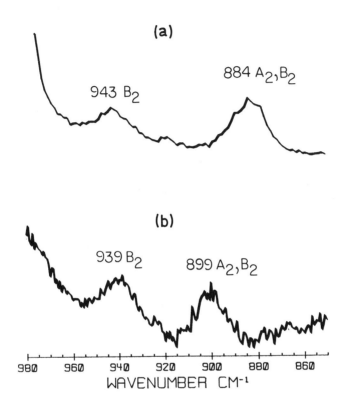

FIGURE 27. *Comparison of the Raman spectra of pyridine 850-980 cm^{-1}: (a) neat pyridine--NRS, laser power = 50 mW at 514.5 nm; (b) 0.05 M pyridine Ag 0.1 M/KCl/H_2O--SERS, laser power = 50 mW at 514.5 nm, E = -0.6 V vs. SCE.*

reflection from an anodized Ag surface. This large difference between ρ for SERS and NRS has been interpreted by assuming that pyridine was axially (i.e., end-on, N-bound) bound to the Ag electrode and undergoing two-dimensional rotational averaging of the scattering tensor. More detailed modeling of depolarization ratios using Snyder's procedure (106) should be undertaken since ρ could be extremely diagnostic for molecular orientation.

d. Lineshapes. Figure 28 shows the results of a lineshape study (107) comparing the NRS of neat pyridine and
0.05 M pyr/0.1 M KCl/H_2O with the SERS at 0.05 M pyr/Ag/0.1 M
KCl/H_2O. Only ν_8 and $\bar{\nu}_9$ have been carefully studied. The
bands in the SERS are significantly broader than those in
either liquid phase NRS. Furthermore both liquid phase
spectra can be accurately fit by a single Lorentzian lineshape function; whereas, the SERS cannot. The broad, non-
Lorentzian lineshapes in SERS are probably the result of the

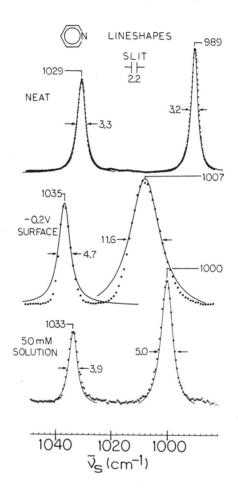

FIGURE 28. *Comparison of the lineshapes of ν_8 and ν_9
for neat pyridine, adsorbed pyridine, and .05 M aqueous
pyridine. Solid lines represent a single Lorentzian fit.
Points represent experimental data.*

simultaneous existence of two or more surface environments for adsorbed pyridine. Support for this interpretation comes from two sources: 1) Pettinger and Wenning (72) have found different linewidths for adsorbed pyridine on Ag(111) and Ag(100) and 2) ν_8 and ν_9 are known to be sensitive to the water content of the environment (108). Figure 29 shows a NR study of ν_8 and ν_9 as a function of the mole ratio of pyr:H_2O. At mole ratios from 1:1 to 2:1, ν_9 can be resolved

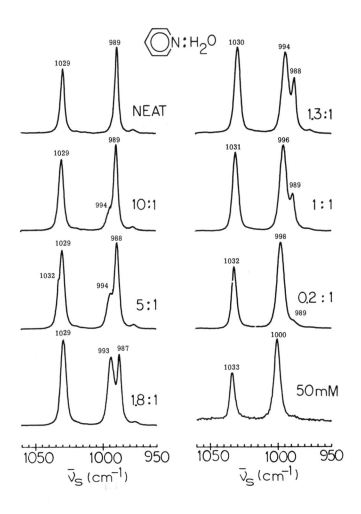

FIGURE 29. Normal Raman spectra of pyridine:water mixtures. Numbers are the mole ratios of pyridine to water.

into two distinct lines. The low frequency member of this doublet is interpreted to be ν_9 for pyridine in a pyridine environment and the high frequency component is ν_9 for pyridine in a water environment. Evidently there are both surface site and environmental contributions to the line broadening of ν_8 and ν_9 in SERS.

The data in Fig. 29 also suggest a possible interpretation for the frequency and linewidth changes observed in the SERPDS experiments (see Fig. 23). The shifts in these spectroscopic parameters with potential probably result from E induced changes in the local molecular environment of adsorbed pyridine and possibly from changes in the population of different surface sites. We view the shift in ν_8 and ν_9 as representing a change in the pyridine environment from one in which pyridine is more surrounded by water at -0.2 V vs. SCE to one in which pyridine is more surrounded by other adsorbed pyridine at -0.8 V vs. SCE. This is consistent with the picture of electrosorption presented in section III.2. where maximum adsorption of neutral organic molecules occurs near $E_z = -0.7$ V vs. SCE.

5. Current Hypotheses Concerning the Origin of SERS. In order to develop a complete picture for the understanding of SERS, both a chemical model for the structure of the Raman active adsorbate/electrode interface and a physical model for the mechanism of the surface Raman enhancement process must be formulated. The chemical structure model for the case of pyridine/silver is based on the experiments cited above and can be summarized as follows.

Neutral pyridine (as shown by the pH dependence) is adsorbed as a monolayer (as shown by the direct chronocoulometric measurement of N_{surf} for 4-acetylpyridine) of a nitrogen bonded, perpendicular or possibly edge bonded, conformation (as shown by the direct observation of a Ag-N vibration at 216 cm^{-1} and by $\rho = 0.6$-0.7).

The molecular and electronic structure of adsorbed pyridine is not highly distorted by the 1-5×10^6 V cm^{-1} interfacial electric field (as shown by the small vibrational frequency shifts between adsorbed and solution phase pyridine).

Chloride ion is specifically coadsorbed with pyridine and probably acts to stabilize the perpendicular conformation of the pyridine (as indicated by the data in Fig. 15).

The SERS observables are sensitive to the crystallinity, crystallographic orientation, surface roughness, and contamination of the metal surface.

The predominant effect of the electrochemical potential is to alter the surface concentrations and molecular environments of the adsorbed species. This is manifested in SERS through the effects of potential on intensity, frequency, and Raman lineshape.

A successful physical model for the origin of the surface Raman enhancement must incorporate these features of the surface chemistry and in addition must account for the magnitude of EF (viz., 10^6), the observation of *both* totally symmetric and non-totally symmetric vibrational modes, and the unusual laser excitation wavelength dependence.

As might be expected, the published as well as the unpublished reports of SERS have generated a reasonable amount of excitement in the surface science community as well as a great deal of theoretical activity designed to explain the origin of SERS. As of this writing the following proposals for the origin of SERS have been made:

Enhancement of Raman scattering by the interfacial electric field (33,70).

RRS induced by coupling of the molecular states of the adsorbate with surface plasmons (56,101) or by excitation of an electronic absorption band in the pyridine surface "complex" that is not existent in the metal alone or in the adsorbate alone (85).

Enhancement of Raman scattering due to large changes in the polarizability derivative with respect to normal coordinate caused by the image field at the adsorbed molecule (109).

Modulation of the electronic polarizability at the metal surface by the vibrating adsorbate giving rise to a strong scattered light signal due to the strong reflectance of the bulk metal (110,111).

Excited electron-hole pairs in the metal substrate, whose wavefunctions penetrate beyond the surface and overlap with the wavefunctions of the adsorbate, are transferred to the adsorbed molecules and back to the metal (112,113) or are scattered by the vibrational modes of the adsorbate (112).

Space in this review does not permit a detailed exposition of
all of these proposals but we will attempt to give an over-
view of the first three which have been published.

a. *Electric field enhancement.* Although originally proposed by this laboratory (for lack of a better explanation at
the time), we now discount the possibility of electric field
enhancement.

b. *RR enhancement.* RR enhancement mechanisms are consistent with much, but not all, of the SERS data. The surface plasmon version is consistent with the ω_L dependence
data (101) and can explain why A_1 modes are preferentially
enhanced over the nontotally symmetric modes. In addition,
since Ag is about the only metal in which surface plasmons
can be excited in external reflection with visible radiation,
it suggests why Ag is the only metal to show high intensity
SERS to date. On the other hand, the surface plasmon model
is not consistent with the electroreflectance experiments of
Pettinger (85) which clearly show that surface plasmons are
not excited by externally reflected, visible radiation (i.e.,
wavelengths longer than ca. 400 nm) via anodization induced
surface roughness. Furthermore, our experiments (92) have
shown that SERS cannot be observed with uv laser excitation
at wavelengths where surface plasmons would be directly
excited.

As an alternative to the surface plasmon RR model,
Pettinger *et al.* (85,88) have recently espoused a RRS model
for SERS based on the excitation of an electronic transition
within a silver (I or II) pyridine surface "complex" formed
during the anodization cycle. The electroreflectance spectrum in Fig. 20b supports this idea. The surface "complex"
RR model is consistent with the special behavior of Ag in
SERS, the preferential enhancement of A_1 vs. nontotally symmetric modes, and the ω_L dependence of the 1037 and 1006 cm^{-1}
vibrations; but, it is not consistent with the ω_L dependence
data for the 1215 cm^{-1} mode (Fig. 21) or the SERS experiment
performed on an unanodized electrode (Fig. 12). In addition
it is not easy to explain the behavior of the 1037 cm^{-1}/
1006 cm^{-1} intensity ratio as a function of ω_L (Fig. 19).
An additional serious problem with RR enhancement models
is that it is very difficult to mathematically formulate
them in such a way as to permit a semi-quantitative estimate
of the EF. So while it is easy to see qualitatively how RR
enhancement can occur, it may turn out that quantitatively
it can only account for factors of 10^1-10^3 enhancement. In
view of the now verified wavelength dependent SERS observations, the safest conclusion to make is that while RR

enhancement, especially via the surface "complex model," is a possible SERS mechanism it may be superimposed on an under-lying SERS mechanism that contributes the major share of the enhancement. Perhaps this review will stimulate theoretical studies on RR mechanisms that will permit calculations of the magnitude of EF.

   c. *Image field enhancement.* Very recently King, Van Duyne, and Schatz (109) have proposed an entirely new SERS mechanism. The basic concept here is that the conduction electrons in free electron gas metals such as silver repre-sent an extremely polarizable medium which can interact with, and be vibrationally modulated by, an appropriately adsorbed molecule. The mathematical formulation of this mechanism is primarily based on classical models of light scattering and electrodynamics and as a result is simple (i.e., crude) enough to permit an order of magnitude estimate of the value of EF to be made. Since the coupling between the conduction electrons and the adsorbate takes place through the image electric field generated in the metal by the adsorbate, we term this the image field enhancement model for SERS.

   Assume the pyridine molecule can be crudely approximated as a point dipole located at the center of the molecule. The induced dipole, μ_{pyr}^{ind}, of pyridine on a metal surface in the presence of the electric field of the laser light, E_L, is:

$$\mu_{pyr}^{ind} = \alpha_{pyr}(E_L + E_{image}) \tag{26}$$

where E_{image} is the image electric field of the pyridine di-pole that is developed in the infinitely polarizable metal substrate and α_{pyr} is the differential polarizability tensor of pyridine. Eq. (26) assumes that α_{pyr} is diagonal and that only the z component need be considered (viz., z is perpen-dicular to the surface and the z subscripts have been omitted everywhere). It will be assumed that α_{pyr} is adequately approximated by the polarizability α_{pyr}^{o} and that higher order contributions such as the hyperpolarizability are small. The z component of the image point dipole electric field evalu-ated at the position of the pyridine dipole is given by:

$$E_{image} = \frac{\gamma \mu_{pyr}^{ind}}{4r^3} \tag{27}$$

where r is the separation of the induced dipole from the
metal surface, and $\gamma = (\varepsilon_M - \varepsilon_A)/(\varepsilon_M + \varepsilon_A)$ with ε_M and ε_A being
the wavelength dependent dielectric constants of the metal
and the adsorbate. The exact meaning of r is illustrated in
Fig. 30. In studies of the response of a metal surface to an
external static charge distribution (114,115), it has been
recommended that some of the imperfections of a point dipole
image field model can be compensated for by assuming that the
effective separation of the induced dipole from the metal
surface is R = r-s. Thus the *effective* image field is obtained from Eq. (27) by replacing r with R.

Combining Eqs. (26) and (27) leads to

$$\mu_{pyr}^{ind} = \alpha_{pyr}^{o}[1-\gamma(\alpha_{pyr}^{o}/4R^3)]^{-1}E_L . \tag{28}$$

Taking the definition of the apparent surface polarizability
of the system adsorbate + metal as (116)

$$\alpha_{pyr,surf} = \mu_{pyr}^{ind}/E_L , \tag{29}$$

we get from Eq. (28):

$$\alpha_{pyr,surf} = \alpha_{pyr}^{o}[1-\gamma(\alpha_{pyr}^{o}/4R^3)]^{-1} \tag{30}$$

Since the intensity per molecule for any Raman process (viz.,
Eq(1)) can be written as (38):

$$I_{if}(\omega_s) = \frac{2^3\pi}{3^2 c^4}\omega_s^4 I_L(\omega_L) \left|Q_{if}\right|^2 \left|\sum_{\rho,\sigma}(\partial\alpha_{\rho\sigma}/\partial Q)\right|^2 \tag{31}$$

the polarizability quantity related to $I_{if}(\omega_s)$ is $\left|\partial\alpha/\partial Q\right|^2$.
By differentiating Eq.(30) we obtain:

$$\frac{\partial\alpha_{pyr,surf}}{\partial Q} = \left[\frac{\alpha_{pyr,surf}}{\alpha_{pyr}^{o}}\right]^2 \frac{\partial\alpha_{pyr}^{o}}{\partial Q} \tag{32}$$

Eq. (25) is the definition of the experimental EF in which
the SERS intensity and the NRS intensity of the internal
standard are appropriately normalized for N_{surf} vs. N_{std} and
A vs. V to put the EF on a per molecule basis. In a comparable manner, combination of Eqs. (30), (31), and (32) leads
to a theoretically defined value of the EF:

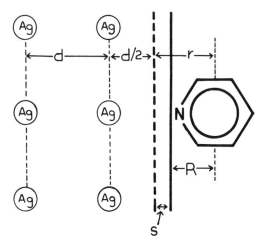

Figure 30. Relevant separation of image dipole from conventionally defined surface, r, and effective surface, R (109).

$$EF = \frac{[I_{if}(\omega_s)]_{surf}}{[I_{if}(\omega_s)]_{soln}} = \left[\frac{\partial\alpha_{pyr,surf}/\partial Q}{\partial\alpha^o_{pyr}/\partial Q}\right]^2$$

$$= [1-(\partial\alpha^o_{pyr}/4R^3)]^{-4}$$

(33)

Further refinements lead to the final expression for the EF (109):

$$EF = \frac{3}{4}(1+\gamma)^2[1 - (\gamma\alpha^o_{pyr}/4R^3)]^{-4}$$

(34)

This simple formula now allows a direct computation of the EF. $\alpha^o_{pyr} = 12$ Å^3 (117) and $\gamma = 1.3-1.6$ for the range of laser excitation wavelengths used in the SERS studies using the known values for ε_{Ag} (118) and assuming $\varepsilon_A = 2$. Figure 31 plots $\log_{10}$ EF as a function of R. The range of experimentally obtained EF = 10^5-10^6 is thus seen to correspond to R = 1.65 Å. The question is: is this reasonable? By using the sum of the covalent radii from the Ag surface to the center of the pyridine molecule, r = 2.10 Å. This value of r implies that s must be 0.45 Å in order for Eq. (34) to give reasonable agreement with experiment. This value of s is in excellent agreement with the estimates of s made in ref. 114. More important, it is seen that for reasonable values of the

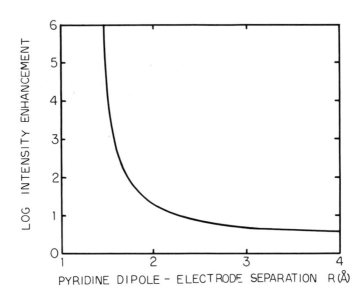

FIGURE 31. Logarithm of the Raman intensity enhancement as a function of the pyridine image-electrode separation R (in Å) (109).

parameters appearing in Eq. (34), *a Raman intensity enhancement of several orders of magnitude is possible at physically obtainable separations between the surface and the adsorbate due to the image field enhancement mechanism.*

How does the IFE model compare with experiment, First, it predicts the right order of magnitude for the EF. Second, it explains why the A_1 modes of adsorbed pyridine are preferentially enhanced over the nontotally symmetric modes since the z component of the image electric field is largest for those vibrations with large z component amplitudes and smallest for vibrations which tend to be in the x-y (i.e., electrode) plane. Third, it predicts that the EF should be approximately ω_L independent (i.e., SERS should follow the ω_L^4 law) which is the observed behavior for the 1215 cm^{-1} pyridine line (Fig. 21). Fourth, it explains the apparently special behavior of silver since it is the best free electron gas metal in the visible region of the spectrum. On the other hand, the IFE model for SERS as presented above is not consistent with the observed ω_L dependence of the 1037 and 1006 cm^{-1} pyridine Raman lines. Recently we became aware of the theoretical studies of Efrima and Metiu (119) concerning the SERS mechanism. These workers have carried through the more detailed calculations involved in extending the IFE model of King, Van Duyne, and Schatz by removing the point dipole and perfectly polarizable, free electron gas metal approximations. Within such a framework Efrima and Metiu have shown that it is possible to explain an EF that decreases slightly with increasing ω_L. It remains to determine experimentally how much of the observed decrease in EF for the 1006 and 1037 cm^{-1} lines is intrinsic to SERS and how much is a laser damage artifact.

Eq. (34) leads to another particularly significant consequence. The value of the EF is extremely sensitive to the adsorbate image-electrode separation, R. As discussed in section II, NRS by itself produces signals that are 10^2-10^3 times smaller than what is required to observe reasonable quality SRS for monolayer quantities of an adsorbate. This implies that if one observes a SERS spectrum at all, an EF of at least 10^2-10^3 is involved. If the IFE model for SERS is correct, this requires that the scattering must originate from adsorbate molecules whose image centers are within ca. 1.8 Å of the effective surface plane. There are two important implications of this consequence: 1) SERS is an intrinsically surface sensitive spectroscopy in exactly the same sense as UPS and XPS (viz., that even though the penetration depth of the input laser beam is a few hundred angstroms, the escape depth of a SERS photon is ≤ 1.8 Å); and 2) this surface sensitivity of SERS signals to the method of surface preparation (i.e., anodization). In order to observe a strong SERS signal, an adsorbate must reside on a clean surface site. Any intervening oxide or carbon contamination layer will drastically reduce the signal strength.

6. *Predictions by and Tests of the IFE Model.* Several of the predictions of the IFE mechanism for SERS are amenable to direct experimental test. In addition these experiments relate to the generality expected for SERS. These predictions are: 1) any molecule that adsorbs on a clean Ag surface should exhibit SERS; 2) the SERS intensity should be sensitive to the orientation of the molecule or substituent group with respect to the electrode surface; 3) SERS should be observable in environments other than the electrochemical one since no electrochemical parameters directly enter Eq. (34); and 4) SERS should be observable for any metal substrate provided that its conduction electrons are highly polarizable (i.e., are a reasonable approximation of a free electron gas) at ω_L. Three of these predictions have been tested experimentally and do seem to add further support to the IFE mechanism.

a. Other adsorbate/Ag systems that exhibit SERS. To date we have observed SERS with EF $\geq 10^5$ for the following adsorbates that are not themselves resonance enhanced at 514.5 nm (33,70,87): piperdine; aniline; N,N-dimethylaniline; benzylamine; N-methylimidazole; sulfanilic acid; 2,4,6-trimethylpyridine; 2-, 3-, and 4-cyanopyridine; 2,4-dicyanopyridine; 4-*tert*-butylpyridine; pyridine-d_5; 4-acetylpyridine; N,N-dimethylcyanamide; and other heterocyclic, aromatic, and aliphatic nitrogen bases. The fact that piperidine (viz., the saturated analog of pyridine) exhibits a strong SERS signal supports our hypothesis that pyridine and its analogs are bonded to the Ag electrode via the N atom in a perpendicular conformation. Piperidine has no π electrons to take part in a flat interaction with the electrode.

In addition to these relatively large molecules, there have been several recent reports showing that small ionic adsorbates exhibit the SERS effect. In particular Hendra *et al.* (120) has published the SERS for SCN^-/Ag and Furtak (121) has recently obtained a beautiful SERS for the CN^-/Ag system. Furtak's data is shown in Fig. 32. Note especially the observation of a low frequency vibration at 226 cm^{-1} which was assigned to the Ag-C stretch. This mode is analogous to the 216 cm^{-1} band observed in the Ag-NC_5H_5 system (Fig. 25).

b. Orientation specificity. The IFE model for SERS implies that one should be able to observe stronger signals for vibrations that generate large image field (i.e., vibrations perpendicular to the metal surface) as compared to vibrations that generate weak image fields (i.e., vibrations in the plane of the metal). This effect is demonstrated in the SERS of 2-, 3-, and 4-cyanopyridine adsorbed on Ag which are shown in Fig. 33 (121). The intensity of the CN stretch (2238 cm^{-1}) is the quantity to be compared. As the CN group is moved from a position where it is perpendicular to the electrode surface (viz., 4-cyanopyridine) to a position where it is approximately parallel to the electrode surface (viz., 2-cyanopyridine), the intensity of CN relative to the ring modes smoothly *decreases*. To demonstrate that the low intensity of CN in the 2-cyanopyridine spectrum is not due to an electrochemical reaction or to some intrinsically low intensity we cite the following: 1) the reduction potential for 2-cyanopyridine is the most negative of the three monocyanopyridines in Fig. 33 (99); and 2) the CN stretch is actually stonger than the ring modes in both a solution of 2-cyanopyridine and in the solid state NRS of the complex Ag(2-CNpyr)$_2^+$NO$_3^-$ (see Fig. 34). This sort of orientation effect would be quite difficult to explain within the context of a RR enhancement mechanism for SERS.

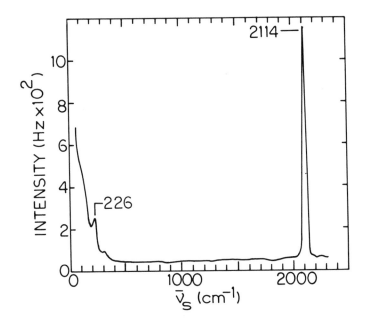

FIGURE 32. SER spectrum of cyanide adsorbed on a silver electrode. Electrolyte solution: 0.1 M Na_2SO_4 + 0.01 M KCN/ H_2O. Laser power = 170 mW at 488.0 nm. E = -0.95 V vs. SCE (121).

c. *SERS in other environments*. In a simple, but elegant experiment, A. Otto has demonstrated that SERS for the CN^-/Ag system can be observed at the Ag/ambient air interface (110) as well as in the more well defined situation of a Ag single crystal in UHV (111). The first SER spectrum at a solid/gas interface rather than at an electrode/electrolyte interface is shown in Fig. 35.

B. *Resonance Raman Spectroscopy of Electrode Surfaces*

Although surface RRS originally looked like the only tractable approach to SRS, when the SERS effect was recognized much effort that otherwise would have been devoted to surface RRS (SRRS) was diverted. Nevertheless there have been several excellent studies which have exploited the RR enhancement effect to examine electrode surfaces *in situ*. Yeager *et al*. have obtained potential dependent SRRS for p-nitroso-dimethyl-aniline adsorbed on a Pt electrode in aqueous 0.1 M KCl (71).

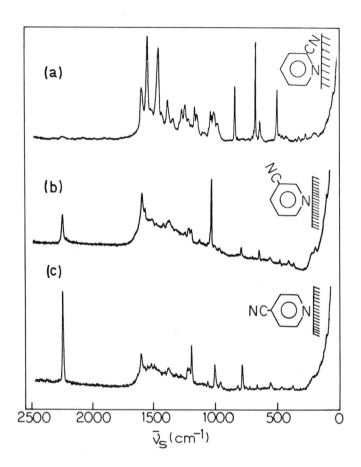

FIGURE 33. *Dependence of SERS on the orientation of the CN group with respect to the electrode surface:* (a) *2-cyano-pyridine;* (b) *3-cyanopyridine; and* (c) *4-cyanopyridine. All pyridines were 0.05 M in a 0.1 M KCl/H₂O electrolyte. Laser power = 50 mW at 514.5 nm. E = -0.6V vs. SCE. (122)*

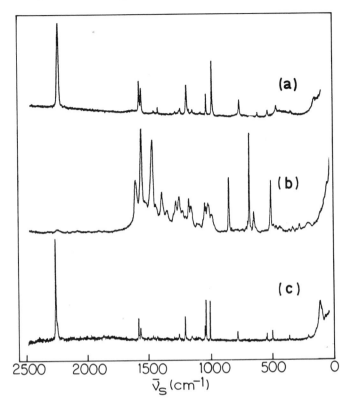

FIGURE 34. *Comparison of the Raman spectra of 2-cyanopyridine in various environments: (a) 0.8 M aqueous solution; (b) adsorbed on Ag electrode as in Fig. 33 (a); (c) solid Ag(2-CNpyr)$_2^+$NO$_3^-$. Laser power = 50 mW at 514.5 nm (122).*

Heitbaum has applied SRRS to observe the adsorbed species formed by the anodic oxidation of hydrazines on Pt in aqueous sulfuric acid solutions (123). Finally the SRRS of the dye Rose Bengal has been observed in the adsorbed state on a ZnO semiconductor electrode (124). In our original SERS paper (33), we found that if the NR adsorbates were replaced by RR adsorbates on Ag electrodes one could effectively add together the RR enhancement and the enhancement due to the SERS effect. Figure 36 shows the RRS + SERS spectra for the dye Crystal Violet adsorbed

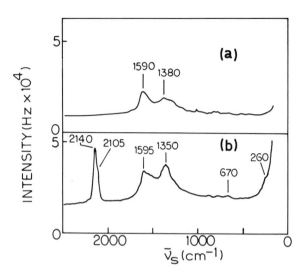

FIGURE 35. Raman spectra of: (a) polycrystalline Ag, freshly polished, exposed to ambient air; and (b) polycrystalline Ag, freshly polished, exposed to CN⁻ in aqueous solution, removed, and air dried. Laser power = ca. 500 mW at 488.0 nm (110).

on a Ag TFE. The remarkable feature about this experiment is that only 5 mW of 514.5 nm laser power was required to obtain these high quality SRS. With such an enormous combined enhancement one can even obtain SRS with a $100 He-Ne laser!

V. PROSPECTS FOR THE FUTURE OF SRS

From the present vantage point it would appear that SRS as implemented in both the SERS and SRRS forms has an extremely bright future. In SERS the key areas for research would seem to be. 1) further efforts to establish the mechanism and to hopefully decide the relative extents to which RR enhancement and IFE contribute to the overall effect; 2) demonstrate the range of applicability of SERS to as wide a variety of adsorbate structures as possible; 3) extend SERS to other metal substrates if possible. For both SRRS and SERS, the main areas of application that would seem to provide the opportunity for exciting research are: 1) the

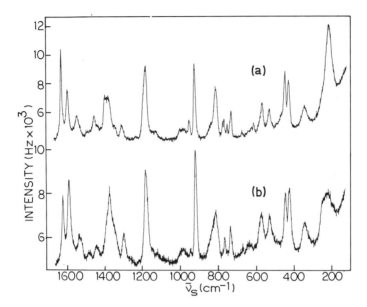

FIGURE 36. Combined *RRS* and *SERS* spectrum of *Crystal Violet adsorbed on a silver thin film electrode:* (a) *-0.6 V vs. SCE and* (b) *0.0 V vs. SCE. Laser power = 5 mW at 514.5 nm.*

ability to study, for the first time, solid/vacuum, solid/ gas, solid/liquid, and solid/transparent solid interfaces with the same technique and 2) the *in situ* molecular characterization of catalytic and/or electrocatalytic reaction mechanisms and kinetics. The application of present and rapidly developing laser technology to surface science may evolve in such a way that it has an enormous impact on many areas of science and technology.

I am deeply grateful for the outstanding efforts of my collaborators Dr. D.L. Jeanmaire, Dr. F.W. King, Dr. G.C. Schatz, Dr. P. Stair, Mr. C.S. Allen, Mr. S. Schultz, Mr. K. Parks, and Mr. J. Lakovits in generating data, interpreting it, and generally contributing to my education concerning surface phenomena. In addition I would like to thank my colleagues around the world who have shared their research results and ideas before publication and have made working in this explosively developing field so exciting and pleasurable. Our work in this area has been generously supported by the National Science Foundation and the Alfred P. Sloan Foundation (1974-1978).

VI.   REFERENCES

1.   Estrup, P.J., *Phys. Today 28*, April 1975, p.33.
2.   Eastman, D.E., and Nathan, M.I., *Phys. Today 28*, April 1975, p. 44.
3.   Plummer, E.W., and Gustafsson, T., *Science 198*, 165 (1977).
4.   Woodruff, D.P., and King, D.A., Eds., Proceedings of the Third Interdisciplinary Surface Science Conference, *Surf. Sci. 68* (1977).
5.   Somorjai, G.A., *Science 201*, 489 (1978).
6.   Fischer, T.E., *Crit. Rev. Solid State Sci. 6*, 401 (1976).
7.   Staehle, R.W., *Corrosion 34*, editorial (1978).
8.   Castle, J.E., *Surf. Sci. 68*, 583 (1977).
9.   Arthur, J.R., Jr., *Crit. Rev. Solid State Sci. 6*, 413 (1976).
10. Hercules, D.M., *Anal. Chem. 50*, 734A (1978) and references therein.
11. McIntyre, J.D.E., *in* "Advances in Electrochemistry and Electrochemical Engineering" (R.H. Muller, ed.), Vol. 9, p. 61. Wiley-Interscience, New York, 1973.
12. McIntyre, J.D.E., *in* "Optical Properties of Solids--New Developments" (B.O. Seraphin, ed.), p. 555.   North Holland, Amsterdam, 1976.
13. Hansen, W.N., *in* "Advances in Electrochemistry and Electrochemical Engineering" (R.H. Muller, ed.), Vol. 9, p. 1. Wiley-Interscience, New York, 1973.
14. Muller, R.H., *in* "Advances in Electrochemistry and Electrochemical Engineering" (R.H. Muller, ed.), Vol. 9, p. 167. Wiley-Interscience, New York, 1973.
15. Kuwana, T., *Ber. Bunsenges. Physik. Chem. 77*, 858 (1973).
16. Kuwana, T., and Winograd, N., *in* "Electroanalytical Chemistry" (A.J. Bard, ed.), Vol. 7, p. 1. Marcel Dekker, New York, 1974.
17. Moskovits, M., and McBreen, P., *J. Chem. Phys. 68*, 4992 (1978).
18. Kolb, D.M., *J. Physique 38*, C5-167 (1977) and references therein.
19. Little, L.H., "Infrared Spectra of Adsorbed Species." Academic Press, New York, 1966.
20. Hair, M.L., "Infrared Spectroscopy in Surface Chemistry." Marcel Dekker, New York, 1967.
21. Krebs, H.-J., and Lüth, H., *Appl. Phys. 14*, 337 (1977).
22. Golden, W.G., Dunn, D.S., and Overend, J., *J. Phys. Chem. 82*, 843 (1978).

23. Ibach, H., Hopster, M., and Sexton, B., *Appl. Surf. Sci.* 1, 1 (1977).

24. Ibach, H., Hopster, M., and Sexton, B., *Appl. Phys. 14,* 21 (1977).

25. Froitzheim, H., Ibach, H., and Lehwald, S., *Phys. Rev.* 14B, 1362 (1976).

26. Simonsen, M., Coleman, R., and Hansma, P., *J. Chem. Phys. 61,* 3789 (1974).

27. Kirtley, J., Scalapino, D.J., and Hansma, P.K., *Phys. Rev. 14B,* 3177 (1976).

28. Greenler, R.G., and Slager, T.L., *Spectrochim. Acta, 29A,* 193 (1973).

29. Fleischmann, M., Hendra, P.J., and McQuillan, A.J.,*Chem. Phys. Lett. 26,* 163 (1974).

30. Chen, Y.J., Chen, W.P., and Burstein, E., *Phys. Rev. Lett. 36,* 1207 (1976).

31. Paul, R.L., and Hendra, P.J., *Miner. Sci. Eng. 8,* 171 (1976).

32. Cooney, R.P., Curthoys, G., and The Tam, Nguyen, *Adv. Catal. 24,* 293 (1975).

33. Jeanmaire, D.L., and Van Duyne, R.P., *J. Electroanal. Chem. 84,* 1 (1977).

34. Albrecht, M.G. and Creighton, J.A., *J. Am. Chem. Soc.* 99, 5215 (1977)

35. Pritchard, J., and Catterick, T., *in* "Experimental Methods in Catalytic Research" (R.B. Anderson and P.T. Dawson, eds.), Vol. III, p. 281. Academic Press, New York, 1976.

36. Cherlow, J.M., and Porto, S.P.S., *in* "Topics in Applied Physics" Vol. 2, "Laser Spectroscopy of Atoms and Molecules" (H. Walther, ed.), p. 255. Springer-Verlag, Berlin, Heidelberg, New York, 1976.

37. Albrecht, A.C., *J. Chem. Phys. 34,* 1476 (1961).

38. Tang, J., and Albrecht, *in* "Raman Spectroscopy" (H.A. Szymanski, ed.), p. 33. Plenum Press, New York, 1970.

39. Spiro, T.G., *in* "Chemical and Biochemical Applications of Lasers" (C.B. Moore, ed.), Vol. I., p. 29. Academic Press, New York, 1974.

40. Skinner, J.G. and Nilsen, W.G., *J. Opt. Soc. Am. 58,* 113 (1968).

41. Kato, Y., and Takuma, H., *J. Opt. Soc. Am. 61,* 347 (1971).

42. Kato, Y., and Takuma, H., *J. Chem. Phys. 54,* 5398 (1971).

43. Greenler, R.G., *Surf. Sci. 69,* 647 (1977).

44. Shriver, D.F., and Dunn, J.B.R., *Appl. Spectrosc. 28,* 319 (1974).

45. Alfano, R.R., and Ockman, N., *J. Opt. Soc. Am. 58,* 90 (1968).

46. Behringer, J., *Mol. Spectrosc. 2,* 100 (1974).
47. Johnson, B.B., and Peticolas, W.L., *Ann. Rev. Phys.Chem. 27,* 465 (1976).
48. Behringer, J., *Mol. Spectrosc. 3,* 163 (1975).
49. Compaan, A., *Appl. Spectrosc. Rev. 13,* 295 (1977).
50. Clark, R.J.H., *Infrared Raman Spectrosc. 1,* 143 (1975).
51. Spiro, T.G., and Stein, P., *Ann. Rev. Phys. Chem. 28,* 501 (1977).
52. Hester, R.E., *Adv. Infrared Raman Spectrosc. 4,* 1 (1978).
53. Spiro, T.G., and Gaber, B.P., *Ann. Rev. Biochem. 46,* 553 (1977).
54. Kitagawa, T., Ozaki, Y., and Kyogoku, Y., *Adv. Biophys. 11,* 153 (1978).
55. Mathies, R., *in* "Chemical and Biochemical Applications of Lasers" (C.B. Moore, ed.), Vol. IV. Academic Press, New York, 1978.
56. Philpott, M.R., *J. Chem. Phys. 62,* 1812 (1975).
57. Inaba, H., *in* "Topics in Applied Physics" Vol. 14 "Laser Monitoring of the Atmosphere" (E.D. Hinckley, ed.), p. 153. Springer-Verlag, Berlin, Heidelberg, New York, 1976.
58. Fleischmann, M., Hendra, P.J., and McQuillan, A.J., *J. Chem. Soc. Chem. Commun.* 80 (1973).
59. McQuillan, A.J., Hendra, P.J., and Fleischmann, M., *J. Electroanal. Chem. 65,* 933 (1975).
60. Paul, R.L., McQuillan, A.J., Hendra, P.J., and Fleischmann, M., *J. Electroanal. Chem. 66,* 248 (1975).
61. Fleischmann, M., Hendra, P.J., McQuillan, A.J., Paul, R.L., and Reid, E.S., *J. Raman Spectrosc. 4,* 269 (1976).
62. Sawyer, D.T., and Roberts, J.L., Jr., "Experimental Electrochemistry for Chemists," Ch. 3. Wiley, New York, 1974.
63. Brown, E. R., Smith, D.E., and Booman, G.L., *Anal. Chem.,* 1411 (1968).
64. Barclay, D.J., and Caja, J., *Croat. Chem. Acta 43,* 221 (1971).
65. Anson, F.C., *Acc. Chem. Res. 8,* 400 (1975).
66. Barradas, R.G., and Conway, B.E., *J. Electroanal. Chem. 6,* 314 (1963).
67. Jaffe, H.H., and Orchin, M., "Theory and Applications of Ultraviolet Spectroscopy", p. 362. Wiley, New York, 1962.
68. Hubbard, A.T., and Anson, F.C., *in* "Electroanalytical Chemistry" (A.J. Bard, ed.), Vol. 4, p. 129. Marcell Dekker, New York, 1970.

69. Van Duyne, R.P., Jeanmaire, D.L., Suchanski, M.R., Wallace, W.L. and Cape, T., Abstracts 149th National Meeting of the Electrochemical Society, Washington, D.C., May 6, 1976, Abs. No. 357.
70. Van Duyne, R.P., *J. Phys. (Paris) 38*, C5-239 (1977).
71. Hagen, G., Simic-Glavaski, B., and Yeager, E., *J. Electroanal. Chem. 88*, 269 (1978).
72. Pettinger, B., and Wenning, U., *Chem. Phys. Lett. 56*, 253 (1978).
73. Mohilner, D.M., *in* "Electroanalytical Chemistry" (A.J. Bard, ed.), Vol. 1, p. 241. Marcel Dekker, New York, 1966.
74. Parsons, R., *in* "Advances in Electrochemistry and Electrochemical Engineering" (P. Delahay, ed.), Vol. 1, p. 1. Interscience, New York, 1961.
75. Delahay, P., "Double Layer and Electrode Kinetics." Interscience, 1965.
76. Gileadi, E., *in* "Electrosorption" (E. Gileadi, ed.), p. 1. Plenum Press, New York, 1967.
77. Payne, R., *J. Electroanal. Chem. 41*, 277 (1973).
78. Reeves, R., *Mod. Aspects Electrochem. 9*, (1974).
79. Yeager, E., *J. Phys. (Paris) 38*, C5-1 (1977).
80. Kolb, D.M., and Kötz, R., *Surf. Sci. 64*, 96 (1977).
81. Paik, W., Genshaw, M.A., and Bockris, J. O'M., *J. Phys. Chem. 74*, 4266 (1970).
82. Trasatti, S., *J. Electroanal. Chem. 53*, 335 (1974).
83. Damaskin, B.B., Petrii, O.A., and Batrakov, V.V., "Adsorption of Organic Compounds on Electrodes." Plenum Press, New York, 1971.
84. Fischer, H., *J. Electroanal. Chem. 62*, 163 (1975).
85. Pettinger,B., Wenning,U., and Kolb,D., *Ber. Bunsenges. Phys. Chem.* in press (1978).
86. Creighton,J., Albrecht,M., Hester,R., and Matthew,J., *Chem. Phys. Lett. 55*, 55 (1978).
87. Allen, C.S., and Van Duyne, R.P., unpublished results, 1977.
88. Pettinger, B., and Wenning, U., in the Proceedings of the Conference on Vibrations in the Adsorbed Layer, Jülich, Germany, 1978.
89. Albrecht, M.G., Evans, J.F., and Creighton, J.A., *Surf. Sci. 75*, L777 (1978).
90. Schultz, S., and Van Duyne, R.P., unpublished results, 1978.
91. Jeanmaire, D.L., and Van Duyne, R.P., *J. Am. Chem. Soc. 98*, 4034 (1976).
92. Allen, C.S., Schultz, S., and Parks, K., *Chem. Phys. Lett.*, submitted, 1978.
93. Schultz, S., and Van Duyne, R.P., to be published.

94. Kiefer, W., *in* "Advances in Infrared and Raman Spectro-
     scopy" (R.J.H. Clark and R.E. Hester, eds.), Vol. 3,
     p. 1. Heyden and Sons, Ltd., London, (1977).
95. Anson, F.C., *Anal. Chem. 38,* 54 (1966).
96. Case, B., and Anson, F.C., *J. Phys. Chem. 71,* 402
     (1967).
97. Christie, J.H., Osteryoung, R.A., and Anson, F.C., *J.
     Electroanal. Chem. 13,* 236 (1967).
98. Lakovits, J., and Van Duyne, R.P., *J. Electroanal.
     Chem.,* submitted, (1978).
99. Dryhurst, D., "Electrochemistry of Biological Mole-
     cules." Academic Press, New York, (1977).
100. Allen, C.S., and Van Duyne, R.P., to be published.
101. Hexter, R.M., and Albrecht, M.G., *Spectrochim. Acta,*
     (in press), (1978).
102. Sverdlov, L.M., Kovner, M.A., and Krainer, E.P.,
     "Vibrational Spectra of Polyatomic Molecules." Wiley,
     New York, (1974).
103. Corrsin, L., Fax, J.B., and Lord, R.C., *J. Chem. Phys.
     21,* 1170 (1953).
104. Long, D.A., and George, W.O., *Spectrochim. Acta 19,*
     1777 (1963).
105. Clark, R.J.H., and Williams, C.S., *Inorg. Chem. 4,* 350
     (1965).
106. Snyder, R.G., *J. Mol. Spectrosc. 37,* 353 (1971).
107. Schultz, S., and Van Duyne, R.P., to be published.
108. Suschinskii, M., "Raman Spectra of Molecules and
     Crystals,", p. 270. Israel Program for Scientific
     Translations, Ltd., New York, (1972).
109. King, F.W., Van Duyne, R.P., and Schatz, G.C., *J. Chem.
     Phys. 69,*   (1978).
110. Otto, A., *Surf. Sci. 75,* L392 (1978).
111. Otto, A., in the Proceedings of the Conference on
     Vibrations in the Adsorbed Layer, Jülich, Germany,
     1978.
112. Burstein, E., Chen, Y.J., and Lundquist, S., *Bull. Am.
     Phys. Soc. 23,* (1), 30 (1978).
113. Aussenegg, F.R., and Lippitsch, M.E., to be published
     in *J. Raman Spectrosc.*
114. Appelbaum, J.A., and Hamann, D.R., *Phys. Rev. B6,*
     1122 (1972).
115. Lang, N.D., and Kohn, W., *Phys. Rev. B7,* 3541 (1973)
116. Antoniewicz, P.R., *J. Chem. Phys. 56,* 1711 (1972).
117. Kielich, S., *in* "Dielectric and Related Molecular
     Processes" (M. Davies, ed.), p. 192. The Chemical
     Society Specialist Periodical Reports, London, (1972).

118. Irani, G.B., Huen, T., and Wooten, F., *Phys. Rev. B3*, 2385 (1971).
119. Efrima, S., and Metiu, H., preprint, 1978.
120. Cooney, R.P., Reid, E., Fleischmann, M., and Hendra, P.J., *J. Chem. Soc. Faraday I 73*, 1691 (1977).
121. Furtak, T., *Solid State Commun.*, in press (1978).
122. Allen, C.S., and Van Duyne, R.P., to be published.
123. Heitbaum, J., *Zeit. für Physik. Chem. 105*, 307 (1977).
124. Yamada, Y., Amamiya, T., and Tsubomura, H., *Chem. Phys. Lett. 56*, 591 (1978).

COHERENT ANTI-STOKES RAMAN SPECTROSCOPY [1]

Sylvie Druet
Jean-Pierre Taran

Office National d'Etudes et de Recherches
Aérospatiales (ONERA)
92320 Chatillon (France)

I. INTRODUCTION

When high power lasers became available, Raman spectroscopy was greatly facilitated. In spite of the weakness of the
scattering process, this spectroscopy was proposed shortly
thereafter as a tool for the nonintrusive in situ chemical
analysis of optically transparent mixtures. Following some
early publications on this subject (1,2), a massive effort was
undertaken in order to evaluate the potential of the method in
the important areas of atmospheric sounding and combustion diagnostics. Two categories of data can be retrieved from the
Raman spectra : species concentration from the intensities of
the relevant Raman bands (vibrational bands in general) and
temperatures from band contours. Detailed accounts of early
experimental work can be found, among other publications, in
several Project SQUID and AIAA workshop proceedings (3,5).
Important instrumental developments were accomplished, with
improved collection efficiencies and signal to noise ratio
enhancement. The fields of applications were rapidly delineated
and it appeared that spontaneous Raman scattering could
prove valuable for the investigation of such easily analyzed
samples as cold or warm aerodynamic jets, but was of limited
potential in low pressure gases, fluorescent samples or in
luminous reactive media.

[1]*Work supported in part by Direction des Recherches,
Etudes et Techniques d'Armement and Service Technique Aéro-
nautique.*

A nonlinear optical technique capable of performing Raman spectroscopy with much improved signal strength was then proposed as a competitor to SRS in the specific area of combustion diagnostics (3-6). This technique is based on a four wave mixing process called Coherent Anti-Stokes Raman Scattering, or CARS. CARS, which is one of many well-known third-order processes, was actually observed as early as 1963 (7,8), and has been since then applied to crystal spectroscopy (9-11) and to the measurement of third-order susceptibilities in gases (12-14). Raman spectroscopy by CARS received a considerable impetus in the early seventies when reliable tunable sources of good optical quality were developed.

The present chapter is devoted to the CARS spectroscopy of molecular species. We review the theory in Part II. A quantum mechanical derivation of the nonlinear source polarization is first presented, together with a diagrammatic representation of all the corresponding density matrix time evolutions. This representation gives novel physical insight into the CARS mechanisms. We then solve the wave equations pertaining to all light waves involved and establish the rate equation for molecular population perturbation.

In part III, the consequences of the theory are deduced. We review the dispersive properties of the CARS susceptibility, both off and on resonance. The spectral content is discussed, and the line contour is analyzed by means of a simple representation of the susceptibility in the complex plane. The wave propagation and the effect of focussing are also considered. Finally the important mechanisms of pump saturation, population perturbation off and on resonance, and Stark broadening (on resonance) are discussed. These processes are important since they restrict the applicability of the technique.

Part IV is devoted to experimental considerations and results. The analysis of mixtures has motivated most of the work performed so far, but special applications like high-resolution Raman spectroscopy are now receiving considerable attention.

II. THEORY

A. *Intuitive Presentation*

The theory of CARS has been discussed in a number of publications (13,15-23) and two good reviews (24,25). These papers are based on general presentations of nonlinear optics by Bloembergen and coworkers (26,27), Maker and Terhune (8,28) and Butcher (29).

Numerous aspects of the theory are now receiving great attention ; these include selection rules (22,30), resonance enhancement and line contours (30-38), linewidth (39,40) and pump focussing (41,42). In our discussion, we shall concentrate on those aspects which are relevant to the measurement, by means of a CARS apparatus, of the properties of matter in the molecular state (i.e. the measurement of temperatures and concentrations).

In gases, CARS is observed when two collinear light beams with frequencies ω_1 and ω_2 (hereafter called laser and Stokes respectively, with $\omega_1 > \omega_2$) , traverse a sample with a Raman active vibrational mode of frequency ω_V such that $\omega_V = \omega_1 - \omega_2$. A new wave is then generated at the anti-Stokes frequency $\omega_3 = \omega_1 + \omega_V = 2\omega_1 - \omega_2$ in the forward direction, and collinear with the pump beams (Fig.1). This new wave results from the inelastic scattering of the wave at ω_1 by the molecular vibrations, which are coherently driven by the waves at ω_1 and ω_2 (hence the name of the effect). We note that the same mechanism creates, for reasons of symmetry, a similar wave at $2\omega_2 - \omega_1$ (CSRS for Coherent Stokes Raman Scattering). This wave has been observed (43,44) and is sometimes used for spectroscopic purposes (45,46), in spite of the difficulties connected with background light rejection and poorer detector efficiencies.

More precisely, using the intuitive Placzek approximation of molecular polarizabilities, CARS can be presented (20,21,24) as involving two distinct and simultaneous mechanisms. If we let $E = E_1 e^{-i\omega_1 t} + E_2 e^{-i\omega_2 t} + cc$, E being the sum of the pump fields at ω_1 and ω_2 respectively, we then see that :

(i) the molecular vibration amplitude q is driven resonantly

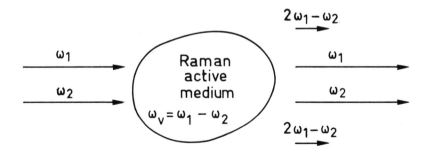

FIGURE 1. CARS and CSRS.

by E_1 and E_2 ; the driving force $\partial W / \partial q$ derives from the molecular energy

$$W = (1/2) \, \alpha \, E^2 \, , \tag{1}$$

where the polarizability α is expanded as

$$\alpha = \alpha_0 + \frac{\partial \alpha}{\partial q} \, q \, , \tag{2}$$

and where the resonant cross term $2 E_1 \, E_2^*$ is retained in E^2 ; the molecular motion is thus driven at frequency $\omega_1 - \omega_2$; (ii) the molecular vibration causes a faint modulation of the macroscopic polarization $P = N \alpha E$, where N is the number density of scatterers, as is evident after expanding α as a function of q . The polarization now contains new frequency components. We are interested in the component at $\omega_1 + (\omega_1 - \omega_2)$, which can generate an electromagnetic wave at this frequency. Efficient coupling to the wave is generally obtained, as we shall see that the phase velocities of the polarization wave and the electromagnetic wave are practically equal.

Our calculation of the CARS intensity and of its spectral behaviour will parallel this classical presentation. To this end, we shall first derive the expression for the polarization, and then use it to calculate the anti-Stokes wave. However, we shall give a quantum–mechanical derivation of the polarization. In addition to being more rigorous than the Placzek model, the quantum picture remains valid when the optical frequencies become resonant with one–photon absorption frequencies (resonant CARS). In estimating the anti–Stokes amplitude, we shall not ignore the extremely important stimulated Raman scattering (SRS) effect ; the latter takes place along with CARS and can cause, if the excitation is too violent, perturbations to the CARS mechanisms (saturation). Therefore, we shall write in full the three wave equations at ω_1 , ω_2 and ω_3 in order to calculate the complete energy exchange between all optical modes, and both CARS and SRS nonlinear polarization terms will be duly accounted for.

B. *Derivation of the CARS Polarization and Susceptibility*

Detailed descriptions of the nonlinear optical polariz-ation created in matter by optical waves have been given by a number of authors (27,29). We recall that the polarization in a medium subjected to large optical fields is written at point $\underline{r}$:

$$\underline{P}(\underline{r}) = \underline{P}^{(1)}(\underline{r}) + \underline{P}^{(2)}(\underline{r}) + \underline{P}^{(3)}(\underline{r}) + \ldots \tag{3}$$

which shows explicitly the linear contribution $P^{(1)}$ and the nonlinear ones. We assume the waves to be plane. Under steady state conditions, we can expand the total field vector $\underline{E}$ as a function of its m distinct frequency components ω_i with wave vector $\underline{k}_i$:

$$\underline{E}(\underline{r},t) = \sum_{i=1}^{m} \left(\underline{E}(\underline{r},\omega_i) e^{-i(\omega_i t - \underline{k}_i \underline{r})} + cc \right) , \tag{4}$$

Due to these applied fields, new frequency components appear in the nonlinear polarization. We write the component at frequency ω_s :

$$\underline{P}^{(n)}(\underline{r},\omega_s) = \underline{\mathcal{P}}^{(n)}(\underline{r},\omega_s) e^{-i(\omega_s t - \underline{k}_s' \underline{r})} + cc , \tag{5}$$

with the phenomenological expansion

$$\underline{\mathcal{P}}^{(n)}(\underline{r},\omega_s) = \underline{\underline{\chi}}^{(n)} \underline{E}_{l_1}(\underline{r},j_1) \underline{E}_{l_2}(\underline{r},j_2) \cdots \underline{E}_{l_n}(\underline{r},j_n), \tag{6}$$

and with $\omega_s = \sum_{i=1}^{n} l_i \omega_{j_i}$, $\underline{k}_s' = \sum_{i=1}^{n} l_i \underline{k}_{j_i}$; here, $\underline{\underline{\chi}}^{(n)}$ is the susceptibility tensor of order n (the rank of this tensor is $n+1$); we also specify $l_i = \pm 1$ and $1 \leqslant j_i \leqslant m$, and

$$\underline{E}_{l_i}(\underline{r},j_i) = \underline{E}(\underline{r},\omega_i) \quad \text{if} \quad l_i = +1$$
$$= \underline{E}^*(\underline{r},\omega_i) \quad \text{if} \quad l_i = -1 \quad .$$

Due to the distinction between local fields and macroscopic fields in dense media, a corrective factor $(\epsilon_i + 2)/3$, where ϵ_i is the optical dielectric constant at frequency ω_i , must be applied to each of the fields in Eq.(6). In gases, we have $\epsilon_i \simeq 1$.

$P^{(1)}$ is associated with linear effects (dispersion and absorption) ; $P^{(2)}$, which is responsible for such effects as frequency doubling or parametric conversion, vanishes in media possessing inversion symmetry, e.g. centrosymmetric crystals, gases and liquids ; all other even order terms also vanish in these media. There are many $P^{(3)}$ effects : third harmonic generation, optical Kerr effect, two photon absorption, as well as Raman scattering and related effects like CARS. Each of these effects is associated with a distinct frequency component, and several distinct effects can contribute to a given frequency component. For instance, there are actually two third-order polarization terms at the frequency ω_3 of the CARS wave :

$$\underline{P}^{(3)}(\underline{r},\omega_3) = \underline{P}^{(3)}{}^{CARS}(\underline{r},\omega_3) + \underline{P}^{(3)}{}^{SRS}(\underline{r},\omega_3) \qquad . \qquad (7)$$

The first one is the CARS component :

$$\underline{P}^{(3)}{}^{CARS}(\underline{r},\omega_3) = \underline{\underline{\chi}}^{(3)}{}^{CARS}(\omega_1,\omega_1,-\omega_2)\, \underline{E}^2(\underline{r},\omega_1)\, \underline{E}^*(\underline{r},\omega_2)$$
$$\times\, e^{i(\underline{k}_3'\cdot\underline{r}\,-\,\omega_3 t)} + cc \qquad , \qquad (8)$$

with $\underline{k}_3' = 2\underline{k}_1 - \underline{k}_2$, while the second one reflects the stimulated Raman scattering (inverse Raman scattering) interaction between the waves at ω_3 and ω_1 :

$$\underline{P}^{(3)}{}^{SRS}(\underline{r},\omega_3) = \underline{\underline{\chi}}^{(3)}{}^{SRS}(\omega_1,-\omega_1,\omega_3)\, \underline{E}(\underline{r},\omega_1)\underline{E}^*(\underline{r},\omega_1) \quad (9)$$
$$\times\, \underline{E}(\underline{r},\omega_3)\, \times\, e^{i(\underline{k}_3\cdot\underline{r}\,-\,\omega_3 t)} + cc \;.$$

The latter is negligible, since the susceptibility components are of comparable magnitude and since

$$|\underline{E}(\underline{r},\omega_3)| \ll |\underline{E}(\underline{r},\omega_1)|,\; |\underline{E}(\underline{r},\omega_2)| \qquad . \qquad (10)$$

We note in passing that the third-order polarization terms at ω_1 and ω_2 can also be broken down into equations similar to Eq.(7). For these, however, the SRS term is the stronger.

The spectral properties of the wave generated at ω_3 thus depend essentially on $\underline{\underline{\chi}}^{(3)}{}^{CARS}(\omega_1,\omega_1-\omega_2)$, the SRS polarization component being negligible. This tensor obeys a number of symmetry restrictions ; it must be invariant with respect to certain permutations of its Cartesian components and the corresponding fields. It also transforms according to the symmetry properties of the medium. These properties have been analyzed by Butcher (29) and were also reviewed in (25, 30). Suffice it to say that in the CARS investigation of molecular media, two experimental situations are of interest : those in which the linearly polarized pump fields have parallel polarizations and those in which they have orthogonal polarizations. It can be shown that the anti-Stokes polarization has the same polarization vector as the Stokes field in both cases. Furthermore, the ratio of the corresponding susceptibility terms off resonance permits one to calculate the familiar Raman depolarization ratio (25,46).

In the remainder of this section, we shall examine the quantum mechanical derivation of $\underline{P}^{(3)}{}_{CARS}(\underline{r}, \omega_3)$, from which the spectral properties of the CARS susceptibility will be inferred. Two presentations of this calculation will be given : the usual iterative treatment of density matrix perturbations will first be outlined, then we shall show that this derivation can be done in a straightforward manner by means of a time-ordered diagrammatic representation. The latter brings novel physical insight into nonlinear optical mechanisms, and into CARS in particular (35-37).

1. *Density Matrix.* The quantum state of the scattering molecules is represented, at point $\underline{r}$, as is conventional, by the density operator $\rho(t)$ with the well-known equation of motion :

$$\frac{\partial}{\partial t} \rho(t) = -\frac{i}{\hbar} \left[H_o + V(t) , \rho(t) \right] + \frac{\partial \rho}{\partial t}\Big|_{damp} . \qquad (11)$$

H_o is the free molecule Hamiltonian with a discrete spectrum of eigenstates $|n\rangle$ corresponding to eigenenergies $\hbar\omega_n$; the Hamiltonian describing the interaction of the molecules with the radiation field is $V(t) = -\underline{p} \cdot \underline{E}(\underline{r}, t)$ in the dipolar approximation ; $\underline{p}$ is the dipole moment operator. $\frac{\partial \rho}{\partial t}\Big|_{damp}$ is the damping term, which is determined by stochastic processes such as spontaneous emission of light and collisions between molecules. Several approximations are usually made on the collisional mechanisms, which lead to a simple expression for the damping terms (27,47-49). These approximations are (i) the assumption of stationary stochastic processes, (ii) the impact approximation, (iii) the isolated line approximation. Approximation (ii) is equivalent to assuming that the time interval between two collisions is larger than their duration. It is usually valid in gases; it allows one to replace the fluctuating time-dependent interactions between molecules by an effective time-independent interaction which is not Hermitian (47). With approximation (iii) the coupling by relaxation among the off-diagonal matrix elements $\rho_{nn'}$ can be neglected. $\frac{\partial \rho}{\partial t}\Big|_{damp}$ in Eq.(11) is then written, as shown in references (27,50) :

$$\frac{\partial}{\partial t} \rho_{nn}\Big|_{damp} = -\Gamma_{nn}\, \rho_{nn} + \sum_{k \neq n} W_{nk}\, \rho_{kk}$$

$$\frac{\partial}{\partial t} \rho_{nn'}\Big|_{damp} = -\Gamma_{nn'}\, \rho_{nn'} \qquad , \tag{12}$$

with $\Gamma_{nn'} = (\Gamma_{nn} + \Gamma_{n'n'})/2 + \Gamma_{nn'}^{e}$; Γ_{nn} is related to the finite lifetime π/Γ_{nn} of state $|n\rangle$ due to spontaneous emission and inelastic collisions, and $2\Gamma_{nn'}^{e}$ is the full broadening of the absorption line between states $|n\rangle$ and $|n'\rangle$ due to phase-interrupting elastic collisions ; W_{nk} is the transition probability via inelastic collisions between states $|k\rangle$ and $|n\rangle$. It is seen from Eqs.(12) that the three approximations above (hereafter called approximation (12)) result in Lorentzian line broadening. Note also that approximation (iii) does not imply that the relaxation of states $|n\rangle$ (or $|n'\rangle$) to other states via inelastic collisions is negligible as can be seen in the first of equations (12) ; this approximation is valid if the condition $\Gamma_{nn'} < \omega_{nn'}$ is satisfied (47).

We assume the perturbation $V(t)$ to be weak enough to validate a solution of Eq.(11) expanded in successive powers of $V(t)$. The density operator is then obtained to any order r by the familiar series expansion

$$\rho(t) = \rho^{(0)}(t) + \rho^{(1)}(t) + \cdots + \rho^{(r)}(t) \quad . \tag{13}$$

The rth order term $\rho^{(r)}(t)$ is proportional to $V^{r}(t)$ and is obtained by r iterative applications of Eq.(11). The term responsible for the CARS polarization is of order 3, and the polarization is given by

$$p^{(3)}\genfrac{}{}{0pt}{}{CARS}{}(\omega_3,t) = N\,\text{Tr}\left[\rho^{(3)}\genfrac{}{}{0pt}{}{CARS}{}(\omega_3,t)\, p\right]$$

$$= N \sum_{mn} \rho_{nm}^{(3)}\genfrac{}{}{0pt}{}{CARS}{}(\omega_3,t)\, p_{mn} \qquad , \tag{14}$$

where $\rho^{(3)}\genfrac{}{}{0pt}{}{CARS}{}(\omega_3,t)$ labels the CARS Fourier component of $\rho^{(3)}(t)$ at frequency ω_3 . We call $\underline{e}$ the unit vector in the direction of the fields. The dipole moment operator p has only off diagonal matrix components :

$$P_{mn} = <m \mid \underline{p} \cdot \underline{e} \mid n>$$

Identification between Eqs. (8) and (14) yields the expression for the CARS susceptibility tensor.

In the following, we shall assume for simplicity that all electric field polarization vectors are oriented along the X axis with common unit vector $\underline{e}$, and that their $\underline{k}$ vectors are oriented along the orthogonal Z axis (collinear propagation). Beam crossing, which is necessary in liquids in order to improve phase-matching (see section III.A) only introduces small angles between $\underline{k}$ vectors and between $\underline{E}$ vectors, so that we can still assume the vectors to be collinear. As discussed above, the CARS polarization is then also aligned along the X axis. In order to simplify notations, we now write field envelopes as $E_i = \underline{E}(\underline{r}, \omega_i)$, with $i = 1, 2, 3$, and the susceptibility tensors are treated as scalars. For simplicity one retains only the terms with the Raman resonance denominator $\omega_{ba} - \omega_1 + \omega_2 \simeq 0$ in the third-order perturbation expansion of $\rho(\underline{r})$. These terms give the Raman contribution χ_R to the CARS susceptibility, while the others, which are not calculated here, are small, frequency independent, and can be lumped together in the non-resonant susceptibility χ_{NR} . Including several possible Raman transitions of the molecules one gets :

$$\chi^{(3)}(\omega_1^{CARS}, \omega_1, - \omega_2) = \chi_{NR} + \sum_{a,b} \chi_R^{ba} \quad .$$

Here, χ_R^{ba} is the Raman susceptibility associated with the Raman transition of frequency ω_{ba}:

$$\chi_R^{ba} = \frac{N}{\hbar^3} \times \frac{1}{(\omega_{ba} - \omega_1 + \omega_2 - i\Gamma_{ba})}$$

$$\times \left\{ \sum_{n'} \left(\frac{P_{an'} P_{n'b}}{\omega_{n'a} - \omega_3 - i\Gamma_{n'a}} + \frac{P_{an'} P_{n'b}}{\omega_{n'b} + \omega_3 + i\Gamma_{n'b}} \right) \times \sum_n (\rho_{aa}^{(o)} - \rho_{nn}^{(o)}) \right.$$

$$\times \left(\frac{P_{bn} P_{na}}{\omega_{na} + \omega_2 - i\Gamma_{na}} + \frac{P_{bn} P_{na}}{\omega_{na} - \omega_1 - i\Gamma_{na}} \right) \qquad (15a)$$

$$- \sum_{n'} \left(\frac{P_{an'} P_{n'b}}{\omega_{n'a} - \omega_3 - i\Gamma_{n'a}} + \frac{P_{an'} P_{n'b}}{\omega_{n'b} + \omega_3 + i\Gamma_{n'b}} \right) \times \sum_n (\rho_{bb}^{(o)} - \rho_{nn}^{(o)})$$

$$\left. \times \left(\frac{P_{bn} P_{na}}{\omega_{nb} - \omega_2 + i\Gamma_{nb}} + \frac{P_{bn} P_{na}}{\omega_{nb} + \omega_1 + i\Gamma_{nb}} \right) \right\} \quad ,$$

which we write as

$$\chi_R^{ba} = (A+B)(\beta+\alpha) - (C+D)(\gamma+\delta) \ . \tag{15b}$$

Here :
- $|n\rangle, |a\rangle, |b\rangle$ are discrete eigenstates of the molecules ($|n\rangle$ excited electronic state, $|a\rangle$ and $|b\rangle$ vibrational states coupled by the Raman transition of shift ω_{ba}) ;
- ω_{na} is an absorption frequency with damping Γ_{na} ;
- $N\rho_{aa}^{(o)}, N\rho_{bb}^{(o)}, N\rho_{nn}^{(o)}$ are initial number densities of molecules in states $|a\rangle$, $|b\rangle$ and $|n\rangle$.
- $\rho_{nn}^{(o)}$ is in general negligible with respect to $\rho_{aa}^{(o)}$ and $\rho_{bb}^{(o)}$; however it can become large if the medium is exposed to electrical discharges or resonant radiation. Furthermore, higher order corrections may have to be applied to these initial population factors due to saturation of some transitions by the pump waves.

In mixtures, χ_{NR} receives contributions from the probed molecules and from the non-Raman-resonant molecular species (diluent molecules). It is composed of terms analogous to those of χ_R^{ba} , but with non-resonant two-photon sum or difference denominators in place of the Raman resonance denominator. In the usual case where the number density N of the probed molecules is small compared to that of the diluent molecules, χ_{NR} exhibits mainly contributions from the latter and is therefore a frequency-independent real number (provided that there are no one-or two-photon electronic resonances in the diluent molecules). However, some of the two-photon non-vibrationally resonant terms contributed by the probed molecules, although small, also depend strongly upon the field frequencies in resonant CARS ; they therefore should be included in χ_R^{ba} rather than left in χ_{NR} . As was shown in section II.B.3 of (37), they can be combined with some of the terms in expression (15a) of χ_R^{ba} under certain assumptions on collisional mechanism, leading to the more widely used expression found in Refs. (29,31,33,51).

2. Diagrammatic Representation. Nonlinear optical susceptibilities have usually been obtained from density operator calculations by perturbation theory up to the desired order. The calculation of any specified order of the density operator involves the evaluation of a number of terms, each associated with a specific time sequence of perturbations to the wave function $|\psi\rangle$ and its complex conjugate $\langle\psi|$. The time-ordering of the perturbations to $|\psi\rangle$ with respect to those to $\langle\psi|$ is important in the case of non-negligible homogeneous

broadening. It arises since the density operator is then a
statistical average of the product of the wave function and
its complex conjugate, which undergo different evolutions in
general (36,37,52). Each of these time sequences can be
depicted by means of a time-ordered diagram which is similar
to a Feynman diagram, as discussed recently in (36,37,52). Any
diagram consists of two parts, viz. Figs.2-4 which we shall
use below to visualize and calculate the main contributions
to the susceptibility : on the left hand side is plotted the
time evolution of $|\psi>$, and on the right hand side that of $<\psi|$,
so that one can follow the time evolution of the density
operator contribution along the time axis (vertical axis).

  Such diagrams have been utilized first by Feynman (53)
for the calculation of light scattering cross-sections, in the
limit of negligible lifetime and collisional broadening. In
this limit the perturbations occuring on the wave function
are strictly decoupled from those occuring on its complex
conjugate and can be calculated independently. Then it is
sufficient to visualize and calculate the time evolution
of $|\psi>$ only (Feynman diagrams). Diagrams have also been
utilized for calculations of transition rates (48,49) in
spontaneous scattering and recently for the calculation
of two-photon absorption contributions to third-order
nonlinear susceptibility (52). There are several advantages
to using the diagrammatic representation. The calculation of
the density matrix elements contributing to $P^{(3)CARS}_{(\omega_3)}$ can be
done directly by means of simple rules and without calculating
lower order terms. The diagrams also permit one to visualize
and interpret the process associated with each of these
elements better than do energy level diagrams, which have
created some confusion concerning the parametric nature of
CARS.

  a. Code for using the diagrams. Each term of the CARS
susceptibility is depicted by a time-ordered diagram representing the evolution of a density matrix element as a
function of time (the main ones are those shown in Figs. 2
and 3).We list here the rules permitting one to draw the
diagrams and to calculate the associated terms for the impact
and isolated line approximation in accordance with refs. (36,
37). Any time-ordered contribution to the n'th order of perturbation of the density operator is associated with a diagram
according to the following rules :

  (1) The time axis is drawn vertically.

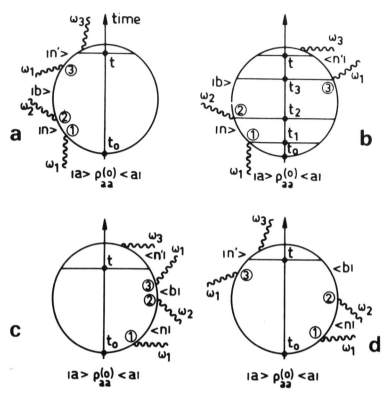

FIGURE 2. Diagrammatic representation of the main contributions from molecules initially in state $|a\rangle$: a,b : Raman resonant contribution ; c,d : Raman antiresonant contribution.

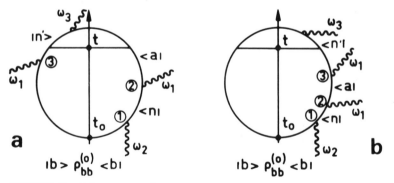

FIGURE 3. Diagrammatic representation of the main contributions to χ_R proportional to $\rho_{bb}^{(0)}$. a : non-parametric diagram ; b : parametric diagram.

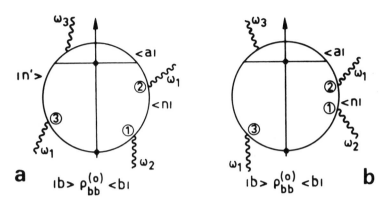

FIGURE 4. Time-ordered diagrams derived from Fig.3a with vertex 3 occuring : a : between vertices 1 and 2 ; b : before vertex 1.

(2) The time evolution of the wave function (ket) is plotted vertically to the left of this axis and that of the complex conjugate (bra) on the right, so that one can follow the time evolution of the density operator contribution along the time axis.

(3) The field interactions are represented in the standard way by vertices that are specified respectively as the photon creation or annihilation operators. Note here that each vertex is drawn at 45 degrees to the tangent to the circle, and not to the time axis as was done in refs. (36).

(4) The number of interactions (or vertices) on the ket side and on the bra side are specified, their total number being equal to the order of the perturbation n.

(5) The time-ordering of the interactions on the ket is specified relative to the interactions on the bra.

We consider the diagram of Fig.2(b) to illustrate the utilization of the rules permitting one to calculate the associated steady state contribution to the polarization. There are three relevant field interactions on this diagram, one of them on the bra side and the other two on the ket side. The first interaction occurs at time t_1 and is specified as a laser photon annihilation operator operating on the ket $|a>$. The 2nd one occurs at time t_2 and is specified as a Stokes photon creation operator operating on the ket $|n>$. The last one at time t_3 is specified as a laser photon creation operator operating on the bra $<a|$. Beginning at t_0 and tracing up the diagram, one introduces factors as follows :

1) The initial density matrix element

$$\rho_{aa}^{(o)}$$

2) The transition moment associated with the first perturbation which results in a change of the molecular wave function from $|a>$ to $|n>$

$$<n|-\underline{p}.\underline{e}|a> E_1 e^{-i(\omega_1 t - k_1 z)}$$

(3) The propagation factor resulting from the simul-taneous evolution of ket $|n>$ and bra $<a|$ between t_1 and t_2 :

$$[\hbar(-\omega_{na} + i\Gamma_{na} + \omega_1)]^{-1}$$

(4) The transition moment associated with the 2nd perturbation resulting in a change of the molecular wave function from $|n>$ to $|b>$:

$$<b|-\underline{p}.\underline{e}|n> E_2^* e^{i(\omega_2 t - k_2 z)}$$

(5) The propagation factor resulting from the simul-taneous evolution of ket $|b>$ and bra $<a|$ between t_2 and t_3 :

$$[\hbar(-\omega_{ba} + i\Gamma_{ba} + \omega_1 - \omega_2)]^{-1}$$

(6) The transition moment associated with the last perturbation which brings the molecular state from $<a|$ to $<n'|$:

$$<a|-\underline{p}.\underline{e}|n'> E_1 e^{-i(\omega_1 t - k_1 z)}$$

(7) The propagation factor resulting from the evolutions of bra $<n'|$ and ket $|b>$ between t_3 and t :

$$[\hbar(-\omega_{bn'} + i\Gamma_{bn'} + \omega_3)]^{-1}$$

(8) If the number of field interactions on the bra side is even (odd) one must put a + (−) sign in front of the contribution of interest (that is a − sign in the case of Fig.2(b)). This gives the density operator matrix element $\rho_{bn'}^{(3)}$, depicted in the diagram at time t. In order to obtain the associated contribution to the total polarization as given by equation (14), one finally has to multiply by

$$<n'|\underline{p}.\underline{e}|b> .$$

The last (4th) vertex on the diagram is of no use in calculating the anti-Stokes polarization. It represents the coupling between the field at ω_3 and the nonlinear polarization, which results in either absorption or emission at ω_3. Whether absorption or emission occurs depends upon the respective phases of all the fields (including E_3) and cannot be determined a priori from the diagram, as will be discussed in Part II.D.

 b. Diagrammatic analysis of $\chi^{(3)}(\overset{CARS}{\omega_1}, \omega_1; \omega_2)$. The contribution to $\chi^{(3)CARS}$ from the molecules initially in vibrational state $|a\rangle$ can be ascertained by considering the four basic diagrams of Fig.2. The terms associated with Figs. 2(a), (b) have a two-photon vibrational resonance (Raman resonance) denominator $(\omega_{ba} - \omega_1 + \omega_2 - i\Gamma_{ba})$, whereas those of Figs.2(c), (d) are Raman anti-resonant $(\omega_{ba} + \omega_1 - \omega_2 + i\Gamma_{ba})$ and hence are smaller. We can associate two more time-ordered diagrams to each of the two initial diagrams of Figs.2(b), (d) as vertex 3 can occur before vertex 1 or between vertices 1 and 2 ; these new possibilities give terms possessing a two-photon sum or difference resonance instead of the vibrational Raman resonance. Each of these resultant eight fundamental diagrams gives another one by interchanging vertices 1 and 2, since emission at ω_2 can occur before absorption at ω_1. On the eight fundamental diagrams, vertices 2 and 3 can also be interchanged, thus providing the last eight terms in χ_{NR}. The diagrammatic analysis thus indicates that there are 24 density matrix terms proportional to the initial population factor $\rho_{aa}^{(o)}$ contributing to $\chi^{(3)CARS}$, four of which are vibrationally resonant. If the three pump waves had distinct frequencies $\omega_1, \omega_1', \omega_2$ (four wave mixing), there would be 48 terms (32). The complete analysis shows that there are as many terms proportional to $\rho_{bb}^{(o)}$, and twice as many to $\rho_{nn}^{(o)}$ whose contributions subtract from the $\rho_{aa}^{(o)}$ and $\rho_{bb}^{(o)}$ terms. Some of the diagrams associated with the terms proportional to $\rho_{bb}^{(o)}$ are shown in Figs.3 and 4.

These time-ordered diagrams have been separated into two categories (37). In the first category (Figs. 2(a), 3(b)) the interactions occur solely on one side (that is either on the wave function or on its complex conjugate) while the other side remains unperturbed during the scattering. *As a consequence, these processes cannot be accounted for by an amplitude squared transition probability.* For simplicity we have called these diagrams "parametric" because they describe the probability for the molecules to be returned to their initial state after the complete succession of interactions. The other type of diagram (Figs.2(b), 3(a)) gives the probability for the molecule initially in state $|a\rangle$ (or $|b\rangle$) to attain state $|b\rangle$ (or $|a\rangle$) and involves two interactions on each side of the diagram. We have therefore called them non-parametric or Raman-type diagrams. If the wave function and its complex conjugate both interact with fields of identical frequencies (as happens in SRS) the processes associated with a non-parametric diagram can be calculated from an amplitude squared transition probability. *When certain simplifying assumptions on collisions are made, this calculation leads to Fermi's Golden Rule.*

In conclusion, one can readily calculate each term in $\chi^{(3)}_{CARS}$ and associate it with the appropriate physical process by means of the diagrammatic representation. In particular, the Raman vibrationally resonant terms in expression (15) of χ^{ba}_R are thereby easily visualized. Term $A\alpha$ is calculated from Fig.2(a) ; it corresponds to a parametric diagram like term $D\gamma$ depicted in Fig.3(b). Terms $B\alpha$ and $C\gamma$ are shown respectively in Figs.2(b) and 3(a) and correspond to a non-parametric diagram. The last four terms $A\beta$, $B\beta$, $C\delta$, $D\delta$ are obtained from the diagrams of Figs.2(a), (b) and 3 by interchanging vertices 1 and 2.

C. Other CARS and Raman Polarization Terms and Molecular Population Changes

We have pointed out in section II.A that the waves at ω_1, ω_2 and ω_3 all undergo both CARS and SRS couplings. In order to fully describe the energy exchange between the waves and the molecules, we need to consider :

(1) CARS polarization components at ω_1 and ω_2 associated with $\chi^{(3)}_{CARS}(-\omega_1,\omega_2,\omega_3)$ and $\chi^{(3)}_{CARS}(\omega_1,\omega_1,-\omega_3)$ respectively (in addition to the ω_3 component just derived).

(2) the SRS component at ω_3 , $\chi^{(3)}_{SRS}(\omega_1,-\omega_1,\omega_3)$, that at ω_2 , $\chi^{(3)}_{SRS}(\omega_1,-\omega_1,\omega_2)$, and the two at $\omega_1,\chi^{(3)}_{SRS}(\omega_3,-\omega_3,\omega_1)$ and $\chi^{(3)}_{SRS}(\omega_2,-\omega_2,\omega_1)$ (we disregard terms like $\chi^{(3)}_{SRS}(\omega_2,-\omega_2,\omega_3)$ which are too far from resonance).

(3) the resulting population changes in $|a\rangle$ and $|b\rangle$, in the form of $(\partial/\partial t)\,\rho^{(4)}_{aa}(r)$ and $(\partial/\partial t)\rho^{(4)}_{bb}(r)$; the latter are best obtained from time-ordered diagrams. We shall disregard all terms associated with the wave at $\omega_4 = 2\omega_2-\omega_1$, since their treatment is identical.

The SRS coupling has no bearing a priori on the generation of the CARS wave. However, this coupling leads to a rapid energy exchange between the wave at ω_1 and that at ω_2, and the molecules. There is also a much weaker SRS coupling involving ω_1, ω_3 and the molecules. The former coupling, if too large, can seriously affect the CARS intensity. We shall calculate the corresponding susceptibility terms $\chi^{(3)}{}^{SRS}_{}(\omega_1,-\omega_1,\omega_2)$ and $\chi^{(3)}{}^{SRS}_{}(\omega_2,-\omega_2,\omega_1)$ first, then those associated with the ω_1, ω_3 interaction.

1. SRS Susceptibility. The basic diagrams for the SRS interaction between ω_1 and ω_2 and associated with $\chi^{(3)}{}^{SRS}(\omega_1,-\omega_1,\omega_2)$ are presented in Fig.5 and 6. The complete set is obtained by appropriate permutations of the first 3 vertices. A total of 24 terms proportional to $\rho^{(o)}_{aa}$ are thus obtained and an additional 24 terms proportional to $\rho^{(o)}_{bb}$. We have, if we write

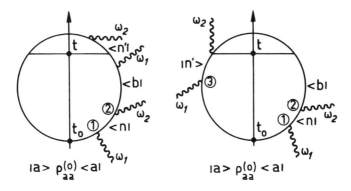

FIGURE 5. *Basic SRS diagrams with contributions proportional to* $\rho^{(o)}_{aa}$.

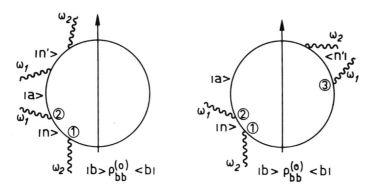

FIGURE 6. *SRS diagrams with contribution proportional to* $\rho^{(o)}_{bb}$.

down only those terms which contain the Raman energy denominator :

$$\chi^{(3)\,\text{SRS}}(\omega_1, -\omega_1, \omega_2) = \chi_{NR} + \frac{N}{\hbar^3} \sum_{a,b} \frac{1}{\omega_{ba} - \omega_1 + \omega_2 + i\Gamma_{ba}}$$

$$\times \sum_{n'} \left(\frac{P_{an'}\, P_{n'b}}{\omega_{n'a} + \omega_2 + i\Gamma_{n'a}} + \frac{P_{an'}\, P_{n'b}}{\omega_{n'b} - \omega_2 - i\Gamma_{n'b}} \right)$$

$$\times \sum_{n} \left[\rho_{aa}^{(0)} \left(\frac{P_{bn}\, P_{na}}{\omega_{na} - \omega_1 + i\Gamma_{na}} + \frac{P_{bn}\, P_{na}}{\omega_{na} + \omega_2 + i\Gamma_{na}} \right) \right. \qquad (16)$$

$$\left. - \rho_{bb}^{(0)} \left(\frac{P_{bn}\, P_{na}}{\omega_{nb} - \omega_2 - i\Gamma_{nb}} + \frac{P_{bn}\, P_{na}}{\omega_{nb} + \omega_1 + i\Gamma_{nb}} \right) \right] \; ;$$

χ_{NR} contains all non-Raman-resonant contributions. Expression (16) is not valid close to electronic resonance. On resonance, a number of terms lumped into χ_{NR} become large, among which are the "hot luminescence" terms described by Shen (50). The latter terms contain energy denominators of the type

$$(\omega_{na} - \omega_1)^2 + \Gamma_{na}^2$$

which are characteristic of the one-photon absorption that[1] takes the molecules through an intermediate diagonal state $|n><n|$. Off resonance, χ_{NR} is small and real. Introducing the expression for the off-resonance spontaneous Raman scattering cross section for excitation at ω_1 and for parallel polarizations (54)

$$\frac{d\sigma}{d\Omega} = \frac{1}{c^4} \, \omega_1 \omega_2^3 \left| \sum_n \left[\frac{P_{bn}\, P_{na}}{\hbar(\omega_{na} - \omega_1)} + \frac{P_{bn}\, P_{na}}{\hbar(\omega_{nb} + \omega_1)} \right] \right|^2 \; , \qquad (17)$$

[1]*Such terms are also found in the expression for the spontaneous Raman scattering cross-section on resonance. They are the source, in both SRS and normal Raman, of the well-known "fluorescence interference". These hot luminescence components are hard to separate spectrally from the resonant Raman ones, are subject to quenching, and can preclude quantitative concentration measurements if quenching rates are not known precisely (see the Chapter by R. Mathies in this volume). The CARS susceptibility also contains similar terms in the χ_{NR}; these terms are small in general and only cause minor correction to the line contours and intensities (37). This difference is fundamental and constitutes one of the key advantages of CARS over SRS and normal Raman for resonant spectral studies.*

and taking advantage of $\omega_{na} - \omega_{nb} \approx \omega_1 - \omega_2$, we arrive at

$$\chi^{(3)\,SRS}(\omega_1, -\omega_1, \omega_2) = s + S \,, \tag{18}$$

with $s = \chi_{NR}$ and

$$S = \frac{N}{\hbar} (\rho_{aa}^{(0)} - \rho_{bb}^{(0)}) \frac{c^4}{\omega_1 \omega_2^3} \frac{d\sigma}{d\Omega} \frac{1}{\omega_{ba} - \omega_1 + \omega_2 + i\Gamma_{ba}} \,.$$

Similar expressions are obtained for the other terms ; neglecting the small variations vs frequency of the one-photon energy denominators, and taking advantage of the relation $\omega_3 - \omega_1 = \omega_1 - \omega_2$, we may write :

$$\chi^{(3)\,SRS}(\omega_2, -\omega_2, \omega_1) = \chi^{(3)\,SRS}(\omega_1, -\omega_1, \omega_3) = \chi^{(3)\,SRS\,*}(\omega_3, -\omega_3, \omega_1)$$
$$= \chi^{(3)\,SRS\,*}(\omega_1, -\omega_1, \omega_2) \,, \tag{19}$$

which can also be expressed as a function of s and S . These results are in agreement with Chapter II of Ref.(27).

2. *Other CARS Susceptibilities.* Susceptibility components at ω_1 and ω_2 are readily obtained from the basic diagrams of Figs.2-4 by redistributing the four vertices, so as to have a final creation operation at ω_1 and ω_2 respectively. We shall not give the detailed result here. Off electronic resonance, we arrive at

$$\chi^{(3)\,CARS}(\omega_1, \omega_1, -\omega_2) = \chi_{NR} + \frac{Nc^4(\rho_{aa}^{(0)} - \rho_{bb}^{(0)})}{\hbar\,\omega_1\,\omega_2^3} \frac{d\sigma}{d\Omega} \frac{1}{\omega_{ba} - \omega_1 + \omega_2 - i\Gamma_{ba}}$$

$$= s + S^* \,, \tag{20}$$

which is in agreement with the result of the derivations performed in the Placzek approximation (21), if we note that Moya's $\gamma = 2\Gamma_{ba}$. We also have :

$$\chi^{(3)\,CARS}(\omega_1, \omega_1, -\omega_3) = \chi^{(3)\,CARS\,*}(\omega_1, \omega_1, -\omega_2) \tag{21}$$

and

$$\chi^{(3)\,CARS}(\omega_2, \omega_3, -\omega_1) = \chi^{(3)\,CARS}(\omega_1, \omega_1, -\omega_2) = \chi^{(3)\,CARS\,*}(\omega_1, \omega_1, -\omega_3) \tag{22}$$

Equation (22) reflects the fact that there are two families of terms possessing the vibrational resonance : one corresponding to the driving frequencies ω_1 and ω_2, the other to the frequencies ω_1 and ω_3.

3. Changes in Vibrational Population. The rate of change of the vibrational population resulting from CARS and SRS can be expressed as a function of $\partial/\partial t\, \rho_{aa}^{(4)}(\omega,t)$ and $\partial/\partial t\, \rho_{bb}^{(4)}(\omega,t)$. In order to calculate these terms, all fields including that at ω_3 must be written explicitly. We retain only the stationary non-oscillating components $\rho_{aa}^{(4)}(0,t)$ and $\rho_{bb}^{(4)}(0,t)$ which give the major contributions. Quite generally, we have for $\rho_{aa}^{(4)}(0,t)$:

$$
\begin{aligned}
\frac{\partial}{\partial t}\rho_{aa}^{(4)}(0,t) = -\frac{1}{i\hbar}\Bigg\{ \sum_n P_{an}\Big[& \rho_{na}^{(3)}(\omega_3,t)\, E_3^*\, e^{i(\omega_3 t - k_3 z)} \\
& + \rho_{na}^{(3)}(\omega_1,t)\, E_1^*\, e^{i(\omega_1 t - k_1 z)} \\
& + \rho_{na}^{(3)}(\omega_2,t)\, E_2^*\, e^{i(\omega_2 t - k_2 z)}\Big] \\
- cc \; \Bigg\} & - \Gamma_{aa}\, \rho_{aa}^{(4)}(0,t).
\end{aligned}
\tag{23}
$$

The three terms within the brackets on the right hand side of (23) are respectively associated with $P^{(3)}(\omega_3,t)$, $P^{(3)}(\omega_1,t)$ and $P^{(3)}(\omega_2,t)$, each of which contains both the CARS and the SRS contributions. A similar relation can be written for $\rho_{bb}^{(4)}(0,t)$. These expressions are extremely complex. However, only a few terms are important near electronic resonance. Then, the rate of growth of the population difference $\Delta = \rho_{aa} - \rho_{bb}$ is given by (37) :

$$
\begin{aligned}
\frac{\partial \Delta}{\partial t} = \frac{\rho_{aa}^{(0)}}{\hbar^4}|E_1|^2|E_2|^2 \Bigg\{ & 2\,\mathrm{Im}\Bigg[\frac{P_{bn}\,P_{na}}{\omega_{ba}-\omega_1+\omega_2+i\Gamma_{ba}} \\
& \times \left(\frac{P_{an}\,P_{nb}}{(\omega_{na}-\omega_1+i\Gamma_{na})^2} + \frac{P_{an}\,P_{nb}}{(\omega_{na}-\omega_1+i\Gamma_{na})(\omega_{nb}-\omega_2-i\Gamma_{nb})} \right)\Bigg] \\
& - \frac{4\Gamma_{na}\Gamma_{nb}}{\Gamma_{nn}} \times \frac{P_{bn}\,P_{na}}{(\omega_{nb}-\omega_2)^2+\Gamma_{nb}^2} \times \frac{P_{an}\,P_{nb}}{(\omega_{na}-\omega_1)^2+\Gamma_{na}^2} \Bigg\} \\
& - \Gamma_{aa}\,\rho_{aa}^{(4)} + \Gamma_{bb}\,\rho_{bb}^{(4)} \quad,
\end{aligned}
\tag{24a}
$$

where the important terms have been kept, and where we have assumed that the temperature was low, so that $\rho_{bb}^{(o)} = 0$. The term between straight brackets is the usual Raman term, whereas the second one corresponds to the hot luminescence part of the scattering (50). The contribution of the CARS terms on resonance is not written here ; some of these terms can be found in Appendix II of (37).

Off resonance, great simplifications occur. Including SRS and CARS contributions, one arrives at

$$\frac{\delta\Delta}{\delta T} = -2\left(\rho_{aa}^{(o)} - \rho_{bb}^{(o)}\right)\frac{c^4}{\hbar^2\omega_1\omega_2^3}\frac{d\sigma}{d\Omega}\frac{\Gamma_{ba}}{(\omega_{ba}-\omega_1+\omega_2)^2 + \Gamma_{ba}^2}$$

$$\times\left(|E_1|^2|E_2|^2 + |E_1|^2|E_3|^2 + E_1^2 E_2^* E_3^* e^{i(k_3'-k_3)z} + cc\right)$$

$$-\Gamma_{bb}\,\rho_{bb}^{(4)} + \Gamma_{aa}\,\rho_{aa}^{(4)} \quad . \tag{24b}$$

The first two terms between brackets correspond to SRS, the last two correspond to CARS.

In conclusion, we have derived the expression for the CARS polarization, and have given diagrams which can be used to visualize all density matrix contributions to this polarization. Our expression is valid even if any of the frequencies (the laser, the Stokes or the anti-Stokes frequency) is in resonance with an allowed dipolar transition ; we shall use this expression in Chapter III to predict the spectral content and line contour properties of the suscepti-bility both on and off-electronic resonance. Our immediate goal, however, is to derive the intensity of the CARS signal and recognize energy exchange processes.

D. CARS Power Density

We are now able to write down the wave equations at ω_1 , ω_2 and ω_3 with source terms in the form of third-order polarizations describing CARS and SRS couplings ; we can also write the expressions giving the changes in the populations of states $|a\rangle$ and $|b\rangle$ which result from these couplings. We shall treat the case off electronic resonance, for which the expressions are simpler and the mechanisms more transparent. First, we shall integrate the wave equation for the anti-Stokes field with given boundary conditions and in the

parametric approximation (i.e. disregarding higher-order
corrections to the polarization due to pump depletion).

   1. *Anti-Stokes Power Density.* We assume the nonlinear
medium to be located between plane boundaries at $z = 0$ and
$z = L$. Two strong waves at ω_1 and ω_2 are present ; their wave
vectors k_1 and k_2 are aligned with the z axis. We take the
anti-Stokes wave in the form of a weak plane wave with
propagation vector k_3 collinear with k_1 and k_2 . The expressions
for these fields are given by Eq.(4). With our assumption of
collinear propagation, each field obeys a wave equation of
the type :

$$\left(\frac{\partial^2}{\partial z^2} - \frac{1}{c^2} \frac{\partial^2}{\partial t^2}\right) \underline{E} = \frac{4\pi}{c^2} \frac{\partial^2}{\partial t^2} \underline{P} \quad , \tag{25}$$

with $\underline{P}$ given by Eq.(3). We use electrostatic units. We have
assumed that $\underline{E}$ vectors are collinear, and we recall that $\underline{P}$
is collinear with them due to the spatial tensor properties
of $\chi^{(1)}$ and $\chi^{(3)}$ (see section II-B). Dropping the vector
notation and expanding P , we arrive at

$$\left(\frac{\partial^2}{\partial z^2} - \frac{1}{c^2} \frac{\partial^2}{\partial t^2}\right) E = \frac{4\pi}{c^2} \frac{\partial^2}{\partial t^2} \left(\chi^{(1)} E + P^{(3)}\right) , \tag{26}$$

where the appropriate frequency components are implicitly
assumed on the right hand side for E and $P^{(3)}$, and where $P^{(3)}$
is given by Eq.(7).
   In order to treat the case of pulsed CARS generation, we
may assume the field envelopes at ω_i , $E_i = E(z, t, \omega_i)$, to be
slowly varying functions of z and t :

$$\frac{\partial}{\partial t}\left|E_i\right| \ll \omega_i\left|E_i\right| ; \frac{\partial}{\partial z}\left|E_i\right| \ll k_i\left|E_i\right| . \tag{27}$$

In order to eliminate consideration of transients, we further
specify

$$\frac{\partial|E_i|}{\partial t} \ll \Gamma\left|E_i\right| \quad , \tag{28}$$

where Γ is the smallest of all molecular rate constants
appearing in the calculation of $\rho^{(3)}(t)$. Using the familiar
definition of the frequency-dependent dielectric constant

$$\epsilon_i = 1 + 4\pi \chi^{(1)}(\omega_i) \quad , \tag{29}$$

we have, at frequency ω_3

$$\left(\frac{\partial^2}{\partial z^2} - \frac{\epsilon_3}{c^2}\frac{\partial^2}{\partial t^2}\right)\left[E_3\, e^{-i(\omega_3 t - k_3 z)} + cc\right] = \frac{4\pi}{c^2}\frac{\partial^2}{\partial t^2} P^{(3)}(z, t, \omega_3) \tag{30}$$

where $k_3 = n_3\omega_3/c$ and $n_3 = \sqrt{\epsilon_3}$. Similar wave equations can be written at ω_1 and ω_2.

The solution of Eq.(30) is composed of a driven wave and of homogeneous solutions in the form of forward-propagating and backward-propagating waves emanating from the medium boundaries (27). The latter are negligible in general in homogeneous mixtures. They will, however, be comparatively large if the driven solution is prevented from growing, e.g. if the medium is thin or if the phase mismatch is large (this situation may be encountered when studying samples placed in thin glass cells).

We use the positive frequency component to carry out our calculations ; the integration of Eq.(30) is best performed by introducing the characteristic variables :

$$\xi = (t + z n_3/c)/\sqrt{2}$$
$$\eta = (t - z n_3/c)/\sqrt{2} \tag{31}$$

and noting that $\partial^2/\partial t^2\, P^{(3)}(z, t, \omega_3) \simeq -\omega_3^2\, P^{(3)}(z, t, \omega_3)$ as a result of conditions (27).

Using $\delta k = k_3' - k_3$ (Fig.7) and neglecting the second order derivatives of E_3 (condition (27)), we arrive at

$$\frac{\partial E_3}{\partial \xi} = \frac{i\pi\sqrt{2}\,\omega_3}{n_3^2}\left[\chi^{(3)\,CARS}_{(\omega_1, \omega_1, -\omega_2)}\, E_1^2\, E_2^*\, e^{\,i\,\delta k c(\xi - \eta)/n_3\sqrt{2}}\right.$$
$$\left. + \chi^{(3)\,SRS}_{(\omega_1, -\omega_1, \omega_3)}|E_1|^2 E_3\right] \quad , \tag{32}$$

where, quite generally, envelopes E_1, E_2 and E_3 depend on ξ and η . Similar equations hold at ω_1 and ω_2. In order to integrate Eq.(32), we note that in the (ξ, η) coordinate system, light pulses of frequency ω_3 undergoing free propagation in the forward direction (i.e. without absorption or nonlinear inter-action) have envelopes that depend on η only. Because $\omega_1 \approx \omega_2 \approx \omega_3$ and because dispersion is small in transparent media, which

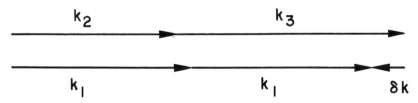

FIGURE 7. Wave vector diagram .

entails $n_1 \simeq n_2 \simeq n_3$, one can assume that light pulses at ω_1 and ω_2 also have envelopes that depend on η only if they propagate without interaction. This approximation applies for E_1 and E_2 in our situation of small pump depletion. One also remarks that the CARS contribution is subject to phase-matching (i.e. depends on δk and z through the exponential) whereas the SRS term does not. This result is classical (27). Since $|E_3|$ is small, the SRS term can be neglected here (but not in the other two equations at ω_1 and ω_2). Given the input boundary condition at $z = 0$, $E_3(\xi = \eta)=0$, integration is straightforward

$$E_3 = \frac{2i\pi\omega_3}{cn_3}\chi^{(3)CARS}(\omega_1,\omega_1,-\omega_2)E_1^2(\eta)\,E_2^*(\eta) \times \frac{\sin\delta k z/2}{\delta k/2}e^{i\delta k z/2}, \quad (33)$$

where we use mixed variables z and η for convenience. The power density $I_3 = (cn_3/2\pi)\,|E_3|^2$ is

$$I_3 = \frac{16\pi^4\omega_3^2}{c^4 n_1^2 n_2 n_3}\left|\chi^{(3)CARS}(\omega_1,\omega_1,-\omega_2)\right|^2 I_1^2(\eta)\,I_2(\eta)$$
$$\times\left[\frac{\sin(\delta k z/2)}{\delta k/2}\right]^2, \quad (34)$$

where I_1 and I_2 are power densities at ω_1 and ω_2 respectively.
The interpretation of Eq.(34) is evident (Fig.8) :

(1) The anti-Stokes pulse travels along with the pump pulses. Its normalized temporal shape at any fixed location z is determined by the product of power densities $I_1^2(\eta)\,I_2(\eta)$.

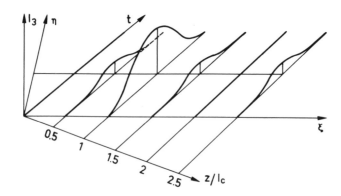

FIGURE 8. Evolution of the anti-Stokes pulse in a medium with negligible dispersion as a function of propagation distance and time ; the pulse retains its temporal shape at each point, but its peak amplitude depends on z/l_c.

As a consequence, if the pump pulses have the same Gaussian dependence with temporal width T_p , the anti-Stokes pulse is also Gaussian, synchronous with the pump pulses, and its width is $T_p / \sqrt{3}$.

(2) This anti-Stokes pulse peak amplitude has a sinusoidal variation as a function of propagation distance z ; the spatial period is $2\pi / \delta k$; the half period l_c , for which the signal reaches its first maximum, is called the coherence length ; the larger l_c , the larger the signal at this maximum. If $\delta k = 0$, the dependence becomes parabolic. The latter situation, if achievable by suitable orientation of k vectors, is favored experimentally, since it gives maximal conversion efficiency for any sample length.

(3) The intensity of the pulse is proportional to the square of the modulus of the susceptibility. The spectroscopic implication   is that, by monitoring the output as a function of $\omega_1 - \omega_2$, one is able to determine the dispersive properties of $|\chi|$, and hence to identify all the Raman resonances within it.

Some aspects of phase-matching and the dispersive properties of $|\chi|$ will be examined in detail in part III.

2. *Energy Exchange.* We now treat the problem of identifying the various modes of energy flow between light waves and molecules. We here make the assumption of a steady state in order to simplify our formalism. This assumption is usually quite valid, except when the laser pulses are short enough for transients to play an important role ; the latter case is complicated and will not be treated here (some considerations on this matter will be found in Chapter IV of Ref.(37)). In the preceding section, we have been able to integrate the wave equation for E_3 in the parametric approximation ; here, we shall assume that all three fields are large so that none of the terms are negligible. Furthermore, we take $|E_3| \neq 0$ at the entrance boundary $z = 0$ of the material. We transform Eq.(32) to obtain the steady state form of the wave equation :

$$\frac{\partial E_3}{\partial z} = \frac{2i\pi\omega_3}{cn_3} \left[\chi^{(3)CARS}(\omega_1, \omega_1, -\omega_2) E_1^2 E_2^* e^{i\delta kz} \right.$$
$$\left. + \chi^{(3)SRS}(\omega_1, -\omega_1, \omega_3) |E_1|^2 E_3 \right] \qquad , - \tag{35}$$

as well as similar obvious equations for E_1 and E_2. We are interested in the rates of exchange of quanta per unit volume at any plane $z = z_0$ within the material. These rates depend upon the phases and amplitudes of the waves in this plane, which parameters are obtained by integrating the set of coupled wave equations for E_1, E_2, E_3. This can only be done numerically and we will not undertake this task here. However, we still can treat the problem by letting

$$E_i(z_0) = |E_i(z_0)| e^{i\varphi_i(z_0)}$$

with $i = 1, 2, 3$ and where φ_i is the phase. Within a thin slab z_0, $z_0 + dz$ of thickness dz, we can take these envelopes for boundary conditions, so that we replace $e^{i\delta k z}$ by $e^{i\delta k(z - z_0)}$ in the wave equations ; since dz is small, we have $e^{i\delta k(z - z_0)} \underset{\sim}{1}$ in the slab. If we introduce $N_i = I_i / \hbar \omega_i$, the photon flux (number of photons per unit time) of energy $\hbar \omega_i$, we have, using Eq.(35) :

$$\frac{\partial N_3}{\partial z} = \frac{cn_3}{\hbar\omega_3}\left(E_3 \frac{\partial E_3^*}{\partial z} + cc\right)$$

$$= \frac{2\pi\hbar}{c^2}\left[i(S^* - S)\frac{N_1 N_3 \omega_1 \omega_3}{n_1 n_3}\right. \tag{36}$$

$$\left. + \left(i(S^* - S)\cos\delta\varphi - (2s + S + S^*)\sin\delta\varphi\right)\frac{N_1\omega_1}{n_1}\left(\frac{N_2 N_3 \omega_2 \omega_3}{n_2 n_3}\right)^{1/2}\right]$$

with $\delta\varphi = 2\varphi_1(z_0) - \varphi_2(z_0) - \varphi_3(z_0)$. We now introduce $a = 2\pi\hbar/c^2$, $W_i = N_i\omega_i / n_i$ ($i = 1, 2, 3$) to obtain the complete set of equations, including that for the change in number of vibrational quanta $(N/2)\partial\Delta/\partial t$:

$$\frac{\partial N_1}{\partial z} = a\left[i(S - S^*)(-W_1 W_2 + W_1 W_3)\right. \tag{37a}$$

$$\left. - 2(2s + S + S^*)(W_1^2 W_2 W_3)^{1/2}\sin\delta\varphi\right]$$

$$\frac{\partial N_2}{\partial z} = a\left[i(S - S^*)W_1 W_2\right. \tag{37b}$$

$$\left. - \left((2s + S + S^*)\sin\delta\varphi - i(S - S^*)\cos\delta\varphi\right)(W_1^2 W_2 W_3)^{1/2}\right]$$

$$\frac{\partial N_3}{\partial z} = a\left[-i(S - S^*)W_1 W_3\right. \tag{37c}$$

$$\left. - \left((2s + S + S^*)\sin\delta\varphi + i(S - S^*)\cos\delta\varphi\right)(W_1^2 W_2 W_3)^{1/2}\right]$$

$$\frac{N}{2}\frac{\partial\Delta}{\partial t} = -a(S - S^*)(W_1 W_2 + W_1 W_3 + 2(W_1^2 W_2 W_3)^{1/2}\cos\delta\varphi) \tag{37d}$$

The interpretation is now transparent, because of the term to term correspondence between the four equations. We first note that $i(S - S^*)$, which is (minus) the imaginary part of the SRS susceptibility $\chi^{(3)SRS}(\omega_1, -\omega_1, \omega_2)$, is positive, while $s + (S + S^*)/2$ is the real part. The first term of Eq.(37a) reflects the familiar SRS coupling to ω_2 ; it corresponds to the first terms in (37b) and (37d). This interaction is usually depicted by the energy level diagram of Fig.9(a) : a photon at ω_1 is split into

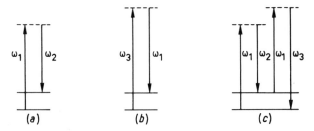

FIGURE 9. Energy level diagrams used for the interpret-ation of energy exchange in SRS and CARS.

a photon at ω_2 and a vibrational quantum at ω_{ba}. This process is the largest in usual CARS experiments and can cause, if the pump is too intense, large changes in ρ_{aa} and ρ_{bb}; these changes in turn reduce $\chi^{(3)CARS}$ and perturb the CARS measurement. A similar process also takes place between ω_3 and ω_1, causing a reduction in I_3 (Fig.9b). In usual CARS experiments, this process is far weaker and can be neglected, as was done in the preceding section.

The last terms of all four equations depend upon the relative phases of the waves and represent CARS per se. Two categories of processes interfering with one-another can be recognized :

(1) a truly parametric process whereby two photons at ω_1 are destroyed to give one at ω_2 and one at ω_3 without energy exchange with the medium. This process, which is associated with the real part of the susceptibility, has been tradition-ally represented by the energy level diagram of Fig.9c. The reverse process is also possible depending on the phases of the waves (e.g. for $I_c < z < 2I_c$ in Fig.8).

(2) a Raman-type process, described by Bloembergen (27) as the interference between the two Raman processes of Figs.9a and b : one photon at ω_3 is destroyed to produce one at ω_2 and two vibrational quanta. This process is associated with the imaginary part of the susceptibility $S - S^*$ and can be reversed by changing the phases of the waves ; it has often been ignored in the literature, where CARS is frequently erroneously presented as non-disturbing to the vibrations. Yet, this process is, at resonance ($\omega_1 - \omega_2 = \omega_{ba}$), the only one actually taking place (the other one then vanishes since the real part, which is equal to S, is very small).

Generally, the boundary conditions encountered in CARS are those of the preceding section II.D.1, i.e. $E_3 = 0$ at $z = 0$. Then, in the parametric approximation, $\delta\varphi$ and the field amplitudes can be calculated explicitly at $z = z_0$; it has been shown (21,37,55) that the rate of the parametric process is proportional to $[Re(\chi^{(3)CARS})]^2$ and that the rate of the Raman-type process proportional to $[Im(\chi^{(3)CARS})]^2$.

3. Energy Level Diagrams vs Time-Ordered Diagrams. We have just hinted at the ambiguities associated with representing the energy exchange in a nonlinear optical process like CARS by means of energy level diagrams : several different diagrams are actually necessary and the phases and frequencies of the waves must be specified in order to fully describe the mechanisms ; furthermore, these diagrams no longer make sense near electronic resonances. As a matter of fact, the representation of energy transfers by means of energy level diagrams is unambiguous only for spontaneous processes like absorption, fluorescence or Raman scattering.

On the other hand, energy level diagrams can also be used to represent how one arrives at a certain polarization frequency in the material by suitably adding or subtracting electric field frequencies, i.e. for a purpose similar to that of time-ordered diagrams. For this usage, the latter are, however, superior in that they fully describe the quantum mechanical sequence of interactions contributing to the creation of the polarization of interest, together with the population changes subsequent to the interaction with the created electromagnetic field. However, energy level diagrams are superior in that molecular states in resonance with the exciting fields are directly visualized. In depicting polarizations, both sorts of diagrams are therefore useful and complementary.

III. PRACTICAL CONSIDERATIONS

The properties of CARS and its applicability as a diagnostics tool will become apparent following a critical evaluation of Eq.(15) (spectral properties) and Eq.(34) (signal strength). The physical mechanisms (Eqs.(37)) must be borne in mind when evaluating saturation.

CARS spectroscopy can be accomplished in various manners depending on the application envisioned (e.g. high-resolution spectroscopy or chemical analysis). Generally, focussed beams are used, which improves spatial resolution and signal strength. The spectra are usually retrieved by holding ω_1 fixed, varying ω_2 so that $\omega_1 - \omega_2$ is swept across the resonances of interest while monitoring the anti-Stokes flux (Fig.10) ; variants of this method can be used, especially for rapid measurements. We begin the discussion with the problems of propagation and beam geometry.

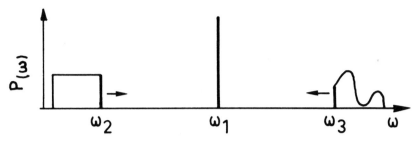

FIGURE 10. Recording of CARS spectra by means of monochromatic lasers (ω_1 fixed, ω_2 tunable).

A. Beam Arrangement

The configuration using unfocussed parallel beams with large diameters is seldom used. It has, as a matter of fact, been used only for a verification of the sinusoidal dependence predicted in Eq.(34)(41) and for the visualization of a small supersonic jet of H_2 (17). Since the power density I_3 is proportional to $I_1^2 I_2$, it seems advantageous to focus the beams to a small diameter and to use high peak power sources. If the condition $\delta k = 0$ is assumed, then it can be shown that :

(1) the anti-Stokes flux is contained within the same cone angle as the pump beams emerging from the focal region ;

(2) this flux is generated for the most part within the focal region (where $I_1^2 I_2$ is large) ;

(3) the total power in the anti-Stokes beam some distance beyond the focus is approximately given by (17)

$$P_3 = \left(\frac{2}{\lambda}\right)^2 \left(\frac{4\pi^2 \omega_3}{c^2}\right)^2 \left| \chi^{(3)CARS}(\omega_1,\omega_1,-\omega_2)\right|^2 P_1^2 P_2 \ , \quad (38)$$

where refractive indices were taken as unity, where $\lambda = 2\pi c / \omega$ with $\omega \simeq \omega_1 \simeq \omega_2 \simeq \omega_3$ and where P_1 and P_2 are the powers at ω_1 and ω_2 respectively. This expression was obtained by assuming that all the signal is generated from a small cylindrical volume about the focus having a length equal to the confocal parameter l of the beams. If Gaussian beams are used, the beam waist at the focus is $\phi = 4\lambda f/\pi d$ where f is the focal length of the lens and d the beam diameter in the plane of the lens ; we also have $l = \pi\phi^2/ 2\lambda$ (56). More accurate calculations were done by Régnier (41) : a derivation similar to that of Ward and New (57) gives a formula which lends itself to numerical integration. It was found that 75% of the anti-Stokes power is generated within a length of $l_r = 6 l$, which we define as the spatial resolution. The study also confirmed that P_3 is approximately independent of the f-number as predicted by Eq.(38). There is, however, a slight decrease in P_3 in going from short to long focal length (a factor of 3 is found from F = 3 cm to F = 50 cm, with d = 3 mm). These results were confirmed recently (25,42,58).

The above results remain valid as long as the condition $l_r < l_c$ is fulfilled (where l_c is the coherence length defined in II.D.1). Régnier's calculations show that, for $l_c < l_r$, P_3 is reduced and becomes inversely proportional to l_c^2 . One can express l_c as a function of the derivatives of n with respect to ω (13,17) :

$$ l_c = \pi c \left[(\omega_1 - \omega_2)^2 \left(2 \frac{\partial n}{\partial \omega} + \omega_1 \frac{\partial^2 n}{\partial \omega^2} \right) \right]^{-1} . \qquad (39) $$

l_c is in the range of 20 to 300 cm in transparent gases at STP. Since l is usually $\approx$ 1–2 mm, dispersion is not expected to reduce conversion efficiencies except at very high densities (100 bar or more).

In liquids and solids, l_c is much shorter (in the range of 0.1 to 1 mm) due to the large dispersion, which results in $|k_2| + |k_3| > 2|k_1|$. One then orients k_1 and k_2 at a small angle (Fig.11). The anti–Stokes radiation is generated in the direction for which δk is zero or minimal. In practice, the angle θ is of the order of 1 to 3° for optimal conversion, and since $|k_1| \simeq |k_2| \simeq |k_3|$, k_3 and k_2 are almost symmetrical with respect to the direction of k_1 . Spatial resolution is then given by the volume common to the pump beams, i.e. of the order of 1 mm.

In gases, we noted that spatial resolution was about 6 l with collinear Gaussian beams. With d = 3 mm and F = 10 cm, the resolution is 8 mm. For some applications, this is insufficient. Eckbreth recently proposed a technique, dubbed BOXCARS (59), based on a distortion of the diagram of Fig.11. The new diagram (Fig.12) also satisfies phase-matching. Experimentally, this arrangement is obtained by splitting the

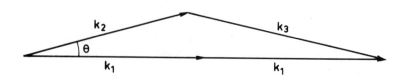

FIGURE 11. Wave vector arrangement for exact phase-matching in condensed, dispersive media.

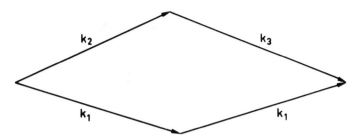

FIGURE 12. *Wave vector diagram for crossed-beams phase-matched CARS or BOXCARS.*

ω_1 pump into two parallel beams (Fig.13). These beams are brought to a common focus with the ω_2 beam. The advantage is that the signal is generated from the focal volume common to all three beams, and that spatial resolution can be made as small as 1 mm. The result of beam crossing is apparent in Fig.14. The curves were obtained by moving a glass platelet of 0.1 mm thickness along the focal regions. The signal is generated by the small, nonresonant susceptibility χ_{NR} of the glass since there is no Raman resonance. For the most part, this signal is composed of the homogeneous solutions of the wave equation at the boundaries, since the plate is too thin for the driven solution to grow appreciably. We note a large improvement in going from collinear (curve 1) to crossed beams (curve 2). Note also that we plotted the *square root* of P_3, i.e. a quantity proportional to the anti-Stokes *field* $E_3(z)$

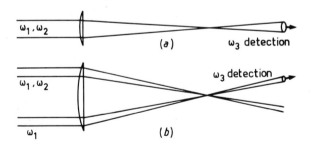

FIGURE 13. *Experimental beam arrangement : (a) conventional CARS ; (b) BOXCARS. In BOXCARS, a conventional CARS beam is also emitted in the direction of ω_2; this beam is actually 10 to 50 times stronger than the BOXCARS beam.*

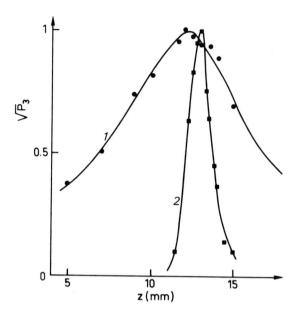

FIGURE 14. *Effect of beam crossing on CARS spatial resolution using near-diffraction-limited ruby laser and dye laser
beams. Curve 1●: collinear beams (conventional CARS of
Fig.13a) ; curve 2■: crossed beams. All beams are 3 mm in dia;
f =10 cm ; distance between parallel beams for BOXCARS : 6 mm.
Profiles have been normalized.*

generated at the platelet[1]. This remark is important since the
P_3 generated in a gas or a liquid from the focal region is
proportional to $|\int E_3(z)\,dz|^2$, i.e. to the square of the
areas under the curves. We have defined spatial resolution l_r
as the length of focal volume over which 3/4 of the total
anti-Stokes power or, equivalently, 87% of the field, are
generated. From the figure, we estimate l_r to be approximately two times the full width at half maximum (FWHM) for
curve 1, i.e. l_r = 20 mm. This estimate is empirical, based
upon the asymptotic behavior of E_1 and E_2 away from the focus
(the behavior is easily seen to be $|z|^{-3}$, with $z = 0$ at the
focus). This experimental l_r is larger than the l_r from the
theory of Régnier by a factor of 1.6, which is excellent
given the fact that the beams diverge slightly more than
the diffraction limit, and that their foci are difficult

---

[1]*Spatial resolutions estimated from the plots of P_3 as a
function of platelet position would appear exceedingly optimistic.*

to align exactly. For curve 2, l_Γ is about one FWHM because of the abrupt decay, i.e. l_Γ = 1.7 mm. Assuming the beams are diffraction-limited and have the same waist ϕ =30 μm, and that λ = 0.7 μm, we calculate that l_Γ should be slightly less than 1 mm given the crossing angle of 50 mrad. Again, the agreement is satisfactory since it is difficult to obtain exact overlap of all the foci and since beam divergence is larger than the diffraction limit. These uncertainties also explain why the maxima of curves 1 and 2 do not coincide.

B. *Signal Strength and Detection of Trace Gases*

The strongest asset of CARS is the considerable magnitude of its signal. Using single frequency pulsed lasers with modest powers (100 kW) and short duration (10 ns), we can collect a respectable number of photons from a diluted gas with collinear focussed beams. Take CO gas, at a dilution of 1% in air at STP, with parameters as given in Table 1. We assume the lasers to be tuned to one of the strongest Q lines at 2143 cm^{-1}. The occupation of the lower level is about 0.15, we have $\rho_{bb} \approx 0$, and the FWHM spontaneous Raman line width is typically $2\Gamma_{ba}/2\pi c$ =0.15 cm^{-1}. We estimate the susceptibility at 1.8 x 10^{-18} cm^3/erg and P_3 at 140 ergs/s, which amounts to 4 x 10^5 photons in 10 ns. Using 3 mm-dia beams and f =5 cm, the spatial resolution is of the order of 2 mm. If one were to perform a spontaneous Raman scattering experiment on the same 2mm-long sample, each shot would produce only 25 photons per steradian in the *entire Stokes sideband* using the laser pulse at ω_1. In calculating this result, we used the formula for the Stokes power pertaining to level $|a>$

$$P_{St} = N \rho_{aa}^{(0)} \frac{d\sigma}{d\Omega} l \Omega P_1 \, , \tag{40}$$

Table 1. Numerical Values for Calculation of CARS Power in a Gas Mixture

$N(cm^{-3})$	$c(cm\,s^{-1})$	$\rho_{aa}^{(0)}$	$h(ergs\,s)$	$\lambda(cm)$	$\frac{d\sigma}{d\Omega}(cm^2/sr)$	$\Gamma(s^{-1})$
3 x 10^{17}	3 x 10^{10}	0.15	6.6x10^{-27}	5 x 10^{-5}	1.5x10^{-31}	0.15πc

and integrated over all $|a\rangle$ levels contributions, using $\sum_a \rho_{aa} \simeq 1$ since $\rho_{bb}=0$ at 300 K. The advantage is clearly in favor of CARS, especially since all of the signal is contained in a solid angle of a few 10^{-3} sr, and since much larger pump powers could be used.

Our estimates of P_3 are very crude due to the numerous approximations used in the derivation. Therefore, agreement with experiments is not expected to be good ; our own experiments have produced a number of photons consistently lower than expectations by a factor of 1 to 10. Crossing the beams as in the experiment of Fig. 14 resulted in an additional reduction of about 20.

From the above estimates and knowing the scaling laws, extremely low concentrations seem to be detectable in principle. For example, with pump powers of 0.1 GW and a pulse duration of 10 ns, a CO concentration as low as 10 ppb will produce 40 photons per shot, which is ample for detection. In practice, several effects will prevent this :

(1) saturation mechanisms ,
(2) spectral interference effects ;

we shall examine them in turn.

C. Saturation

Saturation relates to such phenomena as the SRS population perturbation given by Eqs. (24), or pump depletion. Near electronic resonance, these phenomena are strongly enhanced ; in addition, one-photon absorption and the dynamic Stark effect become important. These effects have been discussed at length (37, 60), and we shall only summarize the results here.

1. Off Resonance.
The population perturbation is given by Eq. (24b). This perturbation can cause changes in the intensities of the CARS lines, and consequently can also bias the concentration and temperature measurements based on these intensities.

The perturbation is largest near the focus. It can be tolerated if, after passage of the pulse and neglecting relaxation, it remains much less than the initial population. We write

$$\frac{\partial \Delta}{\partial t} T_p < 0.1 \, \Delta \qquad , \qquad (41)$$

which corresponds to an error of less than 5% in the measure-
ment of $|\chi|$ due to the time integration. Neglecting relax-
ation, we arrive at

$$I_1 I_2 < 1 \ (GW/cm^2)^2 \quad , \tag{42}$$

using the parameters for CO as listed in Table 1 and $T_p = 10$ ns,
and exciting at line center. This condition is satisfied using
P_1, $P_2 < 3.5$ kW with $d = 3$ mm and $f = 10$ cm. If one allows
for relaxation, higher powers can be applied. Our own experiments showed that, with this geometry and $P_1 = 1$ MW,
$P_2 = 50$ kW, the perturbation on Δ remains undetected (i.e.
less than 10%) at STP. We obtained the same result with
BOXCARS, using 1 MW in each beam at ω_1 and the same P_2, although
the power densities were higher, and the signal contributions
came only from the high power density regions.

It is important to point out that for a constant product
$P_1 P_2$, it is advantageous to use a small P_2 and a large P_1 in
order to maximize P_3 (60). This important consideration must
be borne in mind when designing a CARS set-up. Finally, if
the total pressure is reduced, the linewidth is reduced and
relaxation is slowed down, so that the saturation threshold
is rapidly lowered. The power densities then have to be
reduced by focussing less tightly, which degrades spatial
resolution.

Another aspect of saturation is pump perturbation. This
occurs only for large N, e.g. near 1 bar or above, and for
powers in excess of 0.1 MW. However, the anti-Stokes signal
is then excessively large, and P_1 and P_2 certainly can be
reduced.

2. *On Resonance.* When ω_1 or ω_2 come into resonance with
one-photon absorptions, the Raman saturation processes just
discussed are enhanced. Furthermore, new phenomena come into
play. These are :
(1) population changes in the ground and electronic
states coupled by the transitions in resonance with the
fields ; for $\omega_1 \simeq \omega_{na}$, , we write

$$\frac{\partial}{\partial F} \left(\rho_{nn}^{(2)} - \rho_{aa}^{(2)} \right) = \frac{4 \rho_{aa}^{(0)}}{\hbar^2} \left| p_{na} E_1 \right|^2 \frac{\Gamma_{na}}{(\omega_{na} - \omega_1)^2 + \Gamma_{na}^2}$$
$$- \Gamma_{nn} \rho_{nn}^{(2)} + \Gamma_{aa} \rho_{aa}^{(2)} . \tag{43}$$

For a resonant gas diluted in air at STP, with an absorption
linewidth of 0.2 cm^{-1} and $|p_{na}| = 0.1$ D (61) we calculate a
maximum steady state power density of $I_1 = 10$ MW/cm^2. Using

this value in Eq.(24a), we then calculate that the upper limit on I_2 is 10 MW/cm^2 for the SRS perturbation to $|a\rangle$ and $|b\rangle$ to be tolerable.

(2) attenuation of the pump. This attenuation is given by

$$I_i(z) = I_i(0) e^{-\alpha_i z} \qquad (44)$$

with $i = 1,2,3$, $\alpha_i = 4\pi k_i \, \mathrm{Im}\left(\chi^{(1)}(\omega_i)\right)$ and

$$\chi^{(1)}(\omega_i) = N \rho_{aa}^{(0)} |P_{an}|^2 / \hbar (\omega_{na} - \omega_i - i\Gamma_{na}) \quad . \qquad (45)$$

From Eqs.(44) and (45), it was estimated that the attenuation was negligible at sub-mb partial pressures in cm-long samples ; for these concentrations, no disruption of the phase-matching condition should take place as a result of the dispersive properties of $\chi^{(1)}$ in the vicinity of the resonance (37).

(3) Stark effect. Lineshifts and line broadenings, which are on the order of $|P_{na} E|/\hbar$ become appreciable at the power levels for which population perturbations reach unacceptable levels (37). In avoiding the latter, we also avoid the former.

3. *Conclusion: Design Considerations.* To summarize the preceding discussions, we here wish to elaborate on the engineering problems one has to solve when designing a CARS set-up for a specific gas phase application. The starting point is the thermodynamic state of the molecular system : temperature, pressure and concentration of the species of interest. From these, we can deduce numerical values for N , ρ_{aa} , ρ_{bb} , Γ , which will be used in the calculation. The procedure then is the following :

(1) calculate $|\chi|$ for the weakest line to detect (compatible with the spectral constraints as discussed in the next section) ;

(2) affix the minimum power figure P_3 for this line, given the stray light level, repetition rate and pulse duration ;

(3) deduce $P_1^2 P_2$ using Eq.(38) ;

(4) determine maximum tolerable power densities I_1 and I_2 compatible with one-photon (if any) and SRS saturation ; if there is no one-photon resonance, it is preferable to use a large I_1 and small I_2 for a given $I_1 I_2$;

(5) from P_1 , I_1 and P_2 , I_2, beam waists at the focus are easily deduced ; the larger waist must be taken for the calculation ; from it the f-number is deduced ;

(6) if the resulting spatial resolution is not adequate, a BOXCARS arrangement can be chosen, but beam powers and focal cross-sections will have to be approximately tripled in order to compensate for the reduction in conversion efficiency and to maintain P_3 at the desired level.

D. *CARS Spectra*

Several authors have examined in detail the selection rules for conventional CARS and established that they are the same as for normal Raman scattering (22,62). Therefore, CARS spectra contain in principle the same information as normal Raman spectra. Indeed, it is clear from Eqs.(15a) and (20) that all the Raman-allowed transitions produce a resonance in the CARS spectrum ; it is also evident that these resonances are found, if the spectrum is displayed vs ω_2, at the same location as the corresponding Stokes lines in a spontaneous Raman spectrum. However, a few distinctive differences can be found as regards the population factors, the presence of another form of resonance (labelled double-electronic resonance) in the case of resonant CARS, and the line contours. We shall first consider the spectral content and the line intensities both on and off resonance. Then, we shall discuss the line contours, which are peculiar to CARS and exhibit the interference between neighbouring lines and non-resonant background.

1. *Spectral Content*

a. *Off resonance*. Off resonance, CARS spectra very strongly resemble normal Raman Stokes spectra. The resemblance is particularly striking if one plots the square root of the CARS signal, which is proportional to $|\chi|$, rather than the signal itself (which biases the spectrum in favor of the stronger resonances). There is, however, one difference which stems from the presence of different population factors in CARS and Raman spectra : the CARS susceptibility is proportional to $\rho_{aa}^{(o)} - \rho_{bb}^{(o)}$, whereas the Stokes power, as given by Eq.(40), is proportional to $\rho_{aa}^{(o)}$. The difference is insignificant in liquids and gases at low temperature, but appreciable differences are found in the line intensities and Q-branch contours of gases at high temperature, due to the presence of a large population in the $|b\rangle$ states.

Apart fron this comment, we shall not discuss further off-resonance CARS spectra, and the reader is referred to general presentations on normal Raman spectra for discussions on selection rules, band intensities and envelopes, etc. (3-5).

b. On resonance. We have pointed out in section II.C.1 that the CARS susceptibility on resonance contains a few terms which are the analog of what Shen (50) calls hot luminescence (and Williams et al. (63) "redistribution") and we stressed the fact that these terms are small. Therefore, their contribution to the spectrum generally amounts to but a small distortion of some of the resonance CARS lines. In resonance Raman scattering, on the contrary, the hot luminescence terms are large and give many intense lines in the spectrum ; furthermore, which lines appear and how intense they are depend upon thermo-dynamic conditions and mixture composition. CARS, not being affected by these factors, is therefore superior.

For this reason, experimental efforts have been under-taken in the area of pure spectroscopic studies. In addition, the applications in analytical chemistry are equally promising since, the CARS lines being strongly enhanced on resonance, one hopes to obtain better detection sensitivities. Resonant CARS spectra in solutions were observed and interpreted for the first time in 1976 by Chabay and coworkers (43,64). Others have also reported similar spectra and given theories for them (34,35). Recently, we were able to observe resonant CARS spectra in Iodine in the gas phase, and to present a coherent interpretation for these spectra (37,38). This theory can be generalized to all gases with discrete absorption lines and is also readily transposed to gases with absorption continua and to the liquids, whose absorption bands are usually quite broad. We shall present it now.

From Eq.(15a), one can see that χ_R^{ba} shows resonant en-hancement whenever ω_3 , ω_1 or ω_2 come close to a discrete absorption frequency. Thus, the CARS spectrum may present a complex structure, since any of these three frequencies may come into resonance with absorption lines as the vibrational-rotational structure is being analyzed. Only the levels associ-ated with these one-photon transitions will contribute strong-ly, so that the spectrum will be very different from the off-resonance case. Moreover, the contribution of the population factor $\rho_{aa}^{(o)}$ in χ_R^{ba} is no longer equal to that of $\rho_{bb}^{(o)}$ and their respective weights vary differently as the spectrum is scanned. However, simplifications occur, since :

(1) One of the three frequencies (ω_1 in general) is usually held fixed on resonance with an absorption line for instrumental convenience.

(2) When the Raman spectrum is analyzed, $\omega_1 - \omega_2$ is varied over a finite spectral domain and a limited number of Raman transitions contribute in turn to the susceptibility. For a given Raman transition i from $|a'\rangle$ to $|b'\rangle$, there are only a few types of allowed one-photon transitions to or from each of the rotational sublevels $|a'\rangle$, $|b'\rangle$, e.g. $\Delta J = \pm 1$ for I_2; therefore the number of absorption lines that ω_2 or ω_3 can reach to enhance χ_R^{ba} is finite.

(3) One can often assume $T_V \ll \hbar \omega_{ba}/k$ where T_V is the vibrational temperature and k the Boltzmann constant, so that $\rho_{bb}^{(o)} \ll \rho_{aa}^{(o)}$. Similarly, $\rho_{nn}^{(o)} \ll \rho_{aa}^{(o)}$, $\rho_{bb}^{(o)}$.

(4) Then, only term $A\alpha$ in Eq.(15b) is large, the other electronic terms containing anti-resonance denominators.

For these conditions, disregarding all electronic terms but $A\alpha$, we shall show that the dominant characteristics are provided by double resonances : these involve combinations of any two of the three following resonances, i.e. vibrational on ω_{ba}, electronic with ω_1 tuned on ω_{na} or with ω_3 tuned on $\omega_{n'a}$; triple resonances are improbable because level spacings are different in the ground and excited electronic states.

As an example, let us consider the case of I_2 at low temperature (hence $\rho_{bb}^{(o)} = 0$). We shall seek resonance enhancement from one-photon transitions between the $X^1\Sigma_g^+$ state, whose vibration-rotation manifold contains all the $|a\rangle$ and $|b\rangle$ sublevels, and the upper state $B^3\Pi_{ou}^+$ which contains the $|n\rangle$ sublevels. If ω_1 is tuned to a particular, isolated, absorption line ω_{na} i, one selects the particular and unique initial state, which we label $|a'\rangle = (v'',J)$, among all the possible $|a\rangle$ sublevels. Selection rules prescribe $J' = J \pm 1$ for $|n\rangle$, which we label (v',J'). We shall discuss here the case of an R transition, i.e. $J' = J + 1$ (Fig.15). One then observes, as ω_2 is varied, omitting the i superscript in the molecular frequencies :

(a) $(\omega_{na}, \omega_{ba})$ double resonances, i.e. $\omega_1 = \omega_{na}$ and $\omega_1 - \omega_2 = \omega_{ba}$ (Fig.15a). These are made up of fundamental and overtone Raman doublets with spacings equal to those of the various vibrational states in the electronic ground state (Fig.16a). The doublets are composed of one Q and one S Raman line if ω_1 is resonant with an R absorption line (or of an O and a Q line if the absorption line is a P line). These resonances obviously correspond to the allowed Raman transitions originating from level $|a'\rangle$. If collisional broadening dominates, the strengths of these lines are inversely proportional to $\Gamma_{na} \Gamma_{ba} \times |\omega_3 - \omega_{n'a}|$. When exact vibrational resonance is obtained $(\omega_1 - \omega_2 = \omega_{ba})$, ω_2 is also automatically resonant with the allowed transition ω_{nb}. This resonance,

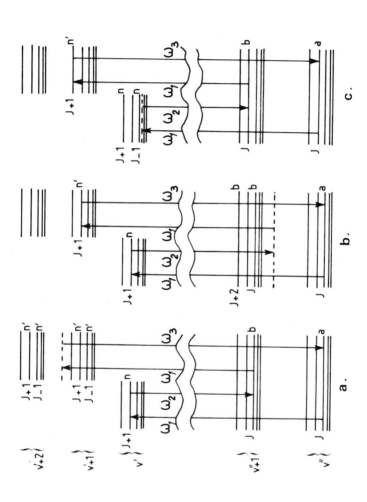

FIGURE 15. Energy level diagrams representing the states contributing to resonance-enhanced CARS in a diatomic molecule :

a : fundamental ($\Delta v = 1$) Raman vibrational resonance enhanced by the resonance condition $\omega_1 = \omega_{na}$;

b : double electronic resonance with $\omega_1 = \omega_{na}$ and $\omega_3 = \omega_{n'a}$;

c : fundamental vibrational resonance enhanced by the resonance condition $\omega_3 = \omega_{n'a}$.

however, only enters into the $C\gamma$ and $D\gamma$ terms, which possess negligible weight in the susceptibility because $\rho_{bb}^{(o)} \simeq 0$. Unless a fortuitous coincidence takes place, ω_3 will not be simultaneously in exact resonance with the $\omega_{n'a}$ allowed transitions from $|a'\rangle$ (triple resonance); actually, its detuning $\omega_{n'a} - \omega_3$ may be as large as 60 cm^{-1} in Iodine, since the spacing between the first vibrational manifolds is about 120 cm^{-1} in the $B\,{}^3\Pi_{ou}^{+}$ state. If a second pump laser at ω_1' were used, then the frequency $\omega_3 = \omega_1 + \omega_1' - \omega_2$ could be tuned independently to one of the $\omega_{n'a}$ resonances, giving additional enhancement of the CARS line.

(b) $(\omega_{na}, \omega_{n'a})$ double resonances (Fig.15b). These are made up of P and R lines occurring in doublets and corresponding to possible resonances between ω_3 and the allowed one-photon transitions from $|a'\rangle$. The positions of these lines in the spectrum are obtained from the condition $\omega_1 - \omega_2 = \omega_{n'a} - \omega_1$. The spacing between the doublets is equal to that between the vibrational levels in the excited electronic state (Fig.16b). The analysis of such lines constitutes a new type of spec-

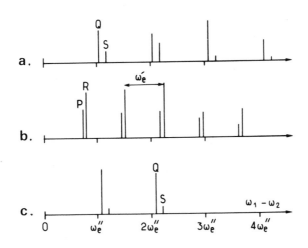

FIGURE 16. Break-up of the CARS spectrum into the three types of double resonances.
a : fundamental and overtone lines resulting from resonances of the type depicted in Fig.15a. Doublet spacing is constant (if anharmonicity is neglected) and is equal to ω_e'' .
b : doublets associated with Fig.15b (spacing is ω_e').
c : doublets associated with Fig.15c, assuming $|b\rangle$ is in the second excited vibrational level, so that only 2 doublets are seen. Here, spacing is ω_e'' but the lines do not coincide with those of Fig.16a in general because J is different.

troscopy which can be of great importance for the precise study of one-photon absorption ; if collisional broadening dominates, the strengths of these lines are inversely proportional to $\Gamma_{na} \Gamma_{n'a} \times |\omega_{ba} - \omega_1 + \omega_2|$. Their intensities are comparable to those of the Raman lines discussed in (a). One can also by this technique select and analyze any weak resonance buried in a continuum by simply tuning ω_1 to one of the other allowed but strong ω_{na} absorptions originating from $|a'\rangle$.

(c) In addition to these two families of double resonances associated with level $|a'\rangle$, one also expects another class of double resonances namely ($\omega_{n'a}, \omega_{ba}$) double resonances pertaining to a different initial level $|a^j\rangle$ (Fig.15c). The location of these resonances can easily be obtained if one realizes that for ω_3 and $\omega_1 - \omega_2$ to be simultaneously resonant, it is necessary for ω_1 to also be resonant with the $\omega_{n'b}$ transitions. The latter usually do not appear in the absorption spectrum because $\rho_{bb}^{(o)} = 0$, but would of course be observed at higher temperatures. Identification of $|a^j\rangle$ immediately gives the doublets to be found (O + Q for ω_1 tuned to an R line, Q + S for a P line) (Fig.16c). If collision broadening dominates, the line strengths are inversely proportional to $\Gamma_{n'a} \Gamma_{ba} \times |\omega_{na} - \omega_1|$.

(d) Whenever a resonance in Fig.16a coincides with one in Fig.16b, one then has a triple resonance and the resulting intensity is strongly enhanced.

In summary : the complete spectrum corresponding to ω_1 fixed to a given ω_{na} contains the two sets of doublets predicted in Figs.16a and 16b. If the absorption spectrum contains several other lines close to or coincident with ω_{na}, each of these will also contribute its two associated sets of doublets. Each one of the possible $\omega_{n'b} = \omega_1$ resonances, if any, will also contribute its associated set of doublets as depicted in Fig.16c. All these strong lines interfere with one another and have their contours distorted as explained in the next section. Furthermore, they emerge on top of, and also interfere with, the rest of the O, Q and S lines which are only weakly enhanced. Finally, the intensities of all these lines also depend on $\rho_{aa}^{(o)}$ and on the transition moments (see section III.C.1 of Ref.(37)).

If the temperature is raised, the resonant CARS spectrum is complicated due to the contribution of the $\rho_{bb}^{(o)}$ terms. These aspects are discussed in detail in Ref.(37).

Tuning the pump frequencies into a vibrational dissociative continuum would eliminate the problems due to multiple interfering sharp resonances associated with the $\rho_{aa}^{(o)}$

and $\rho_{bb}^{(o)}$ terms. The continuum contains many excited rovibronic
intermediate states which give interfering equivalent contributions to χ_R^{ba} , so that the dependence upon the field
frequencies is smoother. Resonance enhancement comes simultaneously from ω_1 and ω_3. Furthermore, we anticipate the CARS
spectrum to only contain the usual vibration-rotation features
(and no longer the double electronic resonance features),
constituting enhanced O, Q and S branches having relative
intensities similar to those encountered off resonance.

2. *Line Contours.* In addition to predicting the features
in the CARS spectrum, one can also describe their spectral
contour. The spectral contour of any CARS line is characteristic. It is the result of the interference between the
complex, frequency dependent, Raman-resonant susceptibility
and the nonresonant background. This interference barely
affects the lineshape if the line is intense, but a pronounced
distortion takes place on weaker lines. The latter, if too
weak, cause but a small ripple on the background and cannot be
detected if the level of fluctuations is too high. This property is the major drawback of CARS in analytical chemistry
applications. In liquid mixtures for instance, if the concentration of the solute is less than 1%, one is not able to
detect its Raman spectrum in the presence of the solvent background χ_{NR}. The same difficulty is encountered in gas mixtures, with detectivities of about 0.1% because the linewidths
are smaller and consequently the lines more intense. This
difficulty has motivated research efforts in the areas of
background reduction or suppression (31,65) and of resonance
enhancement.

a. *Collision broadening.* Collisions are the dominant cause
of line broadening in the majority of CARS experiments performed to date. The expressions in Eqs.(15) and (20) were
derived under the assumption of collision broadening, therefore they can be used for an analysis of line shapes on and
off resonance. These shapes can and have frequently been
obtained from computer calculations. For isolated lines,
however, these shapes can also be predicted in an elegant
manner using the circle model of (55,60). This model was
introduced for off-resonance CARS when it was realized that
the spectrum is just a plot of the modulus of the susceptibility vs ω_2 , and that the susceptibility can be represented as a vector in the complex plane. It was then found that
the extremity of this vector χ describes a circle during the
spectral scan. As matter of fact, χ is the sum of two

vectors, i.e. χ_R whose argument and modulus depend upon ω_1, ω_2 and ω_3, and χ_{NR} which is a constant. The equation for the χ_R component associated with one particular Raman resonance ω_{ba} can be written

$$\chi_R = N\alpha_R / (\omega_{ba} - \omega_1 + \omega_2 - i\Gamma_{ba})$$
$$= X + iY \qquad\qquad (46)$$

where α_R stands for electronic terms, population factors, etc. In the non-resonant case, α_R is a real number proportional to $d\sigma/d\Omega$ and to $\rho_{aa}^{(o)} - \rho_{bb}^{(o)}$; it depends little upon field frequencies and can be taken as a constant. Then the expression for χ_R can be interpreted as the parametric equation for two families of orthogonal circles (Fig.17).

   (1) The main family is described by χ_R as $\omega_1 - \omega_2$ is varied about ω_{ba} with the linewidth Γ_{ba} held fixed. There is a circle for each value of Γ_{ba}. The circle equation is $X^2 + (Y - 1/2\, y)^2 = (1/2y)^2$ with $y = \Gamma_{ba}/N\alpha_R$ as a parameter. For each particular experimental situation, only the circle with the relevant y parameter needs to be drawn. This circle is tangent to the X axis at the origin and lies in the upper half plane. Its diameter is y^{-1}.

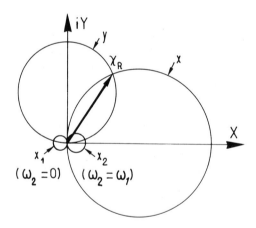

FIGURE 17. Representation of χ_R as a vector in the complex plane. The extremity of this vector describes the circle of parameter y as the detuning x is being varied, and is found at the intersection with circle of parameter x. The vector tip is actually excluded from the small segment of arc contained within the circles of parameters $x_1 = (\omega_{ba} - \omega_1)/N\alpha_R$ and $x_2 = \omega_{ba}/N\alpha_R$.

(2) For a given detuning $\omega_{ba} - \omega_1 + \omega_2$, the position of $\underline{\chi}_R$ on the preceding circle is found at the intersection with the orthogonal circle whose equation is $(X - 1/2x)^2 + Y^2 = (1/2x)^2$ with $x = (\omega_{ba} - \omega_L + \omega_S)/N\alpha_R$ as a parameter. There is a circle for each value of x, and these circles constitute the second family.

It is seen readily that χ_R rotates clockwise as ω_2 is increased from $-\infty$ to $+\infty$; its argument decreases from π to 0 and is equal to $\pi/2$ on resonance, i.e. at $\omega_2 = \omega_1 - \omega_{ba}$. In actuality, ω_2 can only be varied from 0 to ω_1, so that a small segment of arc on either side of the origin is not swept. Adding χ_{NR} results in a translation parallel to the X axis (Fig.18a). The line contour is then drawn readily by plotting $|\underline{\chi}|$ as a function of x (or ω_2). If only an approximate lineshape is to be rendered, it is sufficient to represent χ_{NR} and the circle of parameter y on the diagram, from which the whole contour is directly pictured (this is what we do on the rest of Fig.18).

The lineshape depends upon the relative magnitude of y and χ_{NR} or, in other words, upon the mixing ratio N/N_d where N_d is the diluent density ($\chi_{NR} \propto N_d$). The case depicted in Fig.18a corresponds to $N/N_d > 0.1$ for typical molecular parameters and near STP in gases. Apart from a slight asymmetry caused by the interference with χ_{NR}; the contour of $|\underline{\chi}|$ then closely fits the square root of a Lorentzian. In the other limit, if the concentration is small, the line contour approaches that of $Re(\chi) = \chi_{NR} + X$, since Y gives a higher order correction (Fig.18b), and the line maximum is shifted. In either case, the difference between the maximum and the minimum is strictly equal to the magnitude of the vibrational resonance, i.e. one circle diameter $y^{-1} = N\alpha_R/\Gamma_{ba}$.

When enhancement of χ_R through electronic resonances is produced, a complex situation appears since several of the terms in α_R can become large simultaneously and can vary as $(\omega_1 - \omega_2)$ is varied. The discussion, however, remains simple for the conditions assumed above in the presentation of the spectral content of $\chi^{(3)CARS}(\omega_1, \omega_1, -\omega_2)$, viz. $\rho_{bb}^{(o)}$ negligible, ω_1 set near resonance, ω_3 off resonance. Then, α_R is proportional to $\rho_{aa}^{(o)}$ and independent of ω_2 and ω_3 over the range of interest, and we can write :

$$\alpha_R \propto P_{an'} \, P_{n'b} \, P_{bn} \, P_{na} / (\omega_{na} - \omega_1 - i\Gamma_{na}) = re^{i\theta}. \qquad (47)$$

For ω_1 fixed, it is apparent that $\underline{\chi}$ still describes the two families of circles. The circles are rotated by argument θ about the extremity of χ_{NR}, and their diameters scale as r. The rotation of the circles proceeds counter-clockwise

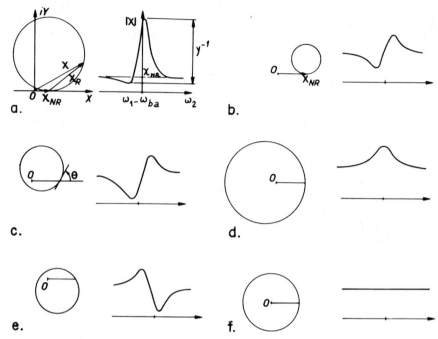

FIGURE 18. *Circle diagrams and corresponding line contours well below electronic resonance (a,b) and in the vicinity of an electronic resonance (c − f), with* P_{bn} P_{na} P_{an}' $P_{n'b}$ *real. The circles and their parameters are defined in the text.*

a − strong isolated Raman resonance ($N/Nd \geqslant 0.1$ *typically) ; the circle described by* $\underline{X}$ *is represented, together with an approximate line contour displayed vs* ω_2 *.*

b − weak Raman resonance ; only $\underline{X}_{NR}$ *and the circle described by* $\underline{X}_R$ *are drawn on the diagram.*

c − same as. (b), but ω_1 *has been increased and tuned close to a one-photon resonance* ω_{na}, *with* $\omega_1 < \omega_{na}$; *as a result the circle has been rotated and its diameter increased.*

d − $\omega_1 = \omega_{na}$; *the circle diameter is at its maximum, and* $\theta = \pi/2$.

e − $\omega_1 > \omega_{na}$ *; the diameter decreases again.*

f − $\omega_1 = \omega_{na}$ *, with the mixing ratio chosen such that the circle radius is equal to* X_{NR} *; the line vanishes.*

as ω_1 increases past the one-photon resonance ω_{na}. If the product of transition moments is real, θ grows from 0 to 180° and is precisely 90° on resonance. Figures 18c, d, e illustrate what happens to the circle of parameter y and the associated vibrational resonance of Fig.18b assuming ω_1 tuned slightly less than ω_{na}, to exactly ω_{na} and slightly greater than ω_{na}, respectively. Cases (c) and (e) produce line shapes which are mirror images of one another with respect to $\omega_2 = \omega_1 - \omega_{ba}$. Case (d) is interesting: depending on the magnitude of the Raman resonance with respect to χ_{NR}, the line, which is symmetrical with respect to $\omega_1 - \omega_{ba}$ appears as a maximum (Fig.18d) or a minimum, or even vanishes completely (Fig. 18f). In the latter case, the circle radius is precisely equal to χ_{NR}. This discussion is actually adequate for most types of double (vibrational + electronic or electronic + electronic) resonances in gases. The lineshapes are in fact more complicated as there is no reason for the product of transition moments in α_R to be real.

The discussion of absorption continua is generally difficult because we possess no closed-form expression for χ_R. However, good insight can be provided by assuming a Lorentzian distribution for the joint density of continuum states $|\alpha\rangle$, with width Γ large compared with ω_{ba}(Fig.19). Then, with the above assumptions ($\rho_{bb}^{(o)} \simeq 0$, ω_1 fixed in resonance into the continuum), it is legitimate to still express α_R for a given Raman line as :

$$\alpha_R \propto (\omega_{\alpha a} - \omega_1 - i\Gamma)^{-1} \times (\omega_{\alpha a} - \omega_3 - i\Gamma)^{-1}$$
$$\simeq (\omega_{\alpha a} - \omega_1 - i\Gamma)^{-2} \quad . \tag{48}$$

As in the previous case, χ_R still describes a circle but the rotation θ is doubled. This analysis provides a simple explanation for the dispersive behaviour of weak resonances

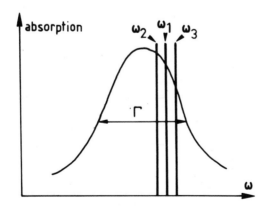

FIGURE 19. Spectral configuration for CARS resonance enhancement in a continuum with a Lorentzian contour. The CARS spectrum is supposed to be of limited extent, so that $\omega_1 - \omega_2 \ll \Gamma$.

reported in   (34,64,66) , since a continuous passage from the
minimum at θ = 90° to a maximum at θ = 270° (Fig.20a, b) can
be directly visualized (for real $\mathcal{X}$'s, θ varies from 0 to 360°).

b. Doppler broadening. High-resolution Raman spectroscopy
of gases is best accomplished by CARS, since :
   (1) the gases must be at low pressure, so that the
collisional broadening is eliminated ; one then takes advantage of the superior signal strength afforded by CARS even at
low densities and with moderate laser powers ;
   (2) the Doppler width of Raman lines is minimal in the
forward direction and one takes advantage of this property in
CARS, since the scattering also takes place in the forward
direction.
   That the Doppler broadening of a CARS resonance should
be small is noteworthy. It is clear that the Doppler shifts
on ω_1 and ω_2 almost cancel one another in the frame of reference of a moving molecule since $\omega_1 \simeq \omega_2$, so that $\omega_1 - \omega_2$
remains in resonance with ω_{ba} whatever the axial velocity
of the molecule. It has been first speculated (40) that the
contour was a Gaussian, like the spontaneous Raman line
contour, with residual width at 1/e intensity of :

$$\Delta \omega_D = 2 \omega_{ba} \sqrt{\frac{2kT}{mc^2}} \quad , \quad (49)$$

where m is the mass of the molecule and T the temperature.
However, Henessian and Byer demonstrated in a recent communication that the real part of the susceptibility is causing a

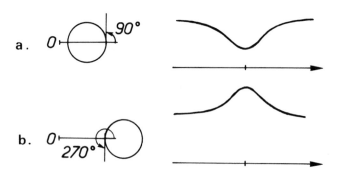

*FIGURE 20. Extension of the diagrams of Fig.18 to specific
situations of resonance with a continuum.*

broadening of the Gaussian contour in the wings, so that the full width of the line at 1/e intensity is actually 1.2 $\Delta\omega_D$ (67).

For resonant CARS, the additional resonance conditions on ω_1 , ω_2 or ω_3 are expected to change the widths and contours of the CARS lines. The theory of this effect is in progress at our laboratory.

E. *Summary*

We have discussed in detail the theory of CARS. We have analyzed the following points :

(1) physical mechanisms involved in the CARS interaction ;

(2) concomitant effects like stimulated Raman scattering, one-photon absorption and Stark broadening, together with their possible interference ;

(3) particularities of CARS spectra off and on resonance. We are now ready to review the recent experimental achievements and to discuss the potential applications of the method.

IV. EXPERIMENTAL RESULTS

Since the first experimental demonstration of CARS in liquids by Maker and Terhune (8), a large number of applications have been found for this effect in liquids, gases and solids. We shall discuss some of the results obtained with liquids and gases especially with regard to concentration and temperature measurements, and we shall examine the state of the art of the technology.

A. *Liquids and Solutions*

It took nearly ten years following Maker and Terhune's original work on benzene and benzene derivatives (the spectra of which were plotted using only a few discrete frequencies at ω_2) before genuine CARS spectroscopy in liquids was reported using tunable lasers (68-70). Begley and coworkers (68) were actually the ones to coin the term CARS, for "Coherent Anti-Stokes Raman Spectroscopy".

1. Experimental Requirements. Liquid densities are typically 10^3 times higher than gas densities ; susceptibilities are correspondingly higher, and laser powers of 1-50 kW are more than enough to generate substantial CARS signals. Furthermore, the beams at ω_1 and ω_2 must be crossed at a few degrees for phase matching (Fig.21), which automatically gives a good spatial resolution, even with beams which are not diffraction-limited. Nitrogen laser-pumped dye lasers, which are the perfect source for this work, have been used in many experiments. Instrumental arrangements and operating procedures are discussed in references (43,71).

2. *Results.*

a. Off resonance. Some of the work in liquids has been oriented towards the measurement of nonlinear susceptibilities and the study of line contours (18,69,70,72-74). Others have reported spectra of liquid N_2 using cw lasers (75,76). Two-photon absorption cross sections have been determined from comparisons between the Raman resonant part and the two-photon terms in the nonresonant part of the susceptibility (69,70,74, 77,31). Higher-order sidebands (HORSES) at $\omega_1 \pm 2 \ (\omega_1 - \omega_2)$ have been reported by Chabay and coworkers (43). Begley et al. discussed the feasibility of using CARS for concentration measurements in liquids (68). The solvent background usually precludes such measurements at mixing ratios less than 1%. In order to improve sensitivity, resonance enhancement can be used.

b. On resonance. Chabay et al. were the first to report resonance enhancement (43). The study was conducted on a 1.4×10^{-3} M solution of diphenyloctatetraene in benzene, and demonstrated appreciable improvements in S/N and fluorescence background rejection. The detectivity limit was then estimated

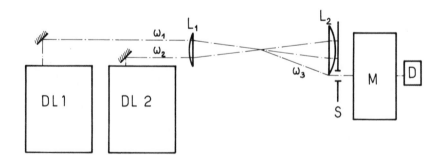

FIGURE 21. Schematic of the optical arrangement for CARS in condensed media : DL : dye laser ; L_1, L_2 : lenses ; S : aperture stop ; M : monochromator ; D : detector.

at 5×10^{-5} M. Nestor et al. reported similar results in
aqueous solutions of ferrocytochrome and vitamin B_{12} (34) ;
they also noted some absorption-like features on the background
level, which they attributed to the inverse Raman process
(associated with the first term of Eq.(37c)). The correct interpretation of these features was given by Hudson and coworkers
(64) and Lynch and coworkers (78). We have presented an interpretation by means of the circle model in section III.D.2. This
effect was also observed in other species and discussed by
several groups (35,79–82).

B. Gases and Plasmas

1. Experimental Requirements. In gases, the Raman linewidths
are usually small (0.1 cm^{-1} or less) and one frequently desires
a good spatial resolution (e.g. in concentration diagnostics
of turbulent flames). This places severe constraints on the
spectral purity and stability of each laser, and also on their
beam divergence and directional stability. A partial discussion
of these constraints can be found in references (55,60) for
off-resonance CARS. The optical schematic of one of the CARS
set-ups in operation at ONERA is shown in Fig.22. A stable

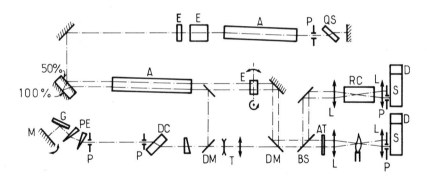

*FIGURE 22. Optical schematic of a CARS set-up for flame
diagnostics : A : ruby amplifier ; P : pinhole ; QS : Q-switch
cell ; E : etalon ; DC : dye laser dye cell ; PE : prism beam
expander ; G : grating ; M : plane mirror for dye laser tuning;
T : telescope ; DM : dichroic mirror ; BS : 10% reflection
beam splitter ; L : achromatic lens ; AT : non fluorescing
neutral density filter ; S : monochromator ; D : detector
(incorporated in monochromator) ; RC : high pressure reference
cell (100 bar of Argon).*

ruby laser is used in this set-up : its cavity is mounted on
a 30cm-long Zerodur bench, and two high refractive index optical flats are used as resonant output reflectors. This
cavity is entirely temperature-controlled to within 0.1°C. The
peak power is 0.7 MW ; the beam is diffraction-limited. The
linewidth is less than 10^{-2} cm^{-1} and the frequency stability
of the same order. The laser beam is split into two parallel
beams by means of an optical flat with 50% and 100% reflective
coatings in order to conduct BOXCARS measurements with 1-2 mm
spatial resolution. These beams are amplified to 1-2 MW, part
of this energy being diverted to pump the dye laser in a collinear arrangement. The latter is tuned by means of a fixed
1200 gr/mm grating and a rotatable mirror, in the fashion described by Shoshan (83). A three-prism beam expander (84) is
used, so that the grating is not operated at grazing incidence,
thus giving better coupling efficiency. The output beam is
diffraction-limited, stable in direction to better than a
tenth of its divergence, and the FWHM of the line is 0.7 cm^{-1}.
The peak power is 10-50 kW. A 3x telescope is inserted to
adapt the beam divergence to that of the ruby laser, and a
dichroic plate is used for combination with the ruby laser
beams. Note that a reference channel is provided in order to
reduce the level of signal fluctuations. The latter are
typically $\pm$ 30% from the mean in stable gases (from shot to
shot). Adding the reference leg and taking the ratio of
sample signal to reference signal reduces the measurement
fluctuations to $\pm$10% or less, which amounts to $\pm$5% in terms of
$\Delta|\chi|/|\chi|$. Incidentally, this level of fluctuations increases
slightly near the spectral lines because of laser frequency
jitter. The use of a reference is also essential as a safeguard
against long term laser power fluctuations, beam wandering,
etc, if any quantitative CARS measurements are to be made. The
pulses are gated electronically, and signal divided by reference is averaged. Due to the large dynamic range on the signal
(1 to 10^{10}), it is necessary to place optical attenuators on
the beam. We prefer to attenuate the pump with calibrated
filters that do not fluoresce rather than place the filters on
the anti-Stokes path. Unless one has to work at high temperatures (2000K and above) and low pressures, there is no need
for more power than quoted for this set-up, provided the
requirements on beam quality and spectral purity are satisfied.
     Multiplex CARS is an interesting development of the method
(85) . It is being pursued at a number of laboratories as a
means for obtaining instantaneous measurements of both temperature and concentration by recording entire spectra in one laser
shot (62,76,86,87). In this method, one uses a broad band dye
laser, but one then has to disperse the anti-Stokes spectrum
and to record it with an optical multichannel analyzer or a TV
camera. This is a very promising area of research.

Another field which is of considerable importance is high-resolution Raman spectroscopy. For some difficult applications, high power pulsed sources have to be used : Moret-Bailly and coworkers were able to obtain 0.005 cm^{-1} resolution on CH_4 at 10mb with a ruby and dye laser assembly (45,88) ; this seems to be the ultimate using such sources. Better resolutions still are possible by using cw lasers. Barrett and Begley were the first to obtain a cw CARS signal from the ν_1 band of CH_4 (89). Shortly thereafter, Hirth reported obtaining large signals from atmospheric N_2 with an intracavity arrangement, in an attempt at following real time fluctuations in a turbulent jet (90). Other cw experiments were also described (40,91—93). In particular, Fabelinsky and coworkers plotted the ν_2 resonance of C_2H_2 with a resolution of 10^{-3} cm^{-1} (92). Resolutions as good as a few kHz are possible in theory with present laser technology.

Resonance CARS work places constraints similar to those for off-resonance spectroscopy, with the added complexities that both lasers must be independently tunable and that one-photon saturation must be carefully avoided. In their work on I_2 vapor, Attal and coworkers used flash-pumped lasers delivering pulses of 10 kW power and 1 μs duration. The beams were diffraction-limited with linewidths of 0.1 cm^{-1} (38).

2. Results. Although CARS has been applied to gases by Rado as early as 1967 (12) and has been used for various measurements including spectroscopy (13,14,94,95) and higher-order studies (HORSES) by Lukasik and Ducuing (96) in 1972, it was not until 1973 that it was proposed for concentration measurements in mixtures. Régnier and Taran (3,6) mapped H_2 concentration profiles in a Bunsen burner flame fueled by town gas and coined the ill-fated "CRAS" (for coherent Raman anti-Stokes scattering"). The same group subsequently measured temperatures from the same flame by recording the rotational structure of the Q-branch of H_2 from Q(0) to Q(5) (21,97,98). Since then, many gases have been studied, either pure or as combustion reactants. The spectrum of the O, Q and S branches of room air N_2 which was reported in (60) is presented in Fig.23. This spectrum exhibits many of the features discussed in section III. One notices, among other things, that the Q-branch is only about 50 times more intense than the nonresonant susceptibility level. The interference between χ_R and χ_{NR} also leads to the asymmetry discussed in III.D, a deep hole being found to the high frequency side of the Q-branch, while there is a slower decay away from line center on the other side. The problem of the detection of trace constituents is also clearly illustrated by the O-branch/S-branch asymmetry. We know that the noise in CARS spectra is caused predominantly by shot to shot fluctuations of the signal to reference ratio ; if both signal and

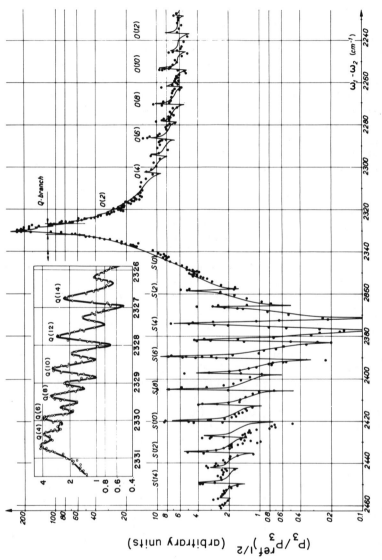

FIGURE 23. Spectrum of atmospheric N_2. The curves are calculated for Boltzmann equilibrium, assuming $(d\sigma/d\Omega)$ $Q = 1.1 \times 10^{-31}$ cm^2/sr. Fitting parameters were $(d\sigma/d\Omega)$ $O,S = 0.55 \times 10^{-32}$ cm^2/sr, $\chi_{NR} = 4 \times 10^{-18}$ cm^3/erg. In the main plot, a width of 0.15 cm^{-1} was assumed for all $O,Q,$ S lines. Note that the vertical scale is roughly in units of 10^{-18} cm^3/erg.

reference are intense enough (which eliminates the influences
of Poisson fluctuations on photocurrent), then there is a
residual level of fluctuations caused by laser instabilities ;
we saw in the preceding section that this level is typically
$\Delta|X|/|X| = \pm 5\%$ (for a state of the art pulsed set-up) and
that is is independent of $|X|$. As a result, the O lines barely
emerge from the noise level of their high-background, whereas
the S lines (which are just as intense) are clearly dis-
tinguished because of lower background level. In practice, the
Q-branch of a trace gas can be detected if the concentration is
larger than 10^{-3} at STP (or between 10^{-2} and 10^{-3} in flames).
Some gases (like H_2 or CO_2) can be detected at lower concentrations (3,6). Note further that in the spectrum of Fig.23 :
    (1) the solid curve was obtained numerically;
    (2) the quantity $\log|X|$ is plotted, which has several
advantages : (i) the error bar is constant whatever $|X|$; (ii)
the slope of the envelope of the Q-branch on its "hot" side
appears as constant over a certain spectral domain and is
inversely proportional to T, which can be used for direct
temperature measurements from the spectra (60);
    (3) the inset plot of the Q-branch alone was obtained at
higher resolution and clearly displays the individual contributions from the Q lines;
    (4) because we plot $|X|$, the linewidths must be measured
at $1/\sqrt{2}$ from the maxima. Numerous spectra have been obtained at
our laboratory in pure samples and in flames, from which concentrations and temperatures have been measured, as close as
50 μm to solid surfaces (55,99). A computer program for data
retrieval has been developed. This program calculates the
modulus of the susceptibility as given by Eq.(20) : the complex
normalized lineshape function for an isolated resonance is
first generated for a given Γ_{ba} , then a summation is carried
out over all the resonances of interest ; this is done by
shifting the central frequency of the normalized lineshape
function to each resonant frequency ω_{ba} and by weighting it
according to the population factors with the assumption of
Boltzmann equilibrium. The nonresonant susceptibility is then
added ; finally, the convolution with the dye laser line contour is calculated (a Gaussian laser line contour gives satisfactory results (21)), and the modulus is taken and plotted.
There are three fitting parameters for any given gas : temperature, concentration and nonresonant susceptibility. Comparison
with experimental spectra usually gives temperatures to within
$\pm 5\%$, and concentrations to within $\pm 10\%$. This program has been
used extensively over the past several years. A systematic
error in N_2 spectra was however found at high temperature in
flames : the concentration of N_2 as calculated with the program
often exceeded 80 and even 100%. This has been attributed to
two causes :

(1) our CARS set-up had insufficient spatial resolution, so that the CARS signal was integrated over regions having vastly different temperatures;
(2) the linewidth Γ_{ba} is a function of temperature. The problem was cured by using a BOXCARS arrangement and by assuming an arbitrary $T^{-1/2}$ dependence for Γ_{ba} ; experimental confirmation of this final assumption is yet to be obtained. An interesting spectrum of pure CO_2 at room temperature was recently obtained by M. Péalat (Fig.24) with the set-up described in Fig.22. One recognizes the following lines : 1 corresponds to the $(11101) \rightarrow (01101)$ transition, 2 to $(10001) \rightarrow (00001)$, 3 is the same as 2 but for the $O^{16}C^{13}O^{16}$ isotope, 4 corresponds to $(10002) \rightarrow (00001)$, 5 to $(11102) \rightarrow (01101)$ and 6 is the same as 4 for the $O^{16}C^{12}O^{18}$ isotope ; feature 5 also contains the $(10002) \rightarrow (00001)$ transition for isotope $O^{16}C^{13}O^{16}$. The O and S lines are also clearly seen. We notice that the stronger lines (2 and 4) are 10^3 to 10^4 times more intense than the nonresonant background of pure CO_2 (which is itself of the same order as that of air). Therefore, the contrast is far stronger than on the N_2 spectrum of Fig.23 ; this fact is

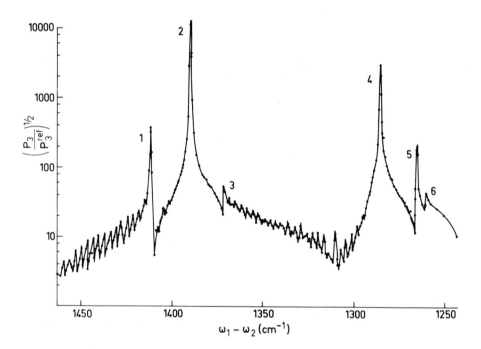

FIGURE 24. *Spectrum of pure CO_2 at room temperature, taken with 0.6 cm^{-1} resolution ; 4 laser shots are averaged per data point.*

attributed to all the rotational lines being grouped in a
structure less than 0.6 cm^{-1} wide. A recent study, conducted
with a commercial CARS unit designed for combustion work and
capable of 0.03 cm^{-1} resolution, actually revealed an intricate
structure of lines with widths of 0.03 cm^{-1} or less. Finally,
Figure 25 is a medium resolution (0.5 cm^{-1}) plot of the atmospheric susceptiblity in the vicinity of line 2 of CO_2, obtained
with the latter set-up ; this plot clearly shows that trace
concentrations of CO_2 as low as 300 ppm can be detected (the
detectivity is better still for H_2, viz. 200 ppm, but is only
1000 ppm for other gases).

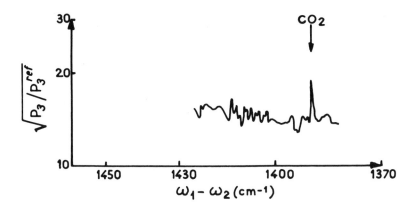

*FIGURE 25. Spectrum of atmospheric CO_2 taken with a com-
mercial CARS set-up using a doubled yag and a tunable dye
laser; spectral resolution is 0.5 cm^{-1}.*

Other groups have conducted spectroscopy work in discharges. The spectrum of Fig.26 was obtained by Harvey and
coworkers on D_2 at 48 Torr (100) ; this spectrum shows a large
difference between the rotational and vibrational temperatures
(400 and 1050 K respectively). This work illustrates the potential of CARS in the diagnostics of gaseous laser media and
plasmas. The same group has also presented evidence of the
existence of photochemical fragments produced by the excitation
of C_6H_6 at 266 nm, and has conducted work on combustion (25,
100). Work on combustions is also being pursued by Schreiber
and coworkers (39,85), Eckbreth and coworkers (59,86,87), Black
and coworkers (101).

Finally, let us mention that resonant CARS in I_2 vapor
at 1 mb diluted in air at 1b has been reported by Attal and
coworkers (38,99). The spectra of Fig. 27 were obtained by
tuning ω_1 at two different frequencies in the vicinity of
the sodium D line at 16.956 cm^{-1}, and scanning the fifth
vibrational overtone. No reference channel was used ; the

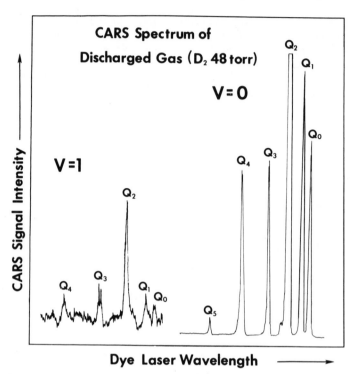

FIGURE 26. CARS spectrum of D_2 in an electrical discharge (from Ref. 100, with permission).

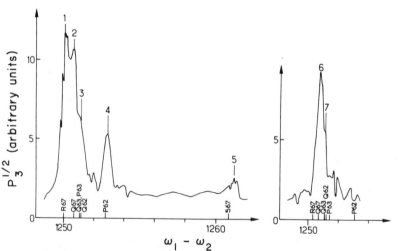

FIGURE 27. Resonant CARS spectra of I_2 with $\omega_1 \simeq 16956$ cm^{-1} (left), then downshifted by 0.3 cm^{-1} (right). The positions of the strongest double resonances are indicated. Horizontal scales are in cm^{-1}.

spectral resolution was about 0.3 cm^{-1}. The stronger features have been interpreted using the theory exposed in III.D.1. We have recently confirmed these spectra, with improved resolution of 0.1 cm^{-1} and using a reference leg. The susceptibility at line center is nearly 1000 times larger than that of any isolated Q line of a nonresonant gas maintained in the same thermodynamic state. This indicates that I_2 can in principle be detected at a concentration level of a few ppm. Similar detectivity improvements are expected for other gases. Clearly, resonant CARS in gases has great potential for chemical diagnostics.

V. CONCLUSION

We have examined in detail the theory of CARS, we have presented the key instrumental requirements and have reviewed some of the most important experimental results obtained to date. In brief, CARS :
(1) gives the same spectroscopic information as spontaneous Raman scattering,
(2) is insensitive to fluorescence interference in resonance Raman work,
(3) gives excellent spatial resolution (1 mm),
(4) is capable of excellent spectral resolution (0.03 to 1 cm^{-1} on ordinary set-ups, 10^{-3} cm^{-1} in special applications using cw sources),
(5) is extremely luminous (10^5 to 10^{10} more intense than normal Raman scattering).
CARS has a few disadvantages :
(1) the major disadvantage is the presence of the non-resonant background, which sets the detectivity limit between 0.1 and 1% ;
(2) it is sensitive to laser instabilities ;
(3) it is subject to saturation at the higher power levels (1 MW or more).
For these reasons, CARS will often be preferred to normal Raman scattering for the following measurements :
(1) study of reactive media, plasmas, gas laser amplifiers, etc... ;
(2) analysis of media containing particulate matter (e.g. sooting flames) or investigation of flows near solid obstacles (as close as 50 μm to the surface) ;
(3) high resolution spectroscopy ;
(4) resonance-enhanced Raman spectroscopy of liquids and gases.
Due to the cost of a state of the art CARS set-up (100,000 to 150,000 $ for flame diagnostics), the decision to prefer CARS

to normal Raman should be carefully weighted. However, fundamental research in all these areas is likely to be actively
pursued.

In addition, there is considerable interest in techniques
capable of reducing or eliminating the background, like
cancellation using three input beams (31,76,102) or using
different input polarizations or RIKES (44,103,105), or like
stimulated Raman gain spectroscopy (106). These new techniques
or variants of CARS are promising, especially for liquid phase
applications. Their usefulness in gases is yet to be demonstrated.

Finally, CARS in principle can be applied to the field of
hyper-Raman spectroscopy with much improved signal to noise
ratio. Our attempts at seeing this effect in CO, ethylene and
other gases have failed thus far, due to stimulated Brillouin
scattering interference (99) ; this spectroscopy is certainly
possible with 10 MW yag lasers using optical isolators, and
will undoubtedly be demonstrated in the near future.

ACKNOWLEDGMENTS

The authors wish to thank Dominique Kalemba for typing
the manuscript with impressive skill and patience and Claire
Laroche for drawing the illustrations with exacting care.

REFERENCES

1. Widhopf, G.F., and Lederman, S., *AIAA J.*, *9*, 309 (1971)
2. Lapp, M., Goldman, L.M., and Penney, C.M., Science, *175*, 1112 (1972).
3. "Laser Raman Gas Diagnostics", Proceedings of the Project SQUID Laser Raman Workshop on the Measurement of Gas Properties, May 10-11, 1973, Schenectady, edited by M. Lapp and C.M. Penney, Plenum Press, New York, London, (1974).
4. Proceedings of Project SQUID Workshop on Combustion Measurements in Jet Propulsion Systems, Ed. by R. Goulard, Purdue University, Lafayette, Indiana, (1975).
5. "Experimental Diagnostics in Gas Phase Combustion Systems", Progress in Astronautics and Aeronautics, Vol. *53*, Edited by B.T. Zinn, Martin Summerfield Series Editor (1977).
6. Régnier, P.R., and Taran, J.P.E., *Appl. Phys. Letters*, *23*, 240 (1973).
7. Terhune, R.W., *Bull. Amer. Phys. Soc.*, *8*, 359 (1963).
8. Maker, P.D., and Terhune R.W., *Phys. Rev.*, *137*, A801 (1965).
9. Yablonovitch, E., Bloembergen, N., and Wynne, J.J., *Phys. Rev.*, *B3*, 2060 (1971).
10. Akhamanov, S.A., Dmitriev, V.G., Kovrigin, A.I., Koroteev, N.I., Tunkin, V.G., and Kholodnykh, A.I., *JETP Letters*, *15*, 425 (1972).
11. Levenson, M.D., Flytzanis, C., and Bloembergen, N., *Phys. Rev.*, *6*, B3962 (1972).
12. Rado, W.G., *Appl. Phys. Letters*, *11*, 123 (1967).
13. Hauchecorne, G., Kerhervé, F., and Mayer G., *J. de Physique*, *32*, 47 (1971).
14. De Martini, F., Guiliani, G.P., and Santamato, E., *Optics Comm.*, *5*, 126 (1972).
15. Minck, R.W., Terhune, R.W., and Wang, C.C., *Applied Optics*, *5*, 1595 (1966).
16. Pantell, R.H., and Puthoff, H.E., "Fundamentals of Quantum Electronics", John Wiley and Sons, New York, (1979), Chap. 6,7.
17. Régnier, P., Moya, F., and Taran, J.P.E., *AIAA Journal*, *12*, 826 (1974).
18. Akhamanov, S.A., and Koroteev, N.I. , *Sov. Phys. JETP*, *40*, 650 (1975).
19. Krochik, G.M., and Khronopulo, Yu.G., *Sov. J. Quant. Electron.*, *5*, 917 (1975).
20. Yariv, A., "Quantum Electronics", John Wiley and Sons, Inc. , New York, Second Edition (1975).
21. Moya, F., "Application de la Diffusion Raman Anti-Stokes Cohérente aux Mesures de Concentrations Gazeuses dans les Ecoulements" Thesis Orsay (1976), available as ONERA Technical Report N°1975-13 and as European Space Agency Technical Translation N° ESATT 331 (1976).

22. Yuratich, M.A., and Hanna, D.C., *Molecular Physics, 33,* 671 (1977).
23. Tolles, W.M., Nibler, J.W., MacDonald, J.R., and Harvey, A.B., *Appl. Spectrosc., 31,* 253 (1977).
24. DeWitt, R.W., Harvey, A.B., and Tolles, W.M., "Theoretical Development of Third Order Susceptibility as related to Coherent Anti-Stokes Raman Spectroscopy," NRL Memorandum Report N°3260 (1976).
25. Nibler, J.W., and Knighten, G.V.,"Coherent Anti-Stokes Raman Spectroscopy", in Topics in "Current Physics", Ed. by A. Weber, Springer Verlag, New York, (1977).
26. Armstrong, J.A., Bloembergen, N., Ducuing, J., and Pershan, P.S., *Phys. Rev., 127,* 1918 (1962).
27. Bloembergen, N. "Nonlinear Optics", W.A. Benjamin, Inc. New York, (1965). Third printing : 1977.
28. Terhune, R.W. and Maker, P.D., in "Nonlinear Optics", Ed. by A.K. Levine, Marcel Dekker, Inc., New York, (1968),Vol. *2,* p. 295.
29. Butcher, P.N., "Nonlinear Optical Phenomena", Ohio State University Engineering Publications, Columbus, Ohio, (1965).
30. Flytzanis, C., in "Quantum Electronics – A Treatise", Rabin, H., and Tang, C., Eds., Academic Press, New York, (1975).
31. Lotem, H., Lynch, R.T., and Bloembergen, N., *Phys. Rev., A14,* 1748 (1976).
32. Lynch, R.T., Jr., Lotem, H., and Bloembergen, N., in "Lasers in Chemistry", Conference Digest, London, May 31st-June 2nd 1977, p.1 ; Bloembergen, N., Lotem, H., and Lynch, R.T., Jr., "Lineshapes in Coherent Resonant Raman Scattering" to be published in the Raman 50th Jubilee Volume of the Indian Journal of Pure and Applied Physics ; Lynch, R.T., Jr., unpublished thesis, Harvard University, 1977.
33. Druet, S., "Diffusion Raman Anti-Stokes Cohérente au voisinage des résonances électroniques", thesis, Orsay, 1976; available as ONERA Technical Note N°1976-6, and, in English, as European Space Agency Technical Translation N°ESA TT 371 (1977).
34. Nestor, J., Spiro, T., and Klauminzer, G., *Proc. Natl. Acad. Sci. USA, 73,* 3329 (1976) ; Carreira, L.A., Goss, L.P., and Malloy, T.B., *J. Chem. Phys., 66,* 4360 (1977).
35. Lau, A., Pfeiffer, M., and Werncke, W., *Optics Comm., 23,* 59 (1977).
36. Yee, S.Y., and Gustafson, T.K., *Phys. Rev.,* A, to be published. Yee, S.Y., Gustafson, T.K., Druet, S., and Taran, J.P., *Optics Comm., 23,* 1 (1977).
37. Druet, S.A.J., Attal, B., Gustafson, T.K., and Taran, J.P., *Phys. Rev.,* A, to be published. (September 1978).

38. Attal, B., Schnepp, O., and Taran, J.P., *Opt. Comm.*, *24*, 77 (1978).
39. Roh, W.B., and Schreiber, P.W., *Appl. Optics*, *17*, 1418 (1978).
40. Henessian, M.A., Kulevskii, L., and Byer, R.L., *J. of Chem. Phys.*, *65*, 5530 (1976).
41. Régnier, P.R., "Application of Coherent Anti-Stokes Raman Scattering to Gas Concentration Measurements and to Flow Visualization," thesis, 1973, available as ONERA Technical Note N°215, also available as European Space Agency Technical Translation N° ESA TT 215 (1973).
42. Shaub, W.M., Harvey, A.B., and Bjorklund, G.C., *J. Chem. Phys.*, *67*, 2547 (1977).
43. Chabay, I., Klauminzer, G., and Hudson, B.S., *Appl. Phys. Letters*, *28*, 27 (1976).
44. Levenson, M.D., *Physics Today*, p.44, (May 1977).
45. Boquillon, J.P., Moret Bailly, J., and R. Chaux, *C.R. Acad. Sc. Paris*, B *284*, 205 (1977).
46. Nestor, J.R., *J. Raman Spectrosc.*, *7*, 90 (1978).
47. Baranger, M., *Phys. Rev.*, *111*, 494 (1958).
48. Fiutak, J., and Van Kranendonk, J., *Can. J. Phys.*, *40*, 1085 (1962).
49. Omont, A., Smith, E.W., and Cooper, J., *The Astroph. J.*, *175*, 185 (1972).
50. Shen, Y.R., *Phys. Rev.*, *B9*, 622 (1974) ; Shen, Y.R., *Phys. Rev.*, *B14*, 1772 (1976).
51. De Witt, R.W., Harvey, A.B., and Tolles, W.M., "Theoretical Development of Third-Order Susceptibility as Related to Coherent Anti-Stokes Raman Spectroscopy", NRL Memorandum Report 3260 (1976).
52. Bordé, C.J., *C.R. Acad. Sc. Paris*, B *282*, 341 (1976) ; Bordé, C.J., in "Laser Spectroscopy III, Proceedings of the Third International Conference", Jackson Lake Lodge, USA, July 4-8, 1977, edited by J.L. Hall and J.L. Carlsten (Springer-Verlag, Berlin, Heidelberg, New York,1977), p. 121.
53. Feynman, R.P., Quantum Electrodynamics, Benjamin, New York, (1962), p. 19-22.
54. Jacon, M., "Theoretical study of resonance Raman Scattering by the methods of quantum electronics" in "Advances in Raman Spectroscopy" (Heyden & Son, London, 1973), Vol. *1*, p. 325.
55. Taran, J.P., in "Tunable Lasers and Applications", ed. by A. Mooradian, T. Jaeger, and P. Stokseth, (Springer-Verlag, Berlin, Heidelberg, New York, 1976), p. 378.
56. Kogelnik, H., and Li, T., "Laser Beams and Resonators," *Proc. IEEE*, *54*, 1312 (1966).

57. Ward, J.F., and New, G.H.C., *Phys. Rev.*, *185*, 57 (1969).
58. Bjorklund, G.C., *IEEE J. Quantum El.*, *QE-11*, 287 (1975).
59. Eckbreth, A.C., *Appl. Phys. Letters*, *32*, 421 (1978).
60. Moya, F., Druet, S., Péalat, M., and Taran, J.P., "Flame Investigation by Coherent Anti-Stokes Raman Scattering" in "Experimental Diagnostics in Gas Phase Combustion Systems", Progress in Astronautics and Aeronautics, *53*, ed. by B.T. Zinn, (Princeton, New-Jersey, 1977), p. 549, presented as paper 76-29 at the AIAA 14th Aerospace Sciences Meeting, Jan. 1976, Washington D.C.
61. Chutjian, A., and James, T.C., *J. Chem. Phys.*, *51*, 1242 (1969).
62. Nibler, J.W., Shaub, W.M., McDonald, J.R., and Harvey, A.B., "Coherent Anti-Stokes Raman Spectroscopy" in "Vibrational Spectra and Structure", *6*, ed. by J.R. Durig (Elsevier Scientific Publishing Co., New York, 1977).
63. Rousseau, D.L., and Williams, P.F., *J.Chem. Phys.*, *64*, 3519 (1976).
64. Hudson, B.S., Hetherington, W., Cramer, S., Chabay, I., Klauminzer, G., *Proc. Natl. Acad. of Sci. USA*, *73*, 3798 (1976).
65. Lynch, R.T., Jr., Kramer, S.D., Lotem, H., and Bloembergen, N., *Optics Comm.*, *16*, 372 (1976).
66. Carreira, L.A., Goss, L.P. and Malloy, T.B., *J. Chem. Phys.*, *66*, 4360 (1977).
67. Henessian, M.A., and Byer, R.L., "High Resolution CARS Line Shape Function", presented as paper G9 at the 10th International Quantum Electronics Conference, Atlanta, Ga, June 29-30, (1978).
68. Begley, R.F., Harvey, A.B., Byer, R.L., *Appl. Phys. Lett.*, *25*, 387 (1974).
69. Levenson, M.D., and Bloembergen, N., *J. Chem. Phys.*, *60*, 1323 (1974).
70. Levenson, M.D., and Bloembergen, N., *Phys. Rev.*, *10*, 3447 (1974).
71. Rogers, L.B., Stuart, J.D., Goss, L.P., Malloy, T.B., Jr., and Carreira, L.A., *Anal. Chem.*, *49*, 949 (1977).
72. Nabara, A., and Kubota, K., *Japanese J. Appl. Phys.*, *6*, 1105 (1967).
73. Itzkan, I., and Leonard, D.A., *Appl. Phys. Letters*, *26*, 106 (1976).
74. Lynch, R.T., Jr., and Lotem, H., *J. Chem. Phys.*, *66*, 1905, (1977).
75. Akhmanov, S.A., Koroteev, N.I., Orlov, R. Yu., and Shumai, I.L., *JETP Letters*, *23*, 249 (1976).

76. Akhmanov, S.A., Bunkin, A.F., Ivanov, S.G., Koroteev, N.I., Kovrigin, A.I., and Shumay, I.L., in "Tunable Lasers and Applications", ed. by A. Mooradian, T. Jaeger and P. Stokseth, (Springer-Verlag, Berlin, Heidelberg, New York, 1976), p. 389.

77. Anderson, R.J.M., Holtom, G.R., and McClain, W.M., *J. Chem. Phys.*, *66*, 3832 (1977).

78. Lynch, R.T., Jr., Lotem, H., and Bloembergen, N., *J. Chem. Phys.*, *66*, 4250 (1977).

79. Nitsch, W., and Keifer, W., Proc. Fifth Raman Conf., Ed. by E.D. Schmid et al., H.F. Schulz Verlag, Freiburg Germany, p. 740 (1976).

80. Carreira, L.A., Goss, L.P., and Malloy, Thomas B., Jr., *J. Chem. Phys.*, *66*, 2621 (1977).

81. Lau, A., Werncke, W., Klein, J., and Pfeiffer, M., *Optics Comm.*, *21*, 399 (1977).

82. Hertz, J., Lau, A., Pfeiffer, M., and Werncke, W., in "Lasers in Chemistry", Conference Digest, London, May 31st-June 2nd, 1977, p. 399.

83. Shoshan, I., and Oppenheim, U.P., *Optics Comm.*, *25*, 375 (1978).

84. Novikov, M.A., and Tertyshnik, A.D., *Sov. J. Quantum Electron.*, *5*, 848 (1975).

85. Roh, W.E., Schreiber, P., and Taran, J.P.E., *App. Phys. Lett.*, *29*, 174 (1976).

86. Eckbreth, A.C., Bonczyk, P.A., and Shirley, J.A., "Experimental Investigations of Saturated Laser Fluorescence and CARS spectroscopic Techniques for Practical combustion Diagnostics", United Technologies Research Center report n° R78-952665-18, East Hartford, Connecticut 06108, Feb. 1978.

87. Eckbreth, A.C., "CARS Investigation in Flames", paper presented at the 17th Internation Symposium on Combustion, Leeds, England, August 1978.

88. Boquillon, J.P., "Spectroscopie à Haute Résolution dans les Gaz Purs par Diffusion Raman Cohérente", unpublished thesis, Dijon, France, June, 19th, 1978.

89. Barrett, J.J., and Begley, R.F., *Appl. Phys. Lett.*, *27*, 129 (1975).

90. Hirth, A., and Volrath, K., *Optics Comm.*, *18*, 213 (1976).

91. Henesian, M.A., Kulevskii, L., Byer, R.L., and Herbst, R.L., *Optics Comm.*, *18*, 225 (1976).

92. Fabelinsky, V.I., Krynetsky, B.B., Kulevsky, L.A., Mishin, V.A., Prokhorov, A.M., Savel'ev, A.D., and Smirnov, V.V., *Optics Comm.*, *20*, 398 (1977).

93. Krynetsky, B.B., Kulevsky, L.A., Mishin, V.A., Prokhorov, A.M., Savel'ev, A.D., and Smirnov, V.V., *Optics Comm.*, *21*, 225 (1977).

94. DeMartini, F., Santamato, E., and Capasso, F., *IEEE J. Quantum El., QE-8*, 542 (1972).
95. DeMartini, F., Simoni, F., and Santamato, E., *Optics Comm., 9*, 176 (1973).
96. Lukasik, J., and Ducuing, J., *Phys. Rev. Letters, 28*, 115 (1972).
97. Moya, F., Druet, S.A.J., and Taran, J.P.E., *Optics Comm., 13*, 169 (1975).
98. Moya, F., Druet, S.A.J., and Taran, J.P.E., in "Laser Spectroscopy," ed. by S. Haroche, J.C. Pebay - Peyroula, T.W. Hansch and S.E. Harris, (Springer-Verlag, New York, 1975), p. 66.
99. Taran, J.P., "Coherent anti-Stokes Raman Spectroscopy", in Laser Spectroscopy III, Proceedings of the Third International Conference, Jackson Lake Lodge, U.S.A., July 4-8, 1977, ed. by J.L. Hall and J.L. Carlsten (Springer-Verlag, Berlin, Heidelberg, New York, (1977)).
100. Nibler, J.W., McDonald, J.R., and Harvey, A.B., *Optics Comm., 18*, 371 (1976).
101. Black, J.J., Gilson, T.R., Greenhalgh, D.A., and Laycock, L.C., *Laser Focus, 84*, March (1978).
102. Lynch, R.T., Jr., Kramer, S.D., Lotem, H., and Bloembergen, N., *Optics Comm., 16*, 372 (1976).
103. Heiman, D., Hellwarth, R., Levenson, M., and Martin, G., *Phys. Rev. Letters, 36*, 189 (1976).
104. Levenson, M.D., and Song, J.J., *JOSA, 66*, 641 (1976).
105. Eesley, G.L., Levenson, M.D., and Tolles, W.M., *IEEE J. Quant. Elect., QE-14*, 45 (1978).
106. Owyoung, A., and Jones, E.D., *Optics Letters, 1*, 152 (1977).

THEORY OF MOLECULAR RATE PROCESSES IN THE PRESENCE OF INTENSE LASER RADIATION[1]

Thomas F. George[2]
I. Harold Zimmerman[3]
Paul L. DeVries
Jian-Min Yuan[4]
Kai-Shue Lam
John C. Bellum[5]
Hai-Woong Lee
Mark S. Slutsky
Jui-teng Lin

Department of Chemistry
University of Rochester
Rochester, New York

I. INTRODUCTION

The interaction between electromagnetic radiation and material systems is an old topic of study. Investigations in this area of physics led directly to the first formulation of

[1]*Research supported by AFOSR Contract No. F49620-78-C-0005, NASA Grant No. NSG 2198 and NSF Grant No. CHE77-27826*
[2]*Camille and Henry Dreyfus Teacher-Scholar, Alfred P. Sloan Research Fellow.*
[3]*Present address: Physics Department, Clarkson Memorial College, Potsdam, New York.*
[4]*Present address: Physics Department, Drexel University, Philadelphia, Pennsylvania.*
[5]*Present address: Fachbereich Physik, Postfach 3049, Universität Kaiserslautern, D-6750 Kaiserslautern, West Germany.*

253

a quantum theory which, as it evolved through successive
stages of sophistication, came to provide a general basis for
explaining virtually all spectroscopic processes associated
with atomic and molecular systems. Of course it also revolu-
tionized the way in which we view the material world. An obvious result of our changed vision is the laser; only in a
quantized environment could it find existence.

In principle, quantum mechanics explains everything that
occurs at the atomic and molecular level. However, there is
left the task of elucidation, and this task has drawn the
attention of skilled workers ever since the general theory
first took on its modern form. It is all very well to have
an explanation in principle, but explanations in particular
are required if useful applications of the theory are to be
made and if the implications of a quantized world are truly
to be appreciated.

Understandably enough the first quantum systems to be
treated were simple models isolated from any outside influences. After a time these models became more realistic. An
external radiation field was introduced that perturbed otherwise stationary particle states. Quantization of the field
put electromagnetic radiation on a formally sound basis while
explaining such subtleties as vacuum polarization, the Lamb
shift and spontaneous emission. Enlarging the material component to include a statistical ensemble of particle systems
led to the discovery of many coherent processes involving interaction with the field. Spin waves and echoes, masers, and
lasers; these discoveries have amply demonstrated the predictive and analytical powers of quantum mechanics and have suggested a number of other effects not mentioned here (1).

In the present paper we wish to report on progress in a
newly-evolved area of research which treats the interaction
of intense laser radiation with particle systems undergoing
dynamical change. A highly descriptive name for this topic
of study is laser assisted chemistry. Its most distinguish-
ing characteristic is the inclusion of dynamical processes
in the particle system even when the radiation field is ab-
sent. For example, the field may illuminate particles in
collision. That a sufficiently strong field can modify the
outcome of such an event is of course well understood. What
we present here are the initial stages of a theory that will
hopefully come to embody that understanding in quantitative
form.

The central point around which the present theory revolves
is this: influences which compete in modifying the behavior
of a molecular system need not act at different times, and
when examining situations in which such influences are in
fact contemporaneous we must include them all together in our

dynamical description. Furthermore, it is entirely reasonable
to suppose that no single one of these competing effects is
dominant, that perturbative modifications to some central process will not properly reflect what happens to the system
under consideration. The problems treated in this chapter are
chosen specifically because they do represent exceptions to
the familiar situation in which a single effect can be isolated and analyzed.

Figure 1 illustrates this point. A representative chemical reaction is shown. In panel A the laser field interacts
with one of the reactant species. This step might represent
preparation of the reactant in some particular excited state,
for example, with the associated dynamics treated separately

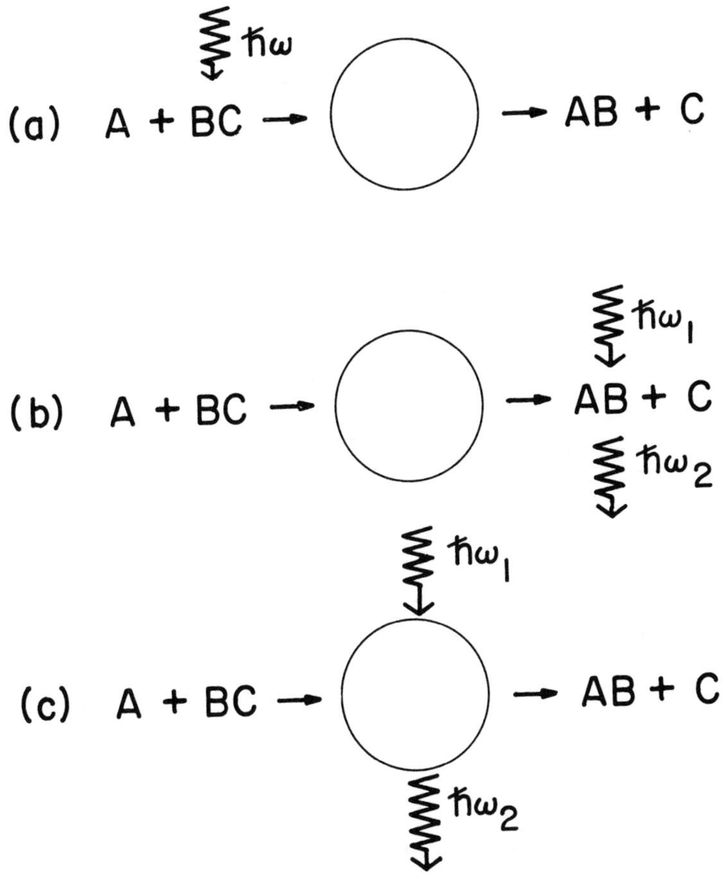

(a) $A + BC \rightarrow \bigcirc \rightarrow AB + C$ with $\hbar\omega$

(b) $A + BC \rightarrow \bigcirc \rightarrow AB + C$ with $\hbar\omega_1$, $\hbar\omega_2$

(c) $A + BC \rightarrow \bigcirc \rightarrow AB + C$ with $\hbar\omega_1$, $\hbar\omega_2$

FIGURE 1. *Three types of laser interaction with a chemical reaction.*

from the subsequent chemical evolution of the system. In
panel B the order is reversed, with chemical reaction leading
to products which are then probed by the field whose quantum
is $\hbar\omega_1$. The output radiation represented by $\hbar\omega_2$ may represent chemiluminescence or Raman-scattered light in spectro-
scopic studies or, if the reaction has yielded a population
inversion among product states, can represent lasing with $\hbar\omega_1$
the stimulating field.

In both of these possibilities the field interacts with
particle systems whose energy levels are stationary. In panel
C, however, the particle energy levels have gone over to
dynamic surfaces among which at least some transitions occur
even in the absence of a field. With the field imposed we may
see otherwise forbidden transitions become allowed; such instances are termed *field assisted* (FA). The field may also
act to change the extent to which observed processes go to
completion, in which case we use the term *field modified* (FM).

Summarizing, we point to three characteristics shared by
all of the systems to be discussed. First, a single-mode
laser field will always be present and will be sufficiently
strong that perturbation theory will not accurately describe
the effects of its interaction with the particle component.
Second, there will always be at least one other competing
disturbance that mixes particle states. Third, there will be
at least one degree of freedom for which, at the energies of
interest, continuum states are accessed. The detail with
which the continuum is treated will depend on the nature of
our interest in each case.

The methods to be applied are described in Section II.
We first take up the radiation field, basing our necessarily
brief presentation on standard references (1-5). The field
may be described in any one of several ways depending on the
problem at hand. The number state representation appears best
suited in cases where few photons are exchanged with the field,
so long as the field is either very intense or is limited to
some small and specifiable number of photons. For intense
fields with the exchange of many photons we find it convenient
to transform to the phase representation (2-6). Under the
approximation that laser beam depletion is negligible we then
recover the familiar "time-dependent" wave equation describing a quantum system exposed to a classical external field.
Beam depletion is certainly negligible if the field is in a
coherent state (7), and this state is described. Finally we
note an interesting approach in which the field is represented
by a classical oscillator in configuration space (8). By this
means a completely classical description of laser-assisted
chemical processes is made feasible. Properly applied, it
should be very useful in helping us to understand the evolu-

tion of various systems subject to the combined influences of field-interactive and molecular dynamics.

We then turn to the interactions whereby the field influences the particle system. As a rule transitions will be generated by the lowest-order particle state transition moments to which the field can couple. In the specific examples to be studied later these will be electric and/or magnetic dipole moments. The restriction to dipole coupling is justified in light of our aim, which is to describe situations in which incidence of the laser field modifies the dynamics of molecular systems (1). Only the strongest possible coupling will bring about this kind of selective interference at reasonable field intensities.

When we discuss molecular dynamics, therefore, we have first to consider the symmetries of the various particle states involved. This will generally be true even in the absence of the field, as we shall most often focus our attention on transitions among quantum states of particle scattering systems. For these we present time-independent coupled equations in a quantum mechanical formulation (9). Parallel semiclassical formulations are also described; these frequently make use of some of the information generated by the quantum treatment, but the semiclassical approach differs fundamentally in its representation of the nuclear degrees of freedom (10).

Several applications of the theory are described in Section III, starting with a series of model calculations designed to illustrate laser-assisted and laser-modified energy transfer processes in non-reactive atomic and molecular collisions. [These examples are drawn from our own work, but theoretical investigation of combined collisional and optical influences on particle systems goes back at least to 1972 (11-23). Experimental research in this area is of somewhat more recent vintage (24-27).] A central ingredient of the calculational method as embodied in the semiclassical theory is the concept of matter-field potential energy surfaces (10, 12,14,28,29), most often associated with coupled electronic and radiative degrees of freedom and thus termed electronic-field surfaces in the cited references. These are described in Section II. Particular versions of this dynamical model are discussed in connection with their corresponding applications.

The sample calculations on non-reactive particle collision systems suggest several further topics that merit discussion. For example, how does the deposition of photon angular momentum in an absorbing species affect the rotational dynamics of the collision complex? What effects will result if the laser field is not perfectly monochromatic, or if it is pulsed in various ways? These points and others are raised and examined with the dual purpose of finding convenient methods to

treat them and developing criteria whereby their importance in
a given situation might be ascertained.

Chemical reaction in a laser field is represented by collinear F + H_2 collisions. Two electronic surfaces are utilized. The lower goes exothermically to HF + H; the upper is
not reactive at the energies of interest, whether or not a
photon is absorbed (multiple photon processes are not included
here). Another interesting problem is that of cooperative
photo- and collision-induced ionization (27,30,31). Harking
back to nomenclature introduced earlier, we call this process
field modified or field assisted collisional ionization depending on whether or not free electrons are produced in the
absence of the field. We might just as easily call it collision assisted photoionization except that we again focus on
single-photon processes.

Photodissociation couples bound particle states to dissociative continuum levels. Two examples concern us, UV and
IR. Absorption of an ultraviolet photon will excite a prop-
erly-chosen molecule from a bound nuclear level on its ground
electronic potential energy surface to a continuum nuclear
state on a dissociative surface (32,33). By way of contrast,
infrared radiation interacts with the vibrational dynamics
associated with a single electronic surface so as to drive the
nuclear energy through higher-lying levels into the continuum.
This usually requires many photons as opposed to the single
photon absorption characteristic of UV photodissociation
(34,35).

Spontaneous emission from (otherwise) stationary excited
particle states has been studied for many years. We have
recently investigated how this process is perturbed if the
particle level separations do not remain constant in time but
vary in a prescribed manner. The method is developed within
the context of a semiclassical approach to nuclear dynamics on
electronic or electronic-field potential energy surfaces.

Finally we look at heterogeneous catalysis, seeking ways
to modify catalytic processes occurring at solid surfaces by
means of laser irradiation. Of interest here are the dynamics
of adsorption and migration on crystal faces, the coupling of
surface to bulk modes via the agency of the adsorbed species,
and laser desorption. This investigation is still in its very
early stages.

A consistent omission in all this is the possibility of
field-induced polarization. Thus Raman and Rayleigh scattering are not among the phenomena being considered even though
they are likely to be observed under the conditions supposed
to be in effect. Eventually we shall have to take these added
competing processes into account. Thus the present theory is
only a beginning. Much remains to be done before it can be

called comprehensive, let alone complete.

II. THEORETICAL METHODS

Representations of the radiation field, the particle system and the interactions involving these two entities are discussed from a general and rather abstract point of view. Our intention in this section is to lay out the various tools that will be utilized in subsequent applications. For that reason our presentation is formal. We assume that the reader is familiar with the basic concepts involved.

A. *Radiation Field*

The Coulomb gauge is most appropriate for describing the propagation of plane electromagnetic waves, since (in the absence of nearby currents and sources) it is equivalent to requiring $\vec{k} \cdot \hat{A} = 0$. Within this gauge we transform the classical field using the correspondence principle to arrive at a quantum mechanical Hamiltonian and associated field operators (2). Since the Hamiltonian is that of a harmonic oscillator we find the stationary states to be $|n\rangle$ (assuming a single-mode field of specified polarization) where n is a non-negative integer such that

$$\hat{H}_f |n\rangle = \hbar\omega\left(n + \frac{1}{2}\right)|n\rangle. \tag{2.1}$$

$\hat{H}_f$ is the Hamiltonian operator for the single-mode field of characteristic quantum $\hbar\omega$. In terms of the photon creation and annihilation operators $\hat{a}^\dagger$ and $\hat{a}$ this Hamiltonian is

$$\hat{H}_f = \hbar\omega\left(\hat{a}^\dagger\hat{a} + \frac{1}{2}\right) \tag{2.2}$$

and the vector potential operator is written as

$$\hat{A} = \frac{i}{\sqrt{V}} \vec{A}_0 [\hat{a}e^{i\vec{k}\cdot\vec{x}} - \hat{a}^\dagger e^{-i\vec{k}\cdot\vec{x}}] \tag{2.3}$$

in the Schroedinger picture. V is the quantization volume. The wave number $\vec{k}$ specifies the mode of the field while $\vec{A}_0$ gives its polarization direction $\vec{\sigma}$ and single-photon magnitude A_0:

$$\vec{A}_0 = \vec{\sigma}A_0 \tag{2.4a}$$

$$= \vec{\sigma}(2\pi\hbar c^2/\omega)^{1/2} \tag{2.4b}$$

in gaussian units. The phase convention in Eq. (2.3) is chosen such that the electric and magnetic field operators,

$$\hat{E} = \frac{-\omega}{c\sqrt{V}} \vec{A}_0 [\hat{a}e^{i\vec{k}\cdot\vec{x}} + \hat{a}^\dagger e^{-i\vec{k}\cdot\vec{x}}], \tag{2.5}$$

$$\hat{H} = -\frac{1}{\sqrt{V}} \vec{k} \times \vec{A}_0 [\hat{a}e^{i\vec{k}\cdot\vec{x}} + \hat{a}^\dagger e^{-i\vec{k}\cdot\vec{x}}], \tag{2.6}$$

will be completely real in the dipole approximation. This is not required, of course, but it is convenient in applications.

The number-state or Fock representation for the field is indicated when only a few photons are exchanged with the particle system. We assume that the occupation number is sufficiently large that the gain or loss of a few photons does not materially affect the strength of the field interaction (exceptions will be noted). However, multiphoton processes are not easily treated in a number-state basis owing to the very large coupling matrices that result. For example, if two electronic states are included in the particle expansion basis, and if there are five vibrational levels for each electronic surface, then we must concern ourselves with ten particle states even without taking rotational degrees of freedom into account. If the field is then imposed we find that the overall system representation includes ten times the number of allowed field states in the overall expansion basis, so that retaining ten Fock states results in a hundred-state basis overall. The situation grows worse very rapidly for more realistic problems in which rotational degrees of freedom are included and in which more precise treatment is made of electronic and vibrational modes. Clearly, an alternative expression for the field is called for.

Converting to the phase representation alleviates much of the formal difficulty associated with the Fock basis. The phase operator $\hat{\phi}$ is defined by writing (2)

$$\hat{a} = (\hat{n} + 1)^{1/2}e^{i\hat{\phi}}, \tag{2.7a}$$

$$\hat{a}^\dagger = e^{-i\hat{\phi}}(\hat{n} + 1)^{1/2}, \tag{2.7b}$$

where the number operator $\hat{n}$ is given by

$$\hat{n} = \hat{a}^\dagger \hat{a}. \tag{2.8}$$

The nonvanishing matrix elements of $e^{\pm i\hat{\phi}}$ in the Fock representation are

$$\langle n-1|e^{i\hat{\phi}}|n\rangle = \langle n+1|e^{-i\hat{\phi}}|n\rangle = 1. \tag{2.9}$$

The operators $e^{\pm i\hat{\phi}}$ do not generally commute:

$$[e^{i\hat{\phi}}, e^{-i\hat{\phi}}] = 1 - \hat{a}^\dagger (\hat{n}+1)^{-1}\hat{a}. \tag{2.10}$$

Thus there is no true eigenstate of the phase except in a limiting sense. To see this, define the state $|\phi\rangle_N$:

$$|\phi\rangle_N = \frac{1}{\sqrt{N+1}} \sum_{n=0}^{N} e^{in\phi}|n\rangle. \tag{2.11}$$

Using the results

$$e^{-i\hat{\phi}}|n\rangle = |n+1\rangle, \tag{2.12}$$

$$e^{i\hat{\phi}}|n\rangle = \begin{cases} 0, & n=0 \tag{2.13a} \\ |n-1\rangle, & n>0, \tag{2.13b} \end{cases}$$

we see that $|\phi\rangle_N$ becomes an eigenfunction of $e^{\pm i\hat{\phi}}$ only in the limit $N \to \infty$. In that limit we also find

$$\langle n|\phi\rangle_{N\to\infty} \to 0 \tag{2.14}$$

which implies that specification of the phase renders the energy state of the field completely indeterminant. We have instead that

$$\hbar\omega\hat{a}^\dagger\hat{a}|\phi\rangle_N = -i\hbar\omega\frac{\partial}{\partial\phi}|\phi\rangle_N. \tag{2.15}$$

Note that this is a time-independent relation, and that it is true for all N. The phase and the photon number are canonically conjugate to one another. If we define a "time" τ by writing

$$\phi = -\omega\tau \tag{2.16}$$

then Eq. (2.15) becomes

$$\hbar\omega\hat{a}^\dagger\hat{a}|\phi\rangle_N = i\hbar\frac{\partial}{\partial\tau}|\phi\rangle_N. \tag{2.17}$$

Finally, define the Hermitian operators

$$\cos\hat{\phi} = \frac{1}{2}[e^{i\hat{\phi}} + e^{-i\hat{\phi}}]$$ (2.18a)

and

$$\sin\hat{\phi} = \frac{1}{2i}[e^{i\hat{\phi}} - e^{-i\hat{\phi}}],$$ (2.18b)

and write Eqs. (2.5) and (2.6) as

$$\hat{\vec{E}} = - \frac{\omega}{c\sqrt{V}} \vec{A}_0 [e^{-i(\hat{\phi}+\vec{k}\cdot\vec{x})} (\hat{n}+1)^{1/2}$$

$$+ (\hat{n}+1)^{1/2} e^{i(\hat{\phi}+\vec{k}\cdot\vec{x})}],$$ (2.19a)

$$\hat{\vec{H}} = - \frac{1}{\sqrt{V}} \vec{k} \times \vec{A}_0 [e^{-i(\hat{\phi}+\vec{k}\cdot\vec{x})} (\hat{n}+1)^{1/2}$$

$$+ (\hat{n}+1)^{1/2} e^{i(\hat{\phi}+\vec{k}\cdot\vec{x})}].$$ (2.19b)

Upon computing $_N\langle\phi|\hat{\vec{E}}|\phi\rangle_N$ and taking the limit $N\to\infty$ we then find $\vec{E}_{ph}$:

$$\vec{E}_{ph} = {}_N\langle\phi|\hat{\vec{E}}|\phi\rangle_N \to$$

$$- \frac{2\omega\sqrt{\frac{N}{2}+1}}{c\sqrt{V}} \vec{A}_0 \cos(\phi+\vec{k}\cdot\vec{x}), \quad N \to \infty$$ (2.20)

and similarly for $_N\langle\phi|\hat{\vec{H}}|\phi\rangle_N$. This expression can be finite only if the number density $N/2V$ is finite, meaning that N can go to infinity and still yield a finite field only if the quantization volume V becomes infinite as well. Calling the average photon number density ρ in this case allows us to recover the classical field amplitude:

$$\vec{E}_0 = - \frac{2\omega}{c} \sqrt{\rho}\, \vec{A}_0, \quad \rho = N/2V,$$ (2.21)

and the oscillating field vector is then given by

$$\vec{E}_{ph}(\tau) = \vec{E}_0 \cos(\omega\tau)$$ (2.22)

in the dipole approximation.

   This is rather an interesting result since it is time independent. For the free field, of course, the interpretation is straightforward: a change $-\Delta\phi$ in the phase ϕ leaves the field exactly as it would be if we changed the time by $\Delta t=-\Delta\phi/\omega$ $=\Delta\tau$. For the coupled field-plus-particle system, however, this is no longer true; the time t is conjugate to the entire Hamiltonian while τ is conjugate only to H_f. Nevertheless we find the approximation t~τ to be sufficiently good that we shall employ it where appropriate.

There are other possible states with which to describe the field. One of the most widely used is Glauber's coherent state (7). It may be defined by writing

$$|\alpha> = e^{-\frac{1}{2}|\alpha|^2} \sum_{n=0}^{\infty} \frac{\alpha^n |n>}{\sqrt{n!}} .$$ (2.23)

This form yields a Poisson distribution over the number states. It may also be expressed as

$$|\alpha> = e^{-\frac{1}{2}|\alpha|^2} e^{\alpha \hat{a}^\dagger} |0> .$$ (2.24)

The averaged energy of this state can be given by

$$\hbar\omega <\alpha | \hat{a}^\dagger \hat{a} |\alpha> = \hbar\omega |\alpha|^2$$ (2.25)

so that $|\alpha|^2$ is seen to be the average value of the number operator on this distribution of number states. Note that the sum in Eq. (2.23) is infinite while the average energy is still finite.

It is of interest to see how the phase operators act:

$$<\alpha |e^{i\hat{\phi}} |\alpha> = \alpha A(\alpha) ,$$ (2.26)

where

$$A(\alpha) = e^{-|\alpha|^2} \sum_{n=0}^{\infty} \frac{|\alpha|^{2n}}{n! \sqrt{n+1}} .$$ (2.27)

$A(\alpha)$ is real and is bounded above by 1, achieving that value only when $\alpha=0$. For large α we find

$$A(\alpha) \to |\alpha|^{-1} , \quad |\alpha| \to \infty ,$$ (2.28)

so that the phase operators return the phases of α and α^* in the classical limit:

$$\alpha \to \sqrt{n_0} \, e^{i\phi} , \quad n_0 \to \infty .$$ (2.29)

The most important property of the coherent state is expressed by writing

$$\hat{a}|\alpha> = \alpha|\alpha> ;\qquad\qquad\qquad\qquad (2.30)$$

i.e., $|\alpha>$ is the eigenstate of the annihilation operator. Thus it is not depleted when photons are absorbed from it, which is consistent with our intuitive picture of the situation that obtains when an intense laser beam illuminates a rarefied gas, for example.

The electric-field operator in the coherent state has the averaged value

$$\vec{E}_{coh} = <\alpha|\hat{E}|\alpha> = -\frac{\omega\vec{A}_0}{c\sqrt{V}} (\alpha e^{i\vec{k}\cdot\vec{x}} + \alpha* e^{-i\vec{k}\cdot\vec{x}})\qquad (2.31)$$

which, upon taking the classical limit of α given by Eq. (2.29), yields

$$\vec{E}_{coh} = \vec{E}_0\cos(\omega\tau)\qquad\qquad\qquad (2.32)$$

in the dipole approximation. This result is identical to the expression found for the phase-state average in Eq. (2.22) with $\vec{E}_0$ defined as in Eq. (2.21). In the present case we have $\rho=n_0/V$ as the average photon density.

The agreement of Eqs. (2.22) and (2.32) indicate the applicability of the classical expression for the field interaction term whenever the average occupation number for the field is large. Given that condition, we find it interesting to consider one more possibility: to express the field operators themselves in classical terms. We define the operators $\hat{P}$ and $\hat{Q}$ by writing (c.f. References 2 and 8)

$$\hat{P} = i\vec{\sigma}\sqrt{\frac{\hbar\omega}{2}} (\hat{a}-\hat{a}^\dagger),\qquad\qquad (2.33a)$$

$$\hat{Q} = -\vec{\sigma}\sqrt{\frac{\hbar}{2\omega}} (\hat{a}+\hat{a}^\dagger).\qquad\qquad (2.33b)$$

The unit vector $\vec{\sigma}$ again gives the polarization. We then find that

$$\hat{H}_f = \frac{1}{2}(\hat{P}^2 + \omega^2\hat{Q}^2),\qquad\qquad\qquad (2.34)$$

$$\hat{E} = \sqrt{\frac{4\pi\omega^2}{V}}\ \hat{Q},\qquad\qquad\qquad (2.35)$$

$$\hat{H} = \sqrt{\frac{4\pi c^2}{V}}\ \vec{k}\times\hat{Q}.\qquad\qquad\qquad (2.36)$$

We then take the classical limits $\hat{P} \to \vec{p}$ and $\hat{Q} \to \vec{q}$, thereby obtaining a completely classical field. The coordinates $\vec{p}$ and $\vec{q}$ are easily related to the phase $\phi = -\omega\tau$ and the average occupation n_0:

$$\vec{q} = - \vec{\sigma} \sqrt{\frac{2\hbar n_0}{\omega}} \cos(\omega\tau) \qquad (2.37a)$$

and

$$\vec{p} = \vec{\sigma}\sqrt{2\hbar\omega n_0} \sin(\omega\tau) \qquad (2.37b)$$

in the classical field. This formulation can be useful in discussing the dynamics of interaction between the field and nonstationary molecular systems when a semiclassical approach is utilized (36).

B. *Matter-Field Interaction*

Our model for the interaction whereby the radiation field couples states of the particle system will depend on the problem at hand. Generally the situation will be sufficiently simple that a single polarized mode may be used to represent the intense field. Additional weak modes can be accommodated with little difficulty. Two general approaches, quantal and semiclassical, are contemplated in treating the particle dynamics; the choice of approach aids in deciding how interaction with the intense field is to be represented. This statement will be enlarged upon when we discuss dynamics. Here we present a few basic points that need to be considered when modeling the interaction mechanism.

Our starting point is the nonrelativistic Schroedinger equation with the particle kinetic energy operator replaced according to the Lagrangian prescription:

$$\frac{\hat{p}^2}{2m} \longrightarrow \frac{1}{2m} (\hat{p} - \frac{q}{c} \hat{A})^2, \qquad (2.38)$$

where q is the charge of the particle whose mass is m. In general there will be a large number of particles which we shall classify as being either electrons or nuclei. We shall adopt the convention that the αth electron, located at $\vec{x}_\alpha$, has spin $\vec{S}_\alpha$ and charge $-e$. Thus we are taking e to be the (positive) magnitude of the unit electronic charge. Similarly, nucleus β is located at $\vec{y}_\beta$, has spin $\vec{S}_\beta$ and charge $Z_\beta e$. The total particle-plus-field Hamiltonian is then

$$\hat{H} = \hat{H}_f + \hat{H}_{s.o.} + \sum_\alpha \{\frac{1}{2m_e} [\frac{\hbar}{i} \vec{\nabla}_\alpha + \frac{e}{c} \hat{A}(\vec{x}_\alpha)]^2$$

$$+ V_\alpha(\vec{x},\vec{y}) + \frac{e}{m_e} \hat{S}_\alpha \cdot [\vec{\nabla}_\alpha \times \hat{A}(\vec{x}_\alpha)]\}$$

$$+ \sum_\beta \{\frac{1}{2m_\beta} [\frac{\hbar}{i} \vec{\nabla}_\beta - \frac{e}{c} Z_\beta \hat{A}(\vec{y}_\beta)]^2$$

$$+ V_\beta(\vec{y}) - \frac{e}{m_\beta} Z_\beta \hat{S}_\beta \cdot [\vec{\nabla}_\beta \times \hat{A}(\vec{y}_\beta)]\}. \qquad (2.39)$$

As usual in nonrelativistic formulations, spin-field and spin-orbit interactions are simply added to the rest of the Hamiltonian. $\hat{S}_\alpha$ is the spin operator for electron α and $\hat{H}_{s.o.}$ is the spin-orbit Hamiltonian. The term $V_\alpha(\vec{x},\vec{y})$ is the potential energy seen by electron α as a function of the positions of all other electrons plus the nuclei. Similarly, $V_\beta(\vec{y})$ represents the potential energy of nucleus β. This term includes no dependence on $\vec{x}$, the collective coordinate for all electrons, in order that we avoid double-counting the electron-nucleus interactions. It is assumed that precautions against double-counting of electron-electron and nucleus-nucleus interactions have also been taken in writing the sums over V_α and V_β.

The Hamiltonian is then expressed in the dipole approximation including just the lowest electric and magnetic moments. We find (9,37)

$$\hat{H} = \hat{H}_f + \hat{H}_{s.o.} + \sum_\alpha \{-\frac{\hbar^2}{2m_e}\nabla_\alpha^2 + V_\alpha(\vec{x},\vec{y})$$

$$+ e[\hat{\vec{E}}\cdot\vec{x}_\alpha + \frac{1}{2m_e} \hat{H}\cdot(\hat{L}_\alpha + 2\hat{S}_\alpha)]\} + \sum_\beta \{-\frac{\hbar^2}{2m_\beta}\nabla_\beta^2 + V_\beta(\vec{y})$$

$$- eZ_\beta[\hat{\vec{E}}\cdot\vec{y}_\beta + \frac{1}{2m_\beta} \hat{H}\cdot(\hat{L}_\beta + 2\hat{S}_\beta)]\}. \qquad (2.40)$$

This is the Hamiltonian expression on which all of our subsequent discussion will rest. The systems whose dynamics will later be analyzed in more detail will be described by Hamiltonians derived from this general form via appropriate simplifying assumptions.

We may separate $\hat{H}$ into two parts,

$$\hat{H} = \hat{H}_{non} + \hat{H}_{int} \, , \tag{2.41}$$

where

$$\hat{H}_{non} = \hat{H}_f + \hat{H}_{s.o.} + \sum_{\alpha} [-\frac{\hbar^2}{2m_e} \nabla_{\alpha}^2 + V_{\alpha}(\vec{x}, \vec{y})]$$

$$+ \sum_{\beta} [-\frac{\hbar^2}{2m_{\beta}} \nabla_{\beta}^2 + V_{\beta}(\vec{y})] \tag{2.42}$$

and

$$\hat{H}_{int} = e \sum_{\alpha} [\hat{\vec{E}} \cdot \vec{x}_{\alpha} + \frac{1}{2m_e} \hat{H} \cdot (\hat{L}_{\alpha} + 2\hat{S}_{\alpha})]$$

$$- e \sum_{\beta} z_{\beta} [\hat{\vec{E}} \cdot \vec{y}_{\beta} + \frac{1}{2m_{\beta}} \hat{H} \cdot (\hat{L}_{\beta} + 2\hat{S}_{\beta})] . \tag{2.43}$$

For now we focus on $\hat{H}_{int}$, which describes the interaction between the field and particle components of the system. Discussion of the noninteracting Hamiltonian $\hat{H}_{non}$ will be deferred until the next part of this section.

The electric field couples particle states of opposite or mixed parity. It is obvious that electronic states of definite symmetry can be constructed for single atoms, at least in principle, if we can disregard effects such as spin-orbit coupling. When more atoms are involved in the molecule, however, we lose our ability to define electronic states of good parity with respect to inversion of electronic coordinates alone. There are exceptions to this observation, most notably in the case of homonuclear diatoms, but generally we find that every conceivable electronic transition becomes electric dipole allowed to a greater or lesser degree when atoms approach one another. In particular, molecular electronic states couple to themselves. To illustrate the significance of this fact, consider a diatom such as HF. If an accurate ground state electronic wave function $\psi_0(\vec{x}, \vec{y})$, written as a function of all relevant electronic and nuclear coordinates, is available for this molecule, then we will find that the moment $\vec{\mu}_{00}(\vec{y})$ defined by

$$\vec{\mu}_{00}(\vec{y}) = e\{\sum_{\beta=1}^{2} z_{\beta} \vec{y}_{\beta} - \sum_{\alpha=1}^{A} \int d^{3A}\vec{x} \psi_0^*(\vec{x}, \vec{y}) \vec{x}_{\alpha} \psi_0(\vec{x}, \vec{y})\}$$

$$\tag{2.44}$$

is not zero until $r = |\vec{y}_1 - \vec{y}_2|$ becomes large (38). This is the source of so-called permanent dipole moments, and it is to the variation of these moments that the field couples during IR emission and absorption on a single electronic surface.

Statements similar to these may also be made in the case of magnetic dipole coupling. The magnetic selection rules in atoms are not quite so stringent as are the electric, and they break down as atoms come together for the same reasons that the electric dipole selection rules do. However, we need not follow the corresponding evolution of the magnetic transition moment if an electric transition dipole comes into being, because effects due to the latter will generally swamp any contribution from the former. If electrical contributions are ignored in such cases we might as well ignore variations in the magnetic dipole as well. Unfortunately we frequently have no expression for the proximity-induced electric dipole transition moments. This lack will be felt in some of the calculations described in Section III, and it represents the most severe shortcoming in current applications of the theory.

C. Dynamics

We consider the combined particle-plus-field system as it evolves. The general quantum theory is displayed first. Semiclassical methods are then outlined and discussed.

1. Quantum Formalism. The time-dependent Schroedinger equation is

$$i\hbar \frac{\partial}{\partial t} \Psi = \hat{H}\Psi, \tag{2.45}$$

with $\hat{H}$ given by Eq. (2.40) and partitioned into $\hat{H}_{non}$ and $\hat{H}_{int}$ as detailed in Eqs. (2.41-43). The noninteractive part of the Hamiltonian may be further divided up into $\hat{H}_f$ and the field-free particle Hamiltonian $\hat{H}_0$. The latter may then be expanded to display electronic and nuclear contributions:

$$\hat{H}_0 = \hat{H}_{el} + \hat{H}_{nuc}, \tag{2.46}$$

where

$$\hat{H}_{el} = \hat{H}_{s.o.} + \sum_\alpha [-\frac{\hbar^2}{2m_e} \nabla_\alpha^2 + V_\alpha(\vec{x},\vec{y})], \tag{2.47a}$$

$$\hat{H}_{nuc} = \sum_\beta [-\frac{\hbar^2}{2m_\beta} \nabla_\beta^2 + V_\beta(\vec{y})]. \tag{2.47b}$$

The solution to Eq. (2.45) is written out in terms of product states:

$$|\Psi(\vec{x},\vec{y},t)> = \sum_{nj\ell} |n>|\psi_j>|\Theta_{j\ell}>f_{nj\ell}.$$ (2.48)

The ket $|n>$ is a number state of the field while $|\psi_j>$ is an electronic basis state and $|\Theta_{j\ell}>$ is a nuclear state. The coefficient $f_{nj\ell}$ will depend on some one of the nuclear separation coordinates and/or time, depending on the problem. We should note that the index j is a composite notation representing all electronic quantum labels; similarly, ℓ is a composite label for the nuclear state.

The electronic basis may be either adiabatic or diabatic (39,40). Assuming completeness, the two are related by writing

$$\psi_j^a(\vec{x},\vec{y}) = \sum_{j'} \psi_{j'}^d(\vec{x})A_{j',j}(\vec{y})$$ (2.49)

where the coefficient matrix $\underset{\sim}{A}$ is a diagonalizing array such that

$$\hat{H}_{el}\psi_j^a(\vec{x},\vec{y}) = U_j^e(\vec{y})\psi_j^a(\vec{x},\vec{y})$$ (2.50)

is an eigenvalue equation for all nuclear configurations $\vec{y}$. The adiabatic states are coupled by the nuclear kinetic energy operator:

$$<\psi_j^a|\nabla_\beta^2|\psi_{j'}^a>_{\vec{x}} = \delta_{jj'}\nabla_\beta^2 + 2\vec{\xi}_{\beta jj'}(\vec{y})\cdot\vec{\nabla}_\beta$$

$$+ \Xi_{\beta jj'}(\vec{y}),$$ (2.51)

with

$$\vec{\xi}_{\beta jj'}(\vec{y}) = <\psi_j^a|\vec{\nabla}_\beta\psi_{j'}^a>_{\vec{x}},$$ (2.52a)

$$\Xi_{\beta jj'}(\vec{y}) = <\psi_j^a|\nabla_\beta^2\psi_{j'}^a>_{\vec{x}}.$$ (2.52b)

By way of contrast, the diabatic basis states do not depend on $\vec{y}$ and so do not couple through ∇_β^2. Rather, they are coupled by $\hat{H}_{el}$:

$$<\psi_j^d|\hat{H}_{el}|\psi_{j'}^d>_{\vec{x}} = H_{jj'}^{el}(\vec{y}).$$ (2.53)

We assume that boundary conditions are chosen such that $H^{el}_{jj'}(\vec{y}) \rightarrow \delta_{jj'} U^e_j(\vec{y})$ as $\vec{y}$ passes into a suitable range of configurations.

For reactive systems, further refinements to the theory are called for. Likely modifications are suggested in Reference 9. Such systems are not examined here, as a working calculation treating chemical reaction in the presence of an intense laser field has yet to be carried out in a quantum mechanical formulation.

The nuclear basis functions $\{\Theta_{j\ell}(\vec{y})\}$ and coefficients $\{f_{nj\ell}\}$ will differ depending on whether the adiabatic or diabatic electronic basis is employed. We shall therefore wish to extend our superscript labeling to these functions. Further decomposition of the nuclear function into vibrational and rotational factors depends on the number of nuclei involved. We shall focus on atom-atom and atom-diatom particle systems. Motion of the overall center of mass will not be taken into account here, even though one of the atoms may be ionized. This approximation is acceptable so long as the particle system remains neutral overall, but it may not be when the particle component carries with it a net charge. This point has yet to be examined quantitatively within the present context.

For atom-atom processes in the center-of-mass frame we are left with just the interatomic vector $\vec{r}$ as illustrated in

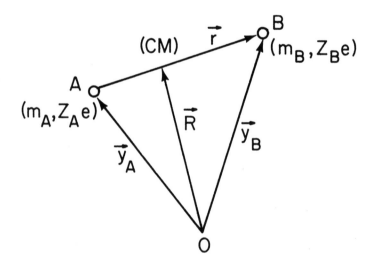

FIGURE 2. The coordinate system for atom-atom processes.

Figure 2. The nuclear Hamiltonian is then

$$\hat{H}_{nuc} = -\frac{\hbar^2}{2m}\nabla^2_{\vec{r}} + Z_A Z_B e^2/r, \tag{2.54}$$

where m is the reduced mass of the diatomic A+B system:

$$m = m_A m_B/(m_A + m_B). \tag{2.55}$$

The full particle Hamiltonian $\hat{H}_0$ will have the matrix representation

$$<\psi^a_j|\hat{H}_0|\psi^a_{j'}>_{\vec{x}} = \delta_{jj'}[-\frac{\hbar^2}{2m}\nabla^2_{\vec{r}} + U^a_j(\vec{r})]$$

$$-\frac{\hbar^2}{2m}[2\vec{\xi}_{jj'}(\vec{r})\cdot\vec{\nabla}_{\vec{r}} + \Xi_{jj'}(r)] \tag{2.56}$$

in the electronically adiabatic basis, where

$$U^a_j(\vec{r}) = Z_A Z_B e^2/r + U^e_j(\vec{r}) \tag{2.57}$$

is the electronically adiabatic potential energy surface governing motion of the two nuclei when the electron state is $|\psi^a_j>$. The elements $\vec{\xi}_{jj'}(r)$ and $\Xi_{jj'}(\vec{r})$ are defined as in Eqs. (2.52) with the subscript β replaced by the label $\vec{r}$.

In the diabatic representation we have

$$<\psi^d_j|\hat{H}_0|\psi^d_{j'}>_{\vec{x}} = -\delta_{jj'}\frac{\hbar^2}{2m}\nabla^2_{\vec{r}} + U^d_{jj'}(\vec{r}) \tag{2.56'}$$

with

$$U^d_{jj'}(\vec{r}) = \delta_{jj'}Z_A Z_B e^2/r + H^{el}_{jj'}(\vec{r}). \tag{2.57'}$$

These may be compared with Eqs. (2.56) and (2.57), respectively. The diabatic representation is much more convenient to use in the matrix mechanical formulation of quantum mechanics than is the adiabatic, unless we are able to ignore the nonadiabatic coupling terms in Eq. (2.56). These terms may not be dropped for several of the problems to be treated. However, it is possible that only adiabatic surfaces and coupling elements are available for a given situation. In such cases we may find it convenient to invert the transformation of Eq. (2.49). A procedure for doing so has been worked out and tested (41). The method evaluates $\underset{\sim}{A}$ using the $\underset{\sim}{\xi}$ coupling arrays, and it has proven useful in the few instances where it was applied. It can be generalized easily to situations involving more than two atoms. We are therefore justified in assuming that the diabatic formulation is given whenever the full adiabatic representation is available for a given problem.

Electric and magnetic dipole matrix elements coupling electronic states are easily calculated in the two representations provided the electronic wave functions are available. Often they are not available, unfortunately, and semiempirical techniques must be called into play. We shall not go into this subject here, choosing rather to defer it to the individual calculations where such methods are employed. However, some basic groundwork can be laid.

The purely nuclear contribution to the electric dipole transition matrix is a diagonal array in either one of the two electronic representations. The diagonal elements are equal to each other, with functional form given by the negative of

$$\vec{\mu}^{nuc}(\vec{r}) = \left(\frac{m_A Z_B - m_B Z_A}{m_A + m_B}\right) e\vec{r} \tag{2.58}$$

for the coordinate system of Figure 2. [We have taken $\vec{R}=0$, which places the origin at the center-of-mass point.] The total dipole array is then given by the formula

$$\vec{\mu}_{jj'}^{a,d}(\vec{r}) = e \sum_\alpha \langle \psi_j^{a,d} | \vec{x}_\alpha | \psi_{j'}^{a,d} \rangle_{\vec{x}} - \delta_{jj'} \vec{\mu}^{nuc}(\vec{r}) \tag{2.59}$$

in the adiabatic or diabatic electronic representation. The only $\vec{r}$-dependence in $\vec{\mu}_{jj'}^d(\vec{r})$ comes from $\vec{\mu}^{nuc}(\vec{r})$ and is therefore confined to the diagonal, while in the adiabatic representation we see that all terms of $\vec{\mu}_{jj'}^a(\vec{r})$ generally vary with $\vec{r}$ owing to the r-dependence of the corresponding electronic basis functions.

The magnetic dipole array can likewise be built up out of two parts, the purely nuclear and the electronic. In the diabatic representation the nuclear contribution is again diagonal, but the adiabatic electronic functions are coupled by the nuclear angular momentum operators and therefore the corresponding nuclear magnetic dipole operator is not diagonal in that basis. We have

$$\frac{e}{2} \sum_{\beta=A,B} \frac{Z_\beta}{m_\beta} \hat{L}_\beta = \frac{e}{2} \left(\frac{m_A Z_B + m_B Z_A}{m_A + m_B}\right) \vec{r} \times \vec{V}_{\vec{r}} \tag{2.60}$$

for the two-atom problem. Contributions due to nuclear spin may also be included but we shall neglect such terms in this paper. [Their inclusion is trivial given the assumption that electronic functions are independent of nuclear spin coordinates in either the diabatic or the adiabatic representation.] We then find the nuclear magnetic dipole operator to be represented by

$$\hat{M}_{jj'}^{d,nuc}(\vec{r}) = \frac{1}{2} Q_{AB} \delta_{jj'} \vec{r} \times \vec{\nabla}_{\vec{r}} \tag{2.60'}$$

in the diabatic representation, and by

$$\hat{M}_{jj'}^{a,nuc}(\vec{r}) = \hat{M}_{jj'}^{d,nuc}(\vec{r}) + \frac{1}{2} Q_{AB} \vec{r} \times \vec{\xi}_{jj'}(\vec{r}) \tag{2.61}$$

in the adiabatic, where

$$Q_{AB} = \left(\frac{m_A Z_B + m_B Z_A}{m_A + m_B} \right) e. \tag{2.62}$$

The total magnetic moment is then written as

$$\hat{M}_{jj'}^{a,d}(\vec{r}) = \frac{e}{2m_e} \sum_\alpha \langle \psi_j^{a,d} | \hat{L}_\alpha | \psi_{j'}^{a,d} \rangle_{\vec{x}} - M_{jj'}^{(a,d),nuc}(\vec{r}). \tag{2.63}$$

Assuming that the necessary dipole elements have been found, we turn to consideration of the nuclear degrees of freedom. There are several possibilities here. If, for example, the two atoms are held together in a well on a single electronic surface, then the two atoms may vibrate and rotate about their common center of mass and, for sufficiently energetic oscillations, may even come apart. The nuclear basis state $|\Theta_{j\ell}\rangle$ will be a product of a vibrational state $|\chi_{jv}\rangle$ and a spherical harmonic $|Y_{LM_L}\rangle$. Note that the spherical harmonic does not have the electronic state index j associated with it. [We also note in passing that here the collective nuclear state index ℓ stands for a triplet of quantum numbers composed of the vibrational index v, the angular momentum label L and the magnetic number M_L.] If the vibrational energy is sufficiently great, then $|\chi_{jv}\rangle$ goes over to a continuum function. The remaining coefficient $f_{nj\ell}$ will depend solely on time in this case or, if Ψ is a stationary state of the system, $f_{nj\ell}$ will be a constant representing the weight of the product $|n\rangle|\psi_j\rangle|\Theta_{j\ell}\rangle$ in the overall wave function. In any case it is the set of f-coefficients on which our attention must eventually focus. All other ingredients of the system wave function

are presumed known.   The various transition matrix elements
are calculable by quadratures, so that matrix equations for
the coefficients can be written down and solved when suitable
boundary conditions are specified.   We shall not carry the detailed analysis any further here, however, as it becomes too
dependent on the particulars of each application.

For atom-diatom processes (in which reaction does not
occur) we employ the coordinate system illustrated in Figure 3.

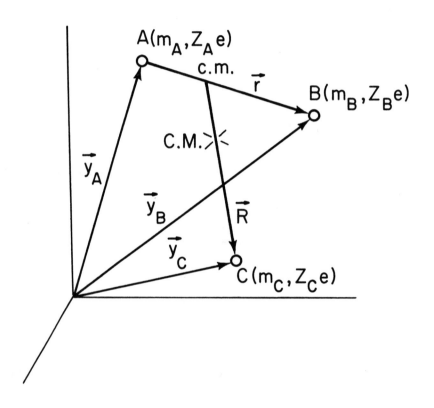

*FIGURE 3.   The coordinate system for atom-diatom processes.
C.M. is the overall center of mass which is stationary.   The
point labeled c.m. on the vector $\vec{r}$ from A to B is the A-B
center of mass.   Note the changed meaning of $\vec{R}$ as compared to
its use in Figure 2.*

The diatom AB appears much as it did in Figure 2, but its
center of mass can no longer be made to coincide with the
origin.   For present purposes we may assume that the diatom

does not dissociate and that the third atom is not captured. The first requirement obviates the need for analysis in Fadeev's terms, at least to a good approximation, while the second further simplifies the task of specifying boundary conditions.

The purely nuclear Hamiltonian is now

$$\hat{H}_{nuc} = -\frac{\hbar^2}{2M}\nabla_{\vec{R}}^2 - \frac{\hbar^2}{2m}\nabla_{\vec{r}}^2 + \frac{Z_A Z_B e^2}{r} + \frac{Z_C(Z_A+Z_B)e^2}{R}. \quad (2.64)$$

The new mass is

$$M = \frac{m_C(m_A+m_B)}{m_A+m_B+m_C}. \quad (2.65)$$

Note that the nuclear coordinates are completely uncoupled from one another in Eq. (2.64). This observation leads us to formulate the atom-diatom problem in such a way that maximal use can be made of the two-atom case discussed above. Toward that end we parallel the earlier development in order to provide a basis for comparing the two cases.

The electronically adiabatic representation of $\hat{H}_0$ is now

$$\langle\psi_j^a|\hat{H}_0|\psi_{j'}^a\rangle_{\vec{x}} = \delta_{jj'}[-\frac{\hbar^2}{2M}\nabla_{\vec{R}}^2 - \frac{\hbar^2}{2m}\nabla_{\vec{r}}^2 + W_j(\vec{r},\vec{R})]$$

$$-\frac{\hbar^2}{2m}[2\vec{\xi}_{\vec{r}jj'}(\vec{r},\vec{R})\cdot\vec{\nabla}_{\vec{r}} + \Xi_{\vec{r}jj'}(\vec{r},\vec{R})]$$

$$-\frac{\hbar^2}{2M}[2\vec{\xi}_{\vec{R}jj'}(\vec{r},\vec{R})\cdot\vec{\nabla}_{\vec{R}}$$

$$+ \Xi_{\vec{R}jj'}(\vec{r},\vec{R})], \quad (2.66)$$

where the adiabatic potential energy surface is

$$W_j(\vec{r},\vec{R}) = \frac{Z_A Z_B e^2}{r} + \frac{Z_C(Z_A+Z_B)e^2}{R} + w_j^e(\vec{r},\vec{R}) \quad (2.67)$$

with $w_j^e(\vec{r},\vec{R})$ defined much as was $U_j^e(\vec{r})$:

$$\hat{H}_{el}\psi_j^a(\vec{x},\vec{r},\vec{R}) = w_j^e(\vec{r},\vec{R})\psi_j^a(\vec{x},\vec{r},\vec{R}). \quad (2.68)$$

The nonadiabatic coupling elements again come from Eqs. (2.52) in an obvious way.

Using the electronically diabatic basis we find that

$$
\langle \psi_j^d | \hat{H}_0 | \psi_{j'}^d \rangle_{\vec{x}} = - \delta_{jj'} [\frac{\hbar^2}{2M} \nabla_{\vec{R}}^2 + \frac{\hbar^2}{2m} \nabla_{\vec{r}}^2] + V_{jj'}(\vec{r},\vec{R}),
$$

$$(2.66')$$

where

$$
V_{jj'}(\vec{r},\vec{R}) = \delta_{jj'} [\frac{Z_A Z_B e^2}{r} + \frac{Z_C (Z_A + Z_B) e^2}{R}]
$$

$$
+ H_{jj'}^{el}(\vec{r},\vec{R}).
$$

$$(2.67')$$

The electric transition dipole is set up in the same way as before, but additional terms appear. Thus we have [c.f. Eq. (2.58)]

$$
\vec{\mu}^{nuc}(\vec{r},\vec{R}) = (\frac{m_A Z_B - m_B Z_A}{m_A + m_B}) e\vec{r}
$$

$$
+ [\frac{(m_A + m_B) Z_C - m_C (Z_A + Z_B)}{m_A + m_B + m_C}] e\vec{R},
$$

$$(2.69)$$

so that

$$
\vec{\mu}_{jj'}^{a,d}(\vec{r},\vec{R}) = e \sum_\alpha \langle \psi_j^{a,d} | \vec{x}_\alpha | \psi_{j'}^{a,d} \rangle_{\vec{x}}
$$

$$
- \delta_{jj'} \vec{\mu}^{nuc}(\vec{r},\vec{R})
$$

$$(2.70)$$

gives the general result for the electric dipole transition element coupling electronic states in either the adiabatic or diabatic representations. Again, the diabatic expression is a constant for all nuclear configurations except for diagonal elements.

Similarly, we find

$$e \sum_{\beta=A,B,C} \frac{Z_\beta}{m_\beta} \hat{L}_\beta = Q_{AB} \frac{\vec{r}\times\vec{V}}{\vec{r}} + Q_{AB,C} \frac{\vec{R}\times\vec{V}}{\vec{R}}$$

$$+ e\left(\frac{Z_B - Z_A}{m_A + m_B}\right) (m \underset{\vec{R}}{\vec{r}\times\vec{V}} - M \underset{\vec{r}}{\vec{R}\times\vec{V}}) , \qquad (2.71)$$

where Q_{AB} is given by Eq. (2.62) and

$$Q_{AB,C} = \left[\frac{(m_A + m_B) Z_C + m_C (Z_A + Z_B)}{m_A + m_B + m_C}\right] e. \qquad (2.72)$$

The contributions to Eq. (2.71) involving the cross terms $\underset{\vec{R}}{\vec{r}\times\vec{V}}$ and $\underset{\vec{r}}{\vec{R}\times\vec{V}}$ will be smaller than the other two terms; indeed, for $Z_A = Z_B$ the cross-terms do not contribute at all. This will always be true of homonuclear diatomics; more generally, it will be approximately true for diatomics treated in a valence-bond or other semiempirical method that leaves the effective core charges equal. In any case we shall not have to worry about the cross terms in the calculations discussed later on. Therefore we write

$$\hat{M}_{jj'}^{a,d}(\vec{r},\vec{R}) = \frac{e}{2m_e} \sum_\alpha \langle\psi_j^{a,d}|\hat{L}_\alpha|\psi_{j'}^{a,d}\rangle_{\vec{x}}$$

$$- \hat{M}_{jj'}^{(a,d),nuc}(\vec{r},\vec{R}) , \qquad (2.73)$$

with

$$\hat{M}_{jj'}^{d,nuc}(\vec{r},\vec{R}) = \frac{1}{2} \delta_{jj'} (Q_{AB} \underset{\vec{r}}{\vec{r}\times\vec{V}} + Q_{AB,C} \underset{\vec{R}}{\vec{R}\times\vec{V}}) \qquad (2.74)$$

and

$$\hat{M}_{jj'}^{a,nuc}(\vec{r},\vec{R}) = \hat{M}_{jj'}^{d,nuc}(\vec{r},\vec{R}) + \frac{1}{2}[Q_{AB} \underset{\vec{r}jj'}{\vec{r}\times\vec{\xi}} (\vec{r},\vec{R})$$

$$+ Q_{AB,C} \underset{\vec{R}jj'}{\vec{R}\times\vec{\xi}} (\vec{r},\vec{R})] . \qquad (2.75)$$

For $R \to \infty$ we assume that $W_j(\vec{r}, \vec{R}) \to U_j^e(r)$, a known function that has no dependence on the direction of the vector $\vec{r}$. We further assume that the nonadiabatic coupling terms $\vec{\xi}_{\vec{r}jj'}(\vec{r}, \vec{R})$ and $\Xi_{\vec{r}jj'}(\vec{r}, \vec{R})$ are negligible at large R. The nuclear basis states $|\Theta_{j\ell}\rangle$ are then products of the diatomic vibrational state $|\chi_{jv}\rangle$, the diatomic rotational state $|Y_{LM_L}\rangle$ and the additional rotational state $|Y_{\lambda m_\lambda}\rangle$ associated with motion about the system center of mass. The coefficient function $f_{nj\ell}$ will depend on R and will always be required to satisfy continuum boundary conditions in the work to be discussed. All of the quantum calculations dealing with atom-diatom processes in the laser field are set up in a time-independent framework.

We summarize our results to this point. In the electronically diabatic representation we have

$$\hat{H}^d_{jj'}(\vec{r}) \equiv \langle \psi^d_j(\vec{x}) | \hat{H} | \psi^d_{j'}(\vec{x}) \rangle_{\vec{x}} \qquad \text{(atom-atom)}$$

$$= \delta_{jj'} [\hat{H}_f - \frac{\hbar^2}{2m} \nabla^2_{\vec{r}}] + U^d_{jj'}(\vec{r})$$

$$+ \hat{E} \cdot \vec{\mu}^d_{jj'}(\vec{r}) + \hat{H} \cdot \hat{\vec{M}}^d_{jj'}(\vec{r}) \qquad (2.76)$$

and

$$\hat{H}^d_{jj'}(\vec{r}, \vec{R}) \equiv \langle \psi^d_j(\vec{x}) | \hat{H} | \psi^d_{j'}(\vec{x}) \rangle_{\vec{x}} \qquad \text{(atom-diatom)}$$

$$= \delta_{jj'} [\hat{H}_f - \frac{\hbar^2}{2M} \nabla^2_{\vec{R}} - \frac{\hbar^2}{2m} \nabla^2_{\vec{r}}]$$

$$+ V_{jj'}(\vec{r}, \vec{R}) + \hat{E} \cdot \vec{\mu}^d_{jj'}(\vec{r}, \vec{R})$$

$$+ \hat{H} \cdot \hat{\vec{M}}^d_{jj'}(\vec{r}, \vec{R}) , \qquad (2.77)$$

while in the adiabatic basis

$$\hat{H}^a_{jj'}(\vec{r}) \equiv <\psi^a_j(\vec{x},\vec{r}) \,|\, \hat{H} \,|\, \psi^a_{j'}(\vec{x},\vec{r}) >_{\vec{x}} \qquad \text{(atom-atom)}$$

$$= \delta_{jj'} [\hat{H}_f - \frac{\hbar^2}{2m} \nabla^2_{\vec{r}} + U^a_j(\vec{r})]$$

$$+ \hat{\vec{E}} \cdot \vec{\mu}^a_{jj'}(\vec{r}) + \hat{\vec{H}} \cdot \hat{\vec{M}}^a_{jj'}(\vec{r})$$

$$- \frac{\hbar^2}{2m} [2\vec{\xi}_{jj'}(\vec{r}) \cdot \vec{\nabla}_{\vec{r}} + \Xi_{jj'}(\vec{r})] \qquad (2.78)$$

and

$$\hat{H}^a_{jj'}(r,R) \equiv <\psi^a_j(\vec{x},\vec{r},\vec{R}) \,|\, \hat{H} \,|\, \psi^a_{j'}(\vec{x},\vec{r},\vec{R}) >_{\vec{x}} \qquad \text{(atom-diatom)}$$

$$= \delta_{jj'} [\hat{H}_f - \frac{\hbar^2}{2M} \nabla^2_{\vec{R}} - \frac{\hbar^2}{2m} \nabla^2_{\vec{r}} + W_j(\vec{r},\vec{R})]$$

$$+ \hat{\vec{E}} \cdot \vec{\mu}^a_{jj'}(\vec{r},\vec{R}) + \hat{\vec{H}} \cdot \hat{\vec{M}}^a_{jj'}(\vec{r},\vec{R})$$

$$- \frac{\hbar^2}{2m} [2\vec{\xi}_{rjj'}(\vec{r},\vec{R}) \cdot \vec{\nabla}_{\vec{r}} + \Xi_{rjj'}(\vec{r},\vec{R})]$$

$$- \frac{\hbar^2}{2M} [2\vec{\xi}_{Rjj'}(\vec{r},\vec{R}) \cdot \vec{\nabla}_{\vec{R}} + \Xi_{Rjj'}(\vec{r},\vec{R})]. \qquad (2.79)$$

Representations of these electronic Hamiltonian operator elements in the number state representation of the field are also easily written down. We display them only for the electronically diabatic forms:

$$\hat{H}^d_{nj,n'j'}(\vec{r}) \equiv <n| H^d_{jj'}(\vec{r}) |n'> \qquad \text{(atom-atom)}$$

$$= \delta_{nn'} \{\delta_{jj'} [\hbar\omega(n-n_o) - \frac{\hbar^2}{2m} \nabla^2_{\vec{r}}] + U^d_{jj'}(\vec{r})\}$$

$$- \frac{1}{\sqrt{V}} (\delta_{n',n+1} \sqrt{n+1} + \delta_{n',n-1} \sqrt{n}) [\frac{\omega}{c} \vec{A}_0 \cdot \vec{\mu}^d_{jj'}(\vec{r})$$

$$+ (\vec{k} \times \vec{A}_0) \cdot \hat{\vec{M}}^d_{jj'}(\vec{r})] \qquad (2.80)$$

and

$$\hat{H}^d_{nj,n'j'}(\vec{r},\vec{R}) \equiv \langle n | \hat{H}^d_{jj'}(\vec{r},\vec{R}) | n' \rangle \qquad \text{(atom-diatom)}$$

$$= \delta_{nn'} \{ \delta_{jj'} [\hbar\omega(n-n_o) - \frac{\hbar^2}{2M} \nabla^2_{\vec{R}} - \frac{\hbar^2}{2M} \nabla^2_{\vec{R}}]$$

$$+ V_{jj'}(\vec{r},\vec{R}) \} - \frac{1}{\sqrt{V}} (\delta_{n',n+1} \sqrt{n+1}$$

$$+ \delta_{n',n-1} \sqrt{n}) [\frac{\omega}{c} \vec{A}_0 \cdot \vec{\mu}^d_{jj'}(\vec{r},\vec{R})$$

$$+ (\vec{k} \times \vec{A}_0) \cdot \hat{M}^d_{jj'}(\vec{r},\vec{R})] \qquad (2.81)$$

Quantum mechanically we continue in this vein, projecting onto the nuclear basis functions so that all bound and periodic nuclear degrees of freedom are integrated out. We end up with a matrix (subscripted with several indices) which is either a constant, if no dynamical degree of freedom remains, or a function of the atom-atom separation r in two-body collisions, or a function of the atom-diatom separation R in atom-diatom collisions. Examples of the various possibilities are treated in section III. Methods of extracting required probabilities and cross-sections from the f-coefficients are presented within the context of each sample calculation.

Semiclassically we proceed as outlined in the next part of this section. As input to the semiclassical method we require classical expressions for the potentials governing nuclear motion. Our procedure is exactly analogous to that followed in the absence of the field, when electronic transitions occur during molecular collisions. In that case the adiabatic electronic potential energy surfaces are employed; the coupling elements $\vec{\xi}_{jj'} \cdot \vec{\nabla}$ and $\Xi_{jj'}$ are discarded and their effects modeled by other means (36). With the field present we find that the electronically adiabatic potential energy surfaces are coupled. The resulting potential matrix is therefore rediagonalized to arrive at the electronic-field representation of the Hamiltonian. The general idea is sketched in Figure 4 for the case where a single photon exchange is allowed. Nuclear dynamics on the electronic-field surfaces are then governed by Hamilton's equations of motion. The details of this approach are our next topic of discussion.

2. *Semiclassical Approach.* The semiclassical approach
treats certain degrees of freedom classically while retaining
a quantum mechanical description for the remaining degrees of
freedom. For the molecular system it is common practice to
designate the nuclear degrees of freedom as classical and the
electronic degrees of freedom as quantum mechanical (36) [al-
though sometimes it is more convenient to quantize one or more
nuclear degrees of freedom, such as vibration when there is
resonance behavior in electronic-to-vibrational energy transfer (42)]. The radiation field can be represented in a
variety of ways as discussed earlier. The semiclassical approach offers two advantages over a fully quantum mechanical
treatment. First, from a computational viewpoint it is generally easier to implement and requires less time on a computer. This is a consequence of the smaller number of
Hamilton's equations of motion which must be integrated as
compared to the number of coupled-channel equations. Second,
it provides a more physical picture of the dynamics of a given
process, since it involves the propagation of classical trajectories which allow one to follow nuclear motion explicitly
in space and time.

In order to obtain as high an accuracy as possible in the
semiclassical approach, it is expedient to work within a representation where classical mechanics is valid over a wide
range of space and time. This can be achieved if we use the
diagonalized form of the electronic-field potential matrix;
namely, we integrate trajectories on the electronic-field surfaces. Typically we expect the nuclear motion to behave
classically except at localized regions where a given pair of
electronic-field surfaces interact, and such regions coincide
with the configurations at which the laser field comes into
resonance with the field-free surfaces.

In summarizing the semiclassical approach we shall assume
a classical limit of the coherent state for the laser field
(10), and we refer the reader to Reference 43 for a corresponding treatment with photon number states. Restricting ourselves to a two-state approximation, we choose the electronic-
field basis states given as (10,12,14,28)

$$\psi_{1,2} = \{{\cos\theta \atop \sin\theta}\}\psi_1^a e^{-in_1\omega t} \mp \{{\sin\theta \atop \cos\theta}\}\psi_2^a e^{-in_2\omega t} \tag{2.82}$$

where ψ_i^a is a field-free electronic state which is an eigenfunction of the electronic Hamiltonian

$$\hat{H}_{el}\psi_i^a = W_i\psi_i^a \tag{2.83}$$

and W_i is a field-free (adiabatic) potential energy surface corresponding to U_i^a (atom-atom) or W_i (atom-diatom) from Section II.C.1. The difference between n_1 and n_2 is the number of photons absorbed or emitted by the molecular system. ψ_i^a depends on nuclear coordinates $(\vec{q})$ as well as electronic coordinates $(\vec{x})$ whereas W_i depends on only nuclear coordinates $(\vec{q})$. The quantity θ is given as

$$\theta(\vec{q}) = \frac{1}{2} \tan^{-1} [2d_{12}(\vec{q}) / \Delta(\vec{q})] \qquad (2.84)$$

where, in the dipole approximation,

$$d_{12}(\vec{q}) = \langle \psi_1^a | \vec{\mu} | \psi_2^a \rangle \cdot \vec{E}_0/2 , \qquad (2.85)$$

$\vec{\mu}$ being the dipole operator and $\vec{E}_0$ the amplitude of the electric field [c.f. Eq. (2.32)], and

$$\Delta(\vec{q}) = W_2(\vec{q}) - W_1(\vec{q}) - \hbar\omega. \qquad (2.86)$$

We shall consider single-photon absorption so that, without loss of generality, we can set $n_1=-1$ and $n_2=0$, which results in the form for the electronic-field surfaces

$$E_{1,2} = \frac{1}{2} \{W_1 + W_2 + \hbar\omega \mp [(W_2 - W_1 - \hbar\omega)^2$$

$$+ 4d_{12}d_{21}]^{1/2}\}. \qquad (2.87)$$

Our goal is to determine the form of an S-matrix element S_{ji} for a transition from some initial state i to some final state j, where i(j) corresponds to $\psi_1(\psi_2)$ and a set of asymptotic vibrational, rotational and electronic states. This can be accomplished by a semiclassical evaluation of the reduced propagator expressed as the path integral (10,36,44,45)

$$\int_{\vec{q}_i}^{\vec{q}_f} D[\vec{q}(t)] K_{kl}[\vec{q}(t)] \exp\left(\frac{i}{\hbar}\int_{t_i}^{t_f} dt\ T\right), \quad \begin{array}{l} t_i \to -\infty \\ t_f \to \infty \\ k=1,2 \end{array} \qquad (2.88)$$

where T is the classical nuclear kinetic energy, and K_{kl} is the electronic-field transition amplitude which depends on the nuclear path $\vec{q}(t)$ and can be written as

$$K_{kl}[\vec{q}(t)] = \int d\vec{x}_k \int d\vec{x}_l \psi_k^*(\vec{x}_k) K[\vec{q}(t)] \psi_l(\vec{x}_l) . \qquad (2.89)$$

In Eq. (2.89) ψ_i is the asymptotic form (i.e., as collision partners separate) of ψ_i as defined in Eq. (2.82). K itself is expressed as a path integral,

$$K[\vec{q}(t)] = \int_{\vec{x}_1}^{\vec{x}_k} D[\vec{x}(t)] \exp\left(\frac{i}{\hbar} \int_{t_i}^{t_f} dt [T_x - V(\vec{x}, \vec{q})\right.$$

$$\left. - H_{int}(\vec{x}, \vec{q}, t)]\right\} \qquad (2.90)$$

where T_x is the electronic kinetic energy, V is the electro-static potential and H_{int} is the field interaction

$$H_{int} = \vec{\mu} \cdot \vec{E}_0 \cos(\omega t) . \qquad (2.91)$$

Assuming a nuclear path $\vec{q}(t)$ is already given, we must solve the time-dependent wave equation

$$i\hbar \frac{\partial \Psi(\vec{x}, \vec{q}(t), t)}{\partial t} = [H_{el}(\vec{x}, \vec{q}(t), t)$$

$$+ H_{int}(\vec{x}, \vec{q}(t), t)] \Psi(\vec{x}, \vec{q}(t), t) \qquad (2.92)$$

where Ψ is given as

$$\Psi(\vec{x}, \vec{q}(t), t) = \sum_{j=1}^{2} a_j(t) \psi_j(\vec{x}, \vec{q}, t) \exp\left(-\frac{i}{\hbar} \int_{t_i}^{t_f} dt E_j(\vec{q})\right) .$$

$$(2.93)$$

In particular, for the boundary conditions $a_1(t_i)=1$ and $a_2(t_i)=0$, we can obtain $K_{kl}[\vec{q}(t)]$ once we have found $a_k(t_f)$:

$$K_{kl}[\vec{q}(t)] = a_k(t_f) \exp\left(-\frac{i}{\hbar} \int_{t_i}^{t_f} dt E_k(t)\right) . \qquad (2.94)$$

Without repeating the steps contained in Reference 10, we simply indicate here that by substituting Eq. (2.93) into (2.92), taking the semiclassical limit $\hbar \to 0$ and dropping highly oscillatory terms as consistent with the rotating wave approximation, we arrive at the form

$$K_{21}[\vec{q}(t)] = \exp\left(-\frac{i}{\hbar}\int_{t_i}^{t_*}dtE_1(t)\right.$$

$$\left.-\frac{i}{\hbar}\int_{t_*}^{t_f}dtE_2(t)\right) \tag{2.95}$$

$$K_{11}[\vec{q}(t)] = (1-p)^{1/2}\exp\left(-\frac{i}{\hbar}\int_{t_i}^{t_f}dtE_1(t)\right) \tag{2.96}$$

where t_* is a (complex) time when $E_1=E_2$ and

$$p = \exp\left(-2\mathrm{Im}\int_{\mathrm{Ret}_*}^{t_*}dtE_1(t) - 2\mathrm{Im}\int_{t_*}^{\mathrm{Ret}_*}dtE_2(t)\right). \tag{2.97}$$

Inserting Eq. (2.95) or (2.96) into (2.88) and using the method of steepest descent to evaluate the nuclear path integral, we obtain the form for S_{fi}

$$S_{fi} = \sum N_{fi}\exp(iA_{fi}/\hbar) \tag{2.98}$$

where the sum runs over all classical trajectories propagating from i to f, N_{fi} is a normalization factor and A_{fi} is the classical action calculated along a given trajectory. We note that A_{fi} gains an imaginary component when a trajectory propagates from E_1 to E_2 by passing through their complex intersection point(s).

To gain some feeling for the role of the electronic-field surfaces in the evaluation of Eq. (2.98), we consider two limiting cases of Eq. (2.87). In the case where $\Delta(\vec{q}) \gg d_{12}$ [see Eq. (2.86)], which we assume to hold in the asymptotic regions, we observe for $\Delta(\vec{q}) > 0$ that

$$E_1 \sim W_1 + \hbar\omega,$$

$$E_2 \sim W_2, \tag{2.99}$$

and for $\Delta(\vec{q}) < 0$ that

$$E_1 \sim W_2,$$

$$E_2 \sim W_1 + \hbar\omega. \tag{2.100}$$

In the case where $\Delta(\vec{q}) \ll d_{12}$ or around the resonance nuclear configuration $[\Delta(\vec{q}) = 0]$, we observe

$$E_{1,2} \sim \frac{1}{2}(W_1 + W_2 + \hbar\omega \mp 2|d_{12}|). \tag{2.101}$$

The behavior of both of the above cases is illustrated in Figure 4, where we consider $\vec{q}$ as a single (translational)

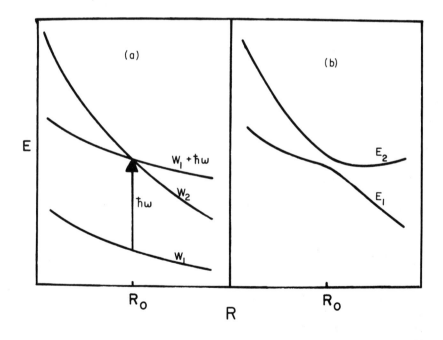

Figure 4. (a) Schematic drawing of two field-free adia-batic surfaces W_1 and W_2. (b) Schematic drawing of the elec-tronic-field surfaces E_1 and E_2. The splitting at R_0 is due to the radiative coupling d_{12}.

coordinate R. We see an avoided crossing around the resonance configuration R_0 [$\Delta(R_0)=0$] due to radiative coupling.
In general, R_0 is approximately equal to the real part of R_*,
where R_* is the branch point of E_1 and E_2 in coordinate space
where $E_1(R_*)=E_2(R_*)$. Integrating Hamilton's equations of
motion in time, we see that the point t_* in time space corresponds to R_* in coordinate space. The noncrossing rules for
the avoidance at R_0 are the same as the usual selection rules
for the nonvanishing of the electric dipole transition element
$\langle\psi_1^a|\vec{\mu}|\psi_2^a\rangle$ at the resonance nuclear configuration. If higher-
order coupling terms are important, selection rules for mag-
netic dipole and/or electric quadrupole transition elements
should be used for the corresponding noncrossing rules.

  If there is no interaction with the field when the molecular collision partners are asymptotic, then the asymptotic
system is in a pure field-free state ψ_i^a. This is always true
when $d_{12}<<|\Delta(\vec{q})|$, which can be accomplished by appropriate
construction of the experiment, for example, by adiabatically
"switching off" the field. One can imagine a beam cell or
crossed-beam experiment where a (third) laser beam is shone
through the collision region; once the collision species move
out of the cross section of the laser beam, the field is automatically switched off. The asymptotic form of K_{kl} is thus

$$K'_{kl}[\vec{q}(t)] = \int d\vec{x}_k \int d\vec{x}_1 \psi_k^{a*}(\vec{x}_k) K[\vec{q}(t)] \psi_1^a(\vec{x}_1).$$
(2.102)

Restricting ourselves to the principal value of 2θ [see Eqs.
(2.82) and (2.84)] from 0 to π, we can relate K'_{kl} and K_{kl} under
two limiting situations: (1) $\Delta>0$, so that $\theta \rightarrow 0$ asymptotically
and hence $\psi_1 \rightarrow \psi_1^a \exp(i\omega t)$ and $\psi_2 \rightarrow \psi_2^a$; (2) $\Delta<0$, so that
$\theta \rightarrow \pi/2$ asymptotically, and hence $\psi_1 \rightarrow -\psi_2^a$ and $\psi_2 \rightarrow \psi_1\exp(i\omega t)$,
where Δ is the asymptotic value of $\Delta(\vec{q})$. As an example, if we
think of R as a reaction coordinate, where $E_1 \rightarrow W_2$ as $R \rightarrow \infty$
and $t_i \rightarrow -\infty$, and $E_2 \rightarrow W_2$ as $R \rightarrow -\infty$ and $t_f \rightarrow \infty$, then we see
that

$$K_{21}[\vec{q}(t)] = - K'_{22}[\vec{q}(t)].$$
(2.103)

Eq. (2.103) indicates that we are looking at a process which
is electronically elastic overall (with respect to field-
free adiabatic states) even though it involves a transition
between electronic-field surfaces. For an overall electroni-
cally inelastic process where $E_1 \rightarrow W_2$ as $R \rightarrow \infty$ and $W_1 + \hbar\omega$ as
$R \rightarrow -\infty$ we have

$$K_{11}[\vec{q}(t)] = - K'_{12}[\vec{q}(t)]e^{-i\omega t}.$$
(2.104)

It is interesting to note that while the process of Eq. (2.104) involves photon absorption, the process of Eq. (2.103) involves no net photon absorption.

A comparison between two field-free surfaces W_1 and W_2 and two electronic-field surfaces E_1 and E_2 suggests that collision dynamics can be dramatically altered by a laser field. For example, a barrier on W_1 of height E_a can be lowered to E_a' on E_1 as seen in Figure 5. Symmetry rules which might be

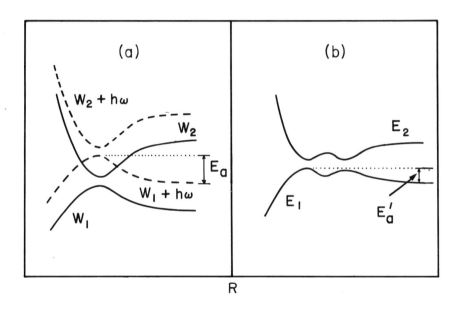

Figure 5. A sketch of two model electronic surfaces W_1 and W_2 plus their shifted images (a). The lower member of each pair is reactive, but interaction with a field of suitable frequency can lead to loss of concertedness along the reaction coordinate as suggested by the well that appears on the lower electronic-field surface E_1 in (b).

valid in the field-free case can be readily violated, and
branching ratios to various product states can be changed. A
special phenomenon which arises due to the laser field is an
interference between radiative coupling and field-free nonadiabatic coupling (this latter coupling occurs between just
W_1 and W_2). An easy way to demonstrate the existence of such
interference is by analyzing the branch-point structure of
the surfaces as utilized in the semiclassical approach. We
can express W_1 and W_2 as solutions of a 2×2 secular determi-
nant

$$W_{1,2} = \frac{1}{2}\{H_{11} + H_{22} \mp [(H_{22}-H_{11})^2$$

$$+ 4H_{12}H_{21}]^{1/2}\}, \tag{2.105}$$

where H_{ij} is a matrix element of the electronic Hamiltonian,
such that H_{11} and H_{22} are field-free diabatic surfaces
coupled through H_{12} and H_{21}. Substituting Eq. (2.105) into
(2.87) we obtain

$$E_{\mp\mp} = \frac{1}{2}\{H_{11} + H_{22} + \hbar\omega \mp \{(\hbar\omega \mp [(H_{22}-H_{11})^2$$

$$+ 4H_{12}H_{21}]^{1/2})^2 + 4d_{12}d_{21}\}^{1/2}\}, \tag{2.106}$$

where E_{--} and E_{+-} corresponds to E_1 and E_2, and we can label
E_{-+} and E_{++} as E_0 and E_3, respectively. We then have two sets
of branch points defined by

$$(H_{22}-H_{11})^2 + 4H_{12}H_{21} = 0 \tag{2.107}$$

$$\{\hbar\omega \mp [(H_{22}-H_{11})^2 + 4H_{12}H_{21}]^{1/2}\}^2$$

$$+ 4d_{12}d_{21} = 0 \tag{2.108}$$

which correspond to two different types of interaction regions
at R_{01} and R_{02}, respectively, in the right side of Figure 6.
The region at R_{01} arises from the field-free interaction between W_1 and W_2 (these follow the dashed lines in the left of
the figure) whereas the regions at R_{02} arise from radiative
coupling. When both types of coupling are comparable, in
evaluating S_{fi} in Eq. (2.98) we must include trajectories
passing through both types of branch points, and hence the
number of terms contributing to the sum in Eq. (2.98) is
higher than if just one type of branch point were used.

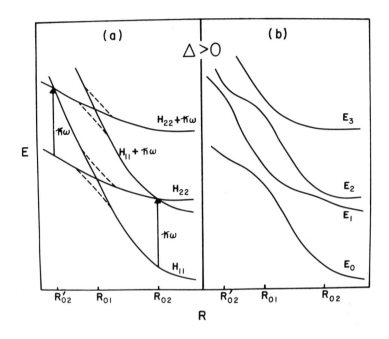

FIGURE 6. (a) Schematic drawing of field-free diabatic (solid lines) and adiabatic (dashed lines) surfaces as functions of the translational coordinate R. The field-free adiabatic surfaces exhibit an avoided crossing at R_{01}, and the field is in resonance with them at R_{02} and R'_{02} (for convenience we assume that the real part of the complex intersection R_{*i} and the resonance configuration are the same, although this is not strictly true. (b) Schematic drawing of the electronic-field surfaces E_0, E_1, E_2, and E_3. The avoided crossings between E_1 and E_2 at R_{02} and R'_{02} are due to the radiative coupling d_{12}.

Since an observable such as a cross section involves the square of S_{fi}, the additional terms can lead to new interference structure.

III. APPLICATIONS

The formalisms introduced in Section II will be illustrated by applications to a variety of specific processes of
interest in the remaining portions of this article. We begin
with quantum mechanical and semiclassical treatments of representative atom-atom and atom-diatom collision processes in the
presence of a field. Examples of bound-continuum processes
and heterogeneous catalysis are then discussed, also within
the framework of both quantum-mechanical and semiclassical
theories.

A. *Energy Transfer*

The most obvious application of the formalism discussed in
the previous sections is to processes involving the transfer
of energy from one collision partner to another. This transfer can be between any of the degrees of freedom of the
collision system (e.g., vibrational-vibrational, electronic-
rotational, etc.), and is a well-studied topic in the absence
of the radiation field. This investigation is thus initiated
with some amount of previously acquired intuition. All the
processes in Section III.A will be treated quantum mechanically,
based on the formalism just presented.

1. Examples. The first example to be considered is the
collinear collision of an atom-diatom particle system in the
presence of a field; specifically, bromine colliding with the
hydrogen molecule (9). In the field-free situation there
exists a strong electronic-to-vibrational resonance, due to
the nearly degenerate energies of the bromine spin-orbit
splitting and the lowest (Morse) vibrational excitation of H_2.
When the translation energy in the collision system is sufficient to make up the defect, excited bromine tends to give
up its energy of excitation while simultaneously H_2 passes from
its ground to its first excited vibrational level. Furthermore, the probability of making an electronic transition
without changing the vibrational excitation is very small.
The formalism for treating the collision system in the
presence of a field is essentially that presented in the previous section. However, some simplification is possible since
the system is collinear and hence the kinetic energy operators
have no angular terms. Beginning with the time-independent
Schroedinger equation

$$(\hat{H}-E)\Psi = 0, \qquad\qquad (3.1)$$

the wavefunction is expanded in the product basis

$$\Psi = \sum_{njv} |n> |\psi_j(\vec{x})> |\chi_{jv}(r)> f_{njv}(R) \tag{3.2}$$

where $|n>$ is the number state of the field, $\psi_j(\vec{x})$ is a diabatic electronic basis state, $\chi_{jv}(r)$ is a diatom vibrational state, and $f_{njv}(R)$ is the coefficient for the scattering coefficient which will ultimately yield transition probabilities. By substituting Eq. (3.2) into the Schroedinger equation and projecting onto the basis states, the close-coupled equations are found to be

$$\sum_{n'j'v'} <n| <\chi_{jv}| <\psi_j |\hat{H}-E|\psi_{j'}> |\chi_{j'v'}> |n'> = 0, \tag{3.3}$$

where $<\psi_j|\hat{H}|\psi_{j'}>$ is just the matrix element $\hat{H}_{jj'}^d$ given in Eq. (2.77), with the angular terms eliminated:

$$<\psi_j|\hat{H}|\psi_{j'}> = \delta_{jj'}[\hat{H}_f - \frac{\hbar^2}{2M}\frac{\partial^2}{\partial R^2} - \frac{\hbar^2}{2m}\frac{\partial^2}{\partial r^2}]$$

$$+ V_{jj'}(r,R) + \hat{E}\cdot\vec{\mu}_{jj'}^d(r,R)$$

$$+ \hat{H}\cdot\hat{M}_{jj'}^d(r,R). \tag{3.4}$$

It is convenient to decompose $V_{jj'}$ as

$$V_{jj'}(r,R) = \delta_{jj'}U_j(r) + W_{jj'}(r,R), \tag{3.5}$$

where

$$[-\frac{\hbar^2}{2m}\frac{\partial^2}{\partial r^2} + U_j(r)]\chi_{jv}(r) = \varepsilon_{jv}\chi_{jv}(r). \tag{3.6}$$

The close-coupled equations are then

$$\sum_{n'j'v'} \{\delta_{nn'}[\delta_{jj'}\delta_{vv'}(-\frac{\hbar^2}{2M}\frac{\partial^2}{\partial R^2} + n\hbar\omega + \varepsilon_{jv} - E)$$

$$+ <\chi_{jv}|W_{jj'}|\chi_{j'v'}>]$$

$$+ <n|\hat{\vec{E}}|n'> \cdot <\chi_{jv}|\vec{\mu}^d_{jj'}|\chi_{j'v'}>$$

$$+ <n|\hat{H}|n'> \cdot <\chi_{jv}|\hat{M}^d_{jj'}|\chi_{j'v'}> \} f_{n'j'v'}(R) = 0 \qquad (3.7)$$

For the Br+H$_2$ system, the electric dipole transition moment is unknown, so that only the magnetic dipole could be used to induce transitions. Truncating the basis to include just two electronic states, the matrix elements are found to be

$$<n|\hat{H}|n'> \cdot <\chi_{vj}|\hat{M}^d_{jj'}|\chi_{j'v'}>$$

$$= \delta_{vv'} [\sqrt{n+1}\, \delta_{n',n+1} + \sqrt{n}\, \delta_{n',n-1}] D_{jj'} \qquad (3.8)$$

where

$$D = -\frac{\alpha}{2} \begin{pmatrix} 4\sqrt{\dfrac{5}{3}} & \dfrac{2}{\sqrt{3}} \\[2ex] \dfrac{2}{\sqrt{3}} & \sqrt{\dfrac{2}{3}} \end{pmatrix} \qquad (3.9)$$

and α is the bromine fine-structure constant (taken to be 1/137). The close-coupled equations could now be solved in the standard manner. However, the electronic matrix element $W_{jj'}$ includes the spin-orbit interaction of bromine and hence is not diagonal asymptotically. A simple change of basis from the pure electronic to the spin-orbit basis is thus performed, enabling boundary conditions to be applied in the usual way.

The calculation was performed with eight states, corresponding to two electronic states, two vibrational states, and two states of the radiation field. The field quantum $\hbar\omega$ was chosen to be about .4571 eV, slightly larger than the spin-orbit splitting of bromine (.4566 eV). Some results are presented in Figure 7, along with the results of a semiclassical calculation based on Section II.C.2 (10b). Perhaps the most interesting result of this calculation is that the probability of making an electronic transition without changing the vibrational state is about 100 times larger in a field (of strength 5.13×10^6 V/cm) than in the field-free situation, and is comparable to the electronic-to-vibrational resonance discussed previously. Thus the radiation field can dramatically affect dynamic processes of the system.

There is reason to question the validity of the interaction

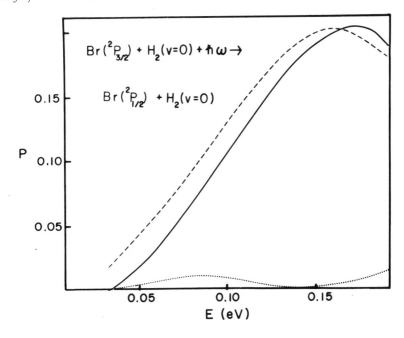

FIGURE 7. Probabilities for the collinear reaction $Br(^2P_{3/2}) + H_2(v=0) + \hbar\omega \rightarrow Br(^2P_{1/2}) + H_2(v=0)$ as functions of initial relative translational collision energy. The value of $\hbar\omega$ is 1.001 times the asymptotic spin-orbit splitting. Shown are the results from the quantum calculations (solid line) and results from semiclassical calculations (dashed line). The quantum probabilities for the field-free transition (dotted line) have been multiplied by 100 and plotted vs. energy above threshold.

mechanism used in this model. As the bromine atom approaches the hydrogen molecule, electric dipole moments will arise coupling both electronic and vibrational states. In an effort to estimate the effects of these proximity-induced moments we have represented them using simple model forms and have repeated the calculation (46). The "permanent" dipole moment for both electronic states was adapted from ground-state HCl results (47). Similarly, the electric transition dipole was based on calculated results obtained for $F+H_2$ (38). The model interactions so obtained are of doubtful accuracy, but they should give some idea of how the actual dipole functions vary during the collision. We find that their presence does not change the original results to any great degree unless the field is overwhelmingly strong ($\geqslant 5 \times 10^7$ V/cm) or the collision

is very energetic. The collision does not otherwise sample
regions where the electric dipole coupling is strong. The
obvious next step is to obtain improved estimates for the
dipole moment functions. This research is still in progress.

Another example of interest is the three-dimensional
treatment of an atom-atom particle system (48). In particular, a two-photon process in the XeF collision system in the
presence of the 248 nm line of the KrF laser has been investigated (49). In this system, the states corresponding to
Xe^++F^- can act as intermediate states in the quenching of the
fluorine atom since these states are coupled to the Xe+F
states by the electric dipole interaction. Both the potential
surfaces and the dipole transition matrix elements were taken
from *ab initio* calculations (50).

The total wavefunction is expanded as

$$\Psi = \sum_{jn} |n> |\psi_j> F_{jn}(\vec{R}) \tag{3.10}$$

where $|n>$ is a number state of the field and ψ_j is a linear
combination of diabatic electronic basis states such that
$|\psi_j>$ is an eigenfunction of the spin-orbit operator. The
close-coupled equations are then easily found to be

$$\sum_{j'n'} \{\delta_{jj'}\delta_{nn'}[-\frac{\hbar^2}{2m}\nabla_{\vec{R}}^2 - \epsilon_{jn}] + \delta_{nn'}V_{jj'}$$

$$+ H_{jn,j'n'}^{int}\}F_{j'n'}(\vec{R}) = 0, \tag{3.11}$$

where ϵ_{jn} is the energy of the system in the jth state with n
photons, $V_{jj'}$ is the potential matrix, and $H_{jn,j'n'}^{int}$ is the interaction matrix between the radiation field and the particle
system. The approximation is now made that both $V_{jj'}$ and
$H_{jn,j'n'}^{int}$ depend only upon the radial coordinate, thus ignoring
angular and Coriolis coupling while greatly simplifying the
collisional dynamics of the system. The nuclear wavefunction
can then be expanded in terms of spherical harmonics,

$$F_{jn}(\vec{R}) = \sum_{\ell m} Y_{\ell m}(\hat{R}) R^{-1} f_{jn}^{\ell m}(R), \tag{3.12}$$

and Eq. (3.11) can be written

$$\sum_{j'n'} \{\delta_{jj'}\delta_{nn'}[\frac{d^2}{dR^2} - \frac{\ell(\ell+1)}{R^2} + k_{jn}^2] - W_{jn,j'n'}\}f_{j'n'}^{\ell m} = 0 \tag{3.13}$$

where we have introduced the notation

$$k_{jn}^2 \equiv \frac{2m}{\hbar^2} (E - \varepsilon_{jn}) \tag{3.14}$$

and

$$W_{jn,j'n'} \equiv \frac{2m}{\hbar^2} (\delta_{nn'} V_{jj'} + H_{jn,j'n'}^{int}). \tag{3.15}$$

Consistent with this approximation, $H_{jn,j'n'}^{int}$ is taken to be

$$H_{jn,j'n'}^{int} = \delta_{n',n\pm1} |E| \mu_{jj'} \tag{3.16}$$

where μ_{ij} is the z component of the transition dipole moment.

For XeF, $W_{jn,j'n'}$ becomes diagonal asymptotically so that boundary conditions can be applied and an S-matrix defined in the usual manner. The total cross section for an inelastic process is then

$$\sigma_{f\leftarrow i} = \frac{\pi}{k_i^2} \sum_{\ell=0}^{\infty} (2\ell+1) |S_{fi}^{\ell}|^2, \tag{3.17}$$

where i(f) is a collective index denoting the initial (final) state.

The wavefunction expansion [Eq. (3.10)] is truncated to include only the six states corresponding to $Xe + F(^2P) + (n+1)$ photons and $Xe^+(^2P) + F^- + n$ photons (see Figure 8). These states can be indexed by the magnitude of the angular momenta projection Ω, which can have the values 1/2 or 3/2. Since $W_{jn,j'n'}$ only couples states having the same value of Ω, the six-state reduces to a two-state problem ($\Omega = 3/2$) and a four-state problem ($\Omega = 1/2$); the quenching process is entirely determined by consideration of the four-state problem. Of the four states, the two covalent are essentially repulsive while the excimer states have substantial wells. The excimer states are shifted by the energy of one KrF photon (248 nm), indicated by the vertical line in the figure, since these states correspond to an absorption of a photon (and hence have $\hbar\omega$ less energy in the radiation field than the covalent states).

The cross section for quenching of $F(^2P_{1/2})$ was calculated for various values of the collision energy and radiation intensity. These results appear in Figure 9. The effect of the field is clearly evident, and is largest at small collision energies. This process is easily described in qualitative terms. As the system, initially in the excited covalent state, collides, it passes through an internuclear separation

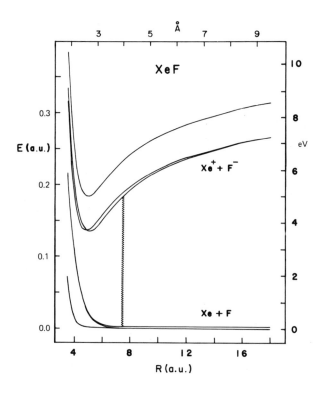

FIGURE 8. Potential energy curves for the XeF system. The vertical line corresponds to the 248 nm line of the KrF laser.

at which it is nearly resonant with the excimer state and hence can absorb a photon. As the collision proceeds, a second near resonance is achieved so that the system emits a photon (by stimulated emission) and makes the transition to the ground covalent state. Thus no photons are lost from the radiation field. Particularly interesting is that the external field is nowhere in resonance with the two covalent states, and resonances with the excimer states are possible only via the collision dynamics. The effect of the field at low collision energies is thus easily understood - for small collision energies the system spends a longer amount of time near the resonant positions, and hence is more likely to make the transitions discussed.

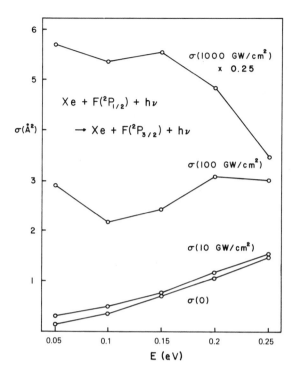

FIGURE 9. Calculated quenching cross-sections for various field strengths as functions of initial collision energy for the process

$$Xe + F(^2P_{1/2}) + \hbar\omega \rightarrow Xe + F(^2P_{3/2}) + \hbar\omega.$$

2. Special Considerations. The examples of the previous section illustrate that the presence of an intense radiation field can substantially alter the collision dynamics of particle systems, and hence serve their designed purpose. However, many of the details of the process were overlooked or simplified by making key assumptions concerning the colli- sion dynamics and the characteristics of the laser itself. In this section we consider a more detailed treatment of these areas.

a. Photon angular momentum. When a photon is absorbed or emitted, angular momentum is transferred between the radia- tion field and the matter system. For a dipole-allowed pro- cess, one unit of angular momentum (the intrinsic spin of the photon) is thus transferred. This transfer of angular momentum was not considered in the examples in Section III.A.1.

As we shall see, photon angular momentum can be included in a straightforward manner, although the resulting formalism is considerably more complex (48).

In the field-free situation, the states used to describe the system are indexed by eigenvalues of the total angular momentum J, and its space-fixed projection, M. Due to the rotational invariance of the Hamiltonian, J and M are conserved. The only allowed transitions are thus within the JM sets, i.e., transitions from a state in the set JM to a state in the set J'M', $J \neq J'$ or $M \neq M'$, are forbidden. Another consequence of the rotational invariance is that the close-coupled equations show no dependence on M so that in the degeneracy-averaged cross section there appears a factor of 2J+1.

Now consider the complications arising from the field. The states used to describe the system are a direct product of the field-free matter states and the states of the radiation field. For simplicity, we will assume that the field is single mode (which we represent by an occupation number state) and interacts with matter only through the electric dipole operator. An appropriate basis set is then indexed by J, M and n, but these quantities need not be conserved in a radiative transition since the presence of the radiation has destroyed the rotational invariance of the system. Consider the system absorbing a photon. Clearly, $n_f = n_i - 1$, and the projection M_f is simply the sum of the initial projection, M_i, and the projection of the photon angular momentum, σ. The final total angular momentum J_f, being the vectorial combination of J_i and the photon angular momentum, satisfies the inequality $|J_i - 1| \leq J_f \leq J_i + 1$. To be fully general, we should consider a second radiative transition, and so on, but the complexity of the problem is already quite clear. The dimensionality has been considerably increased (over the field-free one) since more states are coupled, and because the difficulty of a numerical solution goes as N^3, this is of considerable concern. But of even greater concern is that since the system is no longer rotationally invariant, the close-coupled equations are not independent of M. Thus each set of equations indexed by M must be solved independently; that is, the summation giving rise to the 2J+1 factor in the field-free degeneracy-averaged cross section cannot be performed analytically, but must be performed numerically. (Actually, symmetry considerations do reduce the number of independent M's by one half, but this hardly compares to the 2J+1 reduction in the field-free case.)

For simplicity, we consider an atom-atom collision in the presence of a radiation field. We denote by $|j\Omega\rangle$ a body-fixed diabatic basis for the electronic plus spin-orbit Hamiltonian which is an eigenfunction of $\vec{j}^2$ (where $\vec{j}$ is the sum of all electronic orbital angular momenta and spins) and j_z (the projection of $\vec{j}$ on the internuclear axis). If the internuclear axis is located in the direction (θ,ϕ) with respect to a space-fixed axis (with the origin of both axes taken to be the center of mass of the diatom), a field-free total angular momentum wavefunction can be written as

$$|JMj\Omega> = \sqrt{\frac{2J+1}{4\pi}}\, D^J_{\Omega M}(0,-\theta,-\phi)\,|j\Omega> \qquad (3.18)$$

where $D^J_{\Omega M}$ is a customary rotation matrix. A basis for the particle system in the presence of a field is then $|JMj\Omega>|n>$. The total wavefunction for the system is then a sum of direct products,

$$\Psi^{J\,M\,j\,\Omega\,n} = \sum_{JMj\Omega n} |JMj\Omega>|n>R^{-1}F^{J\,M\,j\,\Omega\,n}_{JMj\Omega n}(\vec{R}), \quad (3.19)$$

which leads in the usual way to the set of close-coupled equations

$$\sum_{J'M'j'\Omega'n'} <n|<JMj\Omega|\hat{H}-E|J'M'j'\Omega'>|n'>R^{-1}$$

$$\times\; F^{J\,M\,j\,\Omega\,n}_{J'M'j'\Omega'n'}(\vec{R}) = 0, \qquad (3.20)$$

where $\hat{H}$ is the total Hamiltonian for the system and E is the total energy of the system. The matrix elements required in this equation (with the exception of the matrix elements of the interaction Hamiltonian) are well known and have been thoroughly discussed elsewhere, so we shall examine only the interaction Hamiltonian matrix elements.

Making the dipole approximation, the interaction Hamiltonian can be written as

$$H^{int} = -\frac{\hbar}{i}\frac{e}{m}\sqrt{\frac{2\hbar\pi}{\omega V}}\sum_{j}[\hat{a}V_{j\sigma} + \hat{a}^\dagger V^*_{j\sigma}], \qquad (3.20)$$

where the j-summation is over all electrons of the system, σ is the photon polarization direction, and the components of $\vec{V}$ are written in spherical coordinates (for convenience, we henceforth define $V_\sigma \equiv \sum_j V_{j\sigma}$). The matrix elements of interest are then of the form

$$<n|<JMj\Omega|\hat{a}V_\sigma|J'M'j'\Omega'>|n'>$$

$$= <n|\hat{a}|n'><JMj\Omega|V_\sigma|J'M'j'\Omega'>$$

$$= \sqrt{n'}\ \delta_{n,n'-1}<JMj\Omega|V_\sigma|J'M'j'\Omega'> \qquad (3.21)$$

and a corresponding matrix element for $\hat{a}^\dagger V_\sigma^*$. Relating V_σ to its body-fixed components, the matrix element of V_σ can be written as

$$<JMj\Omega|V_\sigma|J'M'j'\Omega'> = <JMj\Omega|\sum_\eta D'_{\eta\sigma}V_\eta|J'M'j'\Omega'>$$

$$= \frac{\sqrt{(2J+1)(2J'+1)}}{4\pi}\sum_\eta \int D^{J*}_{\Omega M}D'_{\eta\sigma}D^{J'}_{\Omega'M'}d\hat{R}<j\Omega|V_\eta|j'\Omega'>$$

$$= \left[\frac{\sqrt{(2J+1)(2J'+1)}}{4\pi}\int D^{J*}_{\Omega M}D'_{\Omega-\Omega',\sigma}D^{J'}_{\Omega'M'}d\hat{R}\right]\delta_{M,M'+\sigma}$$

$$\times \quad <j\Omega|V_{\Omega-\Omega'}|j'\Omega'>$$

$$= \left[\sqrt{\frac{2J'+1}{2J+1}}\ C(J'1J;\Omega',\Omega-\Omega')C(J'1J;M'\sigma)\right]\delta_{M,M'+\sigma}$$

$$\times \quad <j\Omega|\Sigma_{\Omega-\Omega'}|j'\Omega'>. \qquad (3.22)$$

It should be clear that quantities such as those in the square bracket of Eq. (3.22) give rise to the complexities discussed earlier.

The close-coupled equations can now be solved in the standard manner, integrating from near the origin to the asymptotic region, transforming to the space-fixed basis, and applying boundary conditions in the usual way. The total, degeneracy-averaged cross section can then be obtained in a straightforward manner and is found to be

$$\sigma(j_o^A j_o^B n_o \rightarrow j^A j^B n) = \frac{1}{(2j^A+1)(2j^B+1)} \frac{\pi}{k^2_{j_o^A j_o^B n_o}}$$

$$\times \sum_{J_o M_o j_o \ell_o} \sum_{JMj\ell} |T^{J_o M_o j_o \ell_o j_o^A j_o^B n_o}_{JMj\ell j^A j^B n}|^2, \qquad (3.23)$$

where j^A, j^B refer to the eigenvalues of the separated atomic total angular momenta ($\vec{j}^A + \vec{j}^B = \vec{j}$), and the summations for J and M are severely restricted. Although this expression is quite similar to the usual field-free one, there are two essential differences. The first is that there is a summation over J as well as J_o, a consequence of the fact that the radiation field couples particle states of different total angular momentum. The second, and perhaps more serious, difference is that there is an explicit summation over M in this expression. In the field-free case, M does not explicitly appear in any matrix element, so that the solutions are M-independent. The M-summation can then be performed trivially, yielding the $2J+1$ factor in the usual expression. In the present situation, the summation must be performed numerically after solving for the various M-dependent T-matrices. We note that in an actual calculation, advantage would be made of the symmetry reducing the number of independent M solutions by one half. Also, the body-fixed basis states $|JMj\Omega\rangle$ would be replaced by linear combinations of states which were eigenfunctions of the parity operator. Use of such a parity basis would reduce the number of equations in the set to be simultaneously solved by one-half, considerably reducing the difficulty of solution. Note that parity is not conserved, however, since the parity (of the particle system) must change when absorbing or emitting a photon via the electric dipole interaction.

The formalism we have discussed is rather complex. Of particular concern is the appearance of M-dependent matrix elements, since this necessitates the solution of the close coupled equations $2J+1$ times for each J, and many J's are necessary to perform the summation for the total degeneracy-averaged cross section. Thus it is desirable to introduce an approximation which uses an averaged radiative matrix element, the average being over the possible M values. Since the projection quantum number represents the orientation of the particle system, the approximation would thus represent an orientational average. The average we choose to use is simply the root-mean-square average of the M-dependent Clebsch-Gordon coefficient in Eq. (3.22). That is, we replace $C(J'1J;M'\sigma)$ by $\bar{C}$,

$$\bar{C} = P\sqrt{\frac{1}{2J'+1} \sum_{M'} C(J'1J;M'\sigma)^2}$$

$$= \frac{P}{\sqrt{3}} \sqrt{\frac{2J+1}{2J'+1}} \tag{3.24}$$

where P is a quantity having unit modulus which is introduced to preserve the parity characteristics of the original Clebsch-Gordon coefficient. By employing this average we have, in essence, imposed rotational invariance upon the system, so that the M-summation in the expression for the total degeneracy-averaged cross section can now be performed trivially, thus accomplishing our stated goal.

To test the validity of this approximation, a model system was chosen for numerical calculations (51,52). Although this model does not represent any specific physical system, it mimics the collision of a halogen with a rare gas atom, i.e., a ^{2}P-state atom colliding with a ^{1}S-state atom. The various parameters appearing in the model were chosen purely for convenience. The electronic states arising in such a collision are of $\sum$ and Π symmetry, and for these potentials we chose (in atomic units)

$$V(\textstyle\sum,R) = e^{-3(R-2.5)},$$

$$V(\Pi^{\pm},R) = e^{-1.5(R-2.5)}. \tag{3.25}$$

Note that a superscript has been added to Π to distinguish between the two degenerate states. We chose the dipole moment to be a Gaussian function peaked at 6 Bohrs, with a maximum value of 2 atomic units. After making the appropriate conversion of units to atomic units and introducing the field intensity (in W/cm^2), the interaction matrix element is given as

$$5.34 \times 10^{-9} \quad \sqrt{I} \, e^{-(R-6)^2}. \tag{3.26}$$

The radiation field is provided by the 10.6 μm line of the CO_2 laser, with a photon energy of .0043 Hartree. The ^{2}P state atom was chosen to have a spin-orbit splitting of .002 Hartree, and the reduced mass of the system was taken to be 2500 a.u.

The transition of interest is the quenching process $^2P_{1/2} \to {}^2P_{3/2}$ (for a halogen the $^2P_{1/2}$ state lies at a higher energy that the $^2P_{3/2}$ state). Rather than the $|JMj\Omega\rangle$ basis, a parity basis was actually used in the calculation, reducing the number of states to nine. As previously discussed, this 9-state problem must be solved repeatedly for various M's, and the sum over M performed numerically. Some savings in effort are realized by recognizing that for linearly polarized radiation, $\sigma=0$, the transition probabilities for $\pm M$ are the same, and so only the positive M's need be numerically considered.

Calculations were performed for several collision energies in the range .1-.2 eV, and for three field intensities. Since the calculations are time consuming, only partial cross sections for three initial J values were considered. Results for J=23/2 are presented in Figure 10. It is evident from these figures that the approximation we have introduced works quite well for intensities as high as 100 GW/cm^2.

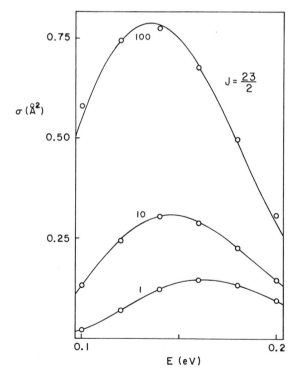

FIGURE 10. *Partial cross sections for quenching (exact results: open circles; approximate results: solid curves) vs. collision energy (in eV). Field intensities in GW/cm^2 are indicated in the figure.*

We are currently in the process of extending our formalism
to atom-diatom collisions in the presence of a field, and
applying it to model calculations. The extension of the formalism, although tedious, is relatively straightforward. Of
course, the extra degrees of freedom present in such systems
complicate the formalism even further than in the atom-atom
case, so that the rigorous treatment of photon angular
momentum in such systems is quite nearly intractable. We are
actively seeking approximations which might alleviate this
problem.

   b. Laser characteristics. The theory of molecular collision processes in the presence of laser radiation has so far
been developed assuming the laser field to be purely monochromatic. This assumption may be questionable in view of experimental reality because even a single-mode field has a
spectral linewidth owing to broadening effects such as Doppler,
collision and radiative lifetime broadening. In other cases
single-mode oscillation may not be practical or even attainable. Therefore, we need to test the assumption of monochromaticity in order to ascertain when it is valid and when it is
not. The results of this study should also be useful in
analyzing experimental data.

   We first consider the effect of a spectral linewidth on
laser assisted collision processes. The question we wish to
answer is whether lasers with a typical linewidth, say 1 GHz,
may properly be assumed to be monochromatic. We next take up
the case where collision occurs in a field of many (typically,
two to twenty) identically polarized, parallel axial modes.
Interference among the different modes gives an oscillatory
time behavior to the field intensity, whose effect on collision
processes we wish to examine. Finally, we shall consider an
extreme case where the number of oscillating modes is large
and the laser pulse duration is very short, say $\sim 10^{-12}$ sec.
The temporal width of such a (sub)picosecond pulse is smaller
than the typical collision time (1-10 psec) for slow collision
processes. The theory described in Section II must be modified in this case because the laser intensity there is assumed
constant over the collision time. We note that, throughout
this section, 'mode' and 'frequency' are used interchangeably
because the propagation and polarization directions of the
laser beam are well-defined.

   At the present time there is no experimental data concerning the importance of the above-named laser characteristics
as they might affect molecular collision processes. However,
these characteristics have been considered in both theoretical
and experimental studies of different processes such as atomic
multiphoton ionization (53-55). Experimental as well as

further theoretical investigation of laser characteristics in molecular collision processes will give a deeper understanding of collisional dynamics taking place in a laser field.

(1) To investigate the effect of the laser linewidth on collision processes, we assume that the laser field consists of N photons and has a Gaussian frequency distribution peaked at ω_0 with the linewidth σ,

$$N(\omega) = \frac{N}{\sqrt{\pi}\,\sigma} \exp[-(\omega-\omega_0)^2/\sigma^2].\tag{3.27}$$

If the linewidth is neglected, i.e., if $\sigma \to 0$, the spectral distribution given by Eq. (3.27) becomes

$$N^{(1)}(\omega) = N\delta(\omega-\omega_0).\tag{3.28}$$

This corresponds to the usual single-mode approximation adopted in almost all theories of collision processes in a laser field. (The superscript 1 in Eq. (3.28) indicates that the field is assumed to consist of one mode.) It is obvious that the approximation breaks down when σ is large. To see whether it is good for typical values of the actual laser linewidth, we use the following approximation scheme. The laser field is represented by a finite number K of frequencies and corresponding weights that best approximate the actual distribution. The choice of the K frequencies and weights is conveniently accomplished by the method of Gaussian quadrature. Details of this K-mode approximation can be found elsewhere (56). If K=1, the approximation gives the distribution of Eq. (3.28) but, for K greater than 1, a distribution results which contains information on the linewidth. For example, if K=2, the actual distribution of Eq. (3.27) is approximated by

$$N^{(2)}(\omega) = \frac{N}{2}\,\delta(\omega-\omega_0-\sigma/\sqrt{2}) + \frac{N}{2}\,\delta(\omega-\omega_0+\sigma/\sqrt{2});\tag{3.29}$$

that is, the field is approximated by one having two modes, of frequencies $\omega_\pm=\omega_0\pm\sigma/\sqrt{2}$, with half of the N photons in each mode. Since the separation between the two frequencies is proportional to the linewidth σ, this two-mode approximation carries at least some information about the linewidth. If K=3, we obtain

$$N^{(3)}(\omega) = \frac{2N}{3}\,\delta(\omega-\omega_0) + \frac{N}{6}\,\delta(\omega-\omega_0-1.2247\sigma)$$

$$+ \frac{N}{6}\,\delta(\omega-\omega_0+1.2247\sigma).\tag{3.30}$$

As K becomes larger, the approximation describes the actual
laser field more closely.

Using the above approximation scheme, we have performed
close-coupling calculations on various model systems for
s-wave collisions and have evaluated the probability P for
making the transition from the lowest to the first excited
electronic state while simultaneously absorbing a photon.
The absorbed photon may be in anyone of the K modes that make
up the K-mode approximation. Figure 11 shows a typical
result. Here, P is plotted against the linewidth σ for the

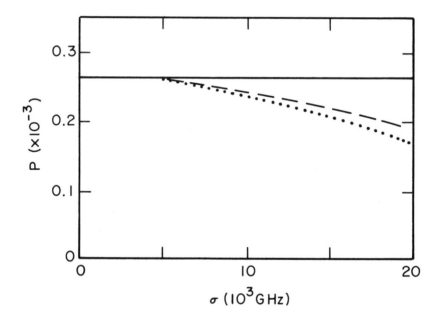

*FIGURE 11. Transition probability P vs. the laser line-
width σ for the case of the single-mode (solid curve), two-
mode (dotted curve), and three-mode (dashed curve) approximations.*

case of the single-mode (solid curve), two-mode (dotted curve)
and three-mode (dashed curve) approximations. For the range
of the values of σ shown in the figure, results using more
than three modes coincide with those obtained in the three-
mode approximation. The adiabatic potential curves W_1 and W_2
for the two electronic states and the electronic transition

moment μ are given by

$$W_1(R) = \exp(-0.8R), \tag{3.31a}$$

$$W_2(R) = 2 \exp(-0.8R) + 0.02, \tag{3.31b}$$

$$\mu(R) = \exp[-(R-4.5)^2] \tag{3.31c}$$

in atomic units, while the central laser wavelength is chosen
to be 9820 Å so that it is tuned to the energy separation
W_2-W_1 at $R \cong 4.5$ Bohrs. The laser power density and the collision energy are taken to be 10^8 W/cm^2 and 0.05 Hartree, respectively. It can be seen from Figure 11 that, if σ is
sufficiently small, all K-mode approximations including the
single-mode approximation give essentially the same value of
the transition probability. However, as σ becomes larger,
($\sigma > 5$THz), the transition probability under the two-mode
approximation begins to deviate from that under the single-
mode approximation. Here, we cannot totally neglect the linewidth and at least the two-mode approximation is necessary to
get accurate results. The value of the linewidth at which the
deviation begins to show up is different for different model
systems. It is typically of the order of a THz for most
systems on which our calculations are performed, and there is
a strong indication that it is at least larger than 200 GHz.
Since a typical laser linewidth is well below 200 GHz, it is
perhaps safe to conclude that the neglect of the laser linewidth in molecular collision processes in a laser field is
justified.

(2) We now consider collision processes which take place
in a multimode laser field. We assume that the laser operates
in M adjacent axial modes and in a single transverse mode
(e.g., TEM$_{00}$). Since two adjacent axial modes are separated by
$\Delta\omega = \pi c/L$, the overall frequency separation is $M\Delta\omega = \pi Mc/L$.
Taking the laser cavity length L to be 1 meter, we obtain
$M\Delta\omega = 3\pi M \times 10^8$ Hz, and as long as M is not too large, this
overall frequency separation is still smaller than 200 GHz
which may be considered to be a safe upper limit for the linewidth to be neglected. Therefore, we may approximate the
actual M-mode field by a single-mode field of average frequency $\bar{\omega}$. The intensity of this approximate field, however,
shows a different time behavior from that of a single-mode
field, i.e., it varies with time because of interference
among the original mode contributions. This time variation is
now the only characteristic that distinguishes between multimode and single-mode fields in the present model. To be more
specific, let us consider collision processes in a two-mode

laser field and a mode-locked ten-mode laser field whose intensities are given, respectively, by

$$I(t) = \bar{I}[1 + \cos(\Delta\omega t + \Delta\phi)] \qquad (3.32)$$

and

$$I(t) = \frac{\bar{I}}{10} \frac{\sin^2[5(\Delta\omega t + \Delta\phi)]}{\sin^2[(\Delta\omega t + \Delta\phi)/2]} , \qquad (3.33)$$

where $\bar{I}$ is the average intensity and $\Delta\omega$ and $\Delta\phi$ are the frequency difference and phase difference between two adjacent modes, respectively. We first note that the intensity given by Eq. (3.32) or (3.33) varies little over a typical collision time. Therefore, we may assume that each pair of colliding atoms experiences a constant intensity during the collision process. However, a different pair samples a different intensity. Thus, the average total cross section is given by

$$\sigma = \frac{1}{T} \int_0^T dt \; \sigma(I(t)), \qquad (3.34)$$

where T is the period of the intensity variation and $\sigma(I(t))$ corresponds to the cross section for a pair of atoms which samples an intensity $I(t)$. If we consider a radiative transition, and if the laser intensity is sufficiently low that the cross section is proportional to the intensity, i.e.,

$$\sigma(I(t)) = \alpha I(t), \qquad (3.35)$$

then Eq. (3.34) yields

$$\sigma = \frac{1}{T} \int_0^T dt \; \alpha I(t) = \alpha\bar{I}, \qquad (3.36)$$

indicating that the cross section in a multimode laser field is the same as that in a single-mode laser field of the same average intensity. However, if the laser intensity is high enough that the cross section shows nonlinear behavior with respect to the intensity, the integration over the intensity profile, Eq. (3.34), in general yields a cross section that depends on the number of modes.

We have performed close-coupling calculations on various model systems and evaluated collision cross sections for different numbers of modes in the laser. Figure 12 shows a typical result in which the cross section σ is plotted vs. the average intensity $\bar{I}$ for the case of a single-mode (solid

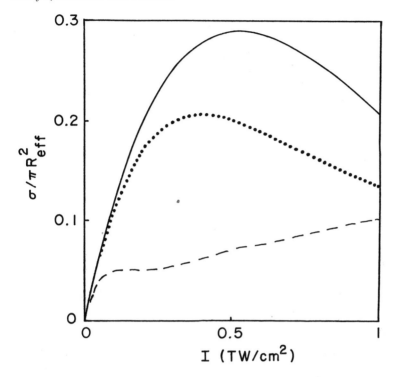

FIGURE 12. *The total cross section σ as a function of the average laser intensity Ī. The solid, dotted, and dashed curves represent the cross section in a single-mode, two-mode, and mode-locked ten-mode laser field, respectively.*

curve), two-mode (dotted curve) and mode-locked ten-mode (dashed curve) laser. The model to which this figure applies is the same as the one defined by Eqs. (3.31). The laser radiation is assumed to be centered at the 10643 Å line and the collision energy is chosen to be 0.03 Hartree. It can be clearly seen that the cross sections are virtually the same regardless of the number of modes if the laser intensity is sufficiently low. However, in the high intensity region, the cross sections are quite different for differing numbers of modes. For the model adopted here, the difference begins to show up when the intensity exceeds ~10^{10} W/cm^2. Although this particular value of the intensity belongs to the model under consideration, the general behavior of the cross section with respect to the number of modes is expected to hold in general because the cross section varies nonlinearly with respect to the laser intensity in a high intensity region regardless of the model. More detailed discussion of collision processes in a multimode laser field is given in Reference 57.

(3) We finally consider molecular collision processes in the presence of a short laser pulse whose temporal width is smaller than or, at most, comparable to the collision time. For the collision time of 1~10 psec, the temporal width of the pulse is required to be of the order of a picosecond or less. Since each pair of colliding atoms experiences the full intensity variation of the pulse during the collision, we need more than just an integration over the intensity profile in order to obtain the total cross section. The variation in the pulse shape must be included in our dynamical description of each individual collision. The same conclusion can be reached by looking at the problem in the frequency domain. For a (sub)picosecond pulse, the linewidth can be 10^3 GHz or larger, and therefore the single-mode approximation, which totally neglects the linewidth, may not be valid; i.e., we need to consider the entire spectral distribution of the pulse. This is another way of saying that the entire pulse shape needs to be considered because the temporal shape and the spectral distribution are Fourier transforms of one another.

Another new feature, in addition to a short temporal width or a large bandwidth that must be taken into account when dealing with collisions in a (sub)picosecond pulse is that we no longer have symmetry about the turning point, unless the pulse shape is symmetric and the center of the pulse exactly coincides with the turning point. If the pulse is sufficiently short, it is possible in principle to illuminate the molecular system during just the incoming or outgoing part of the collision. This suggests that the study of collisions in a short pulse can provide considerable insight into the control of collisional transitions.

A semiclassical theory of collision processes in an ultrashort pulse has been developed in Reference 58. If the laser intensity is sufficiently high that the laser field may be considered coherent, the formal structure of the theory is not much different from that described in Section II.C.2. The essential difference is contained in the expression for adiabatic potential surfaces. If we consider an electronic transition from the lowest to the first excited state under the rotating wave approximation, the two adiabatic potential surfaces are given by

$$E_{1,2}(R) = \frac{1}{2}(\bar{W}_1(R) + \bar{W}_2(R) \pm \{[\bar{W}_2(R) - \bar{W}_1(R)]^2$$
$$+ 4[\vec{\mu}(R) \cdot \vec{E}_0(R)]^2\}^{1/2}) \qquad (3.37)$$

where the + and the − signs in front of the radical correspond to E_2 and E_1, respectively; $\vec{E}_0(R)$ is the electric field

strength of the laser pulse when the internuclear separation
is equal to R, and

$$\bar{W}_1(R) = W_1(R) + N\hbar\omega_0$$

$$\bar{W}_2(R) = W_2(R) + (N-1)\hbar\omega_0.$$

Here, N is the number of photons initially present in the
pulse and ω_0 is the central frequency of the pulse. Were the
intensity variation negligible over the collision time, $\vec{E}_0(R)$
would have been a constant leading to Eq. (2.87) of Section
II.C.2. Here, however, $\vec{E}_0(R)$ is a sharply varying function of
R, and the structure of the adiabatic potential surfaces and,
consequently, the whole collisional dynamics are highly sensitive to the intensity variation.

Eq. (3.37) for adiabatic potential surfaces enables one to
evaluate collision cross sections for various molecular systems (which determines $W_1(R)$ and $W_2(R)$) and for various shapes
of the pulse (which determines $\vec{E}_0(R)$). Calculations on the
Landau-Zener and the Demkov model with a square pulse indicate
that the cross sections are critically dependent upon pulse
parameters such as intensity, temporal width, general shape
and polarization (58). Therefore, by appropriate choice of
parameters, a particular transition may be enhanced, sup-
pressed or limited in time and space. Further investigation
on realistic physical systems is needed if we are to see how
to control actual collisional transitions occurring in the
presence of ultrashort pulses.

B. *Chemical Reactions*

Here we shall study the effects of a laser field on chemical reactions. This field, on the one hand, is of recent academic interest, and on the other hand, has great potential for
practical use. It is easy to visualize these effects by using
the electronic-field representation. As mentioned before, the
electronic-field surfaces, which arise from shifting field-free
(FF) adiabatic surfaces relative to each other by $n\hbar\omega$ and
being coupled through the radiative coupling, may be quite
different from the FF surfaces in a relatively strong radiation
field. Especially when distortions take place around some
important regions of the potential surfaces, such as the
transition state region, the reaction rate, branching ratio
and even the reaction mechanism can be greatly changed. A
lowering of the activation barrier at the transition state of
a reaction in a laser field has been illustrated in Figure 5.
A careful, dynamical study of a chemical reaction system which

may assist in finding new synthetic routes in organic chemistry
is still too laborious a job to perform at the present stage.
To take a modest first step toward this goal, we have carried
out a dynamic study of the triatomic reactive system, $F+H_2 \rightarrow$
$HF+H$, under the irradiation of a laser field. This is one of
the well-known reactions taking place in the HF-laser cavity
(59). Several dynamical calculations have been done for this
reaction in the field-free case (60-62). We shall present
below results of a first calculation of this reaction in a
laser field (63). A semiclassical trajectory method (62) is
used where the three atoms are confined in a collinear configuration. In the next paragraph we shall briefly describe
the method and special features associated with our calculations. Results and some discussions will then follow.

We use the classical trajectory method with a uniform
grid of initial conditions and modified to include transitions
between electronic-field surfaces. Restricting ourselves to
two electronic-field' states, we treat nonadiabatic transitions
according to the decoupling approximation (64,65) developed
for the Miller-George theory (36); that is, a trajectory is
propagated until it reaches the "seam" —— a curve formed by
projecting a proper set of the branch points onto the real
space. Then we decompose the momentum into perpendicular and
parallel components to the "seam", and propagate the trajectory
perpendicularly into the complex configuration space, where it
circles a branch point to pick up a damping factor corresponding to a leakage to the other electronic-field surface.

The FF ground electronically diabatic surface is the semiempirical Muckerman V surface. The first excited FF electronic surface is obtained from fitting parameters to data
based on GRHF-CI calculations (66). FF adiabatic surfaces, W_1
and W_2, are then constructed by coupling diabatic surfaces
through a constant spin-orbit interaction term. W_1 and W_2
plotted against the reaction coordinate s are shown schematically in Figure 13, in which an arrow is drawn to show where
an Nd:glass laser photon is in resonance with the surfaces.
The transition dipole moment between the $^1\Sigma$ and $^1\Pi$ of HF as a
function of interatomic distance has been published by Bender
and Davidson (38). The overall radiative coupling is taken to
be (approximately) proportional to the moment they have calculated, so that the coupling is a function of just the H-F
distance. The electronic-field surfaces, E_1 and E_2, can then
be constructed from W_1, W_2 and d_{12} from Eq. (2.87).

To select initial conditions uniformly, we have transformed
the coordinates of the Morse oscillator of the reactant diatomic molecule into action-angle variables. The action variable of the Morse oscillator is just its vibrational quantum
number. The angle variable is the phase of the oscillator.
All the trajectories that we have considered start on the upper

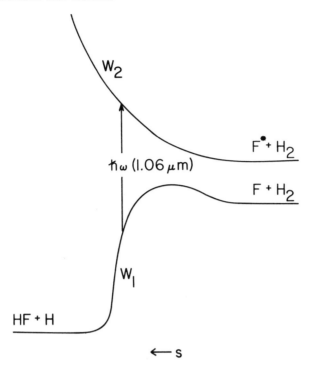

FIGURE 13. *Two field-free potential energy surfaces,*
W_1 *and* W_2, *for the reactive* $F+H_2$ *system, where the vertical*
arrow indicates the resonance between $\hbar\omega(1.06 \ \mu m)$ *and the*
surfaces.

electronic-field surface, E_2, which correlates to $F(^2P_{3/2})$ +
$H_2(v=0)$. Since the Muckerman V potential surface reduces to a
Morse potential plus a free atom in the asymptotic region,
the use of action-angle variables to select the initial conditions is well justified. For a fixed, initial atom-diatom
molecule distance we have uniformly selected the initial
phases of the oscillators. The results reported below, in
general, are obtained from grids of a hundred trajectories.

To use the Miller-George theory we need to locate complex
branch points (i.e., intersection points) between the electronic-field surfaces. For each field strength of the Nd:glass
laser we have different sets of electronic-field surfaces, and
therefore a different set of branch points. We restrict ourselves to collinear collisions, whereby two hydrogen atoms
cannot interchange with each other. We shall label them as
H_a and H_b and write the reaction as

$$F(^2P_{3/2}) + H_aH_b(v=0) \xrightarrow{\hbar\omega(1.06 \ \mu m)} FH_a(v') + H_b, \quad (3.38)$$

where v and v' are the vibrational quantum numbers of H_aH_b and FH_a, respectively. To locate the branch points where $E_1=E_2$ we analytically continue the electronic-field surfaces into complex coordinate space.

Let r_1 be the interatomic distance between H_a and H_b, and r_3 the distance between F and H_a. r_2 is the distance between F and H_b which is equal to r_1+r_3. The corresponding translational coordinates R_1 and R_3 associated with r_1 and r_3 are from the center of mass of H_aH_b to F and center of mass of FH_a to H_b, respectively. For a fixed r_1 we find the complex branch points (r_1,r_3^*) by the Newton-Raphson iterative method. The real part of r_3^* satisfies approximately the relation $r_3=r_1-$ 0.30 Bohr. As mentioned before, in the decoupling approximation (64,65) one needs to integrate the trajectory perpendicularly (67,68) to the "seam" into the complex space to find the transition probability. To effectively separate the nuclear kinetic energy term into two components, one containing the perpendicular momentum and the other the parallel one, we project the set of branch points onto the (r_3,R_3)-plane. The "seam" lies around approximately a straight axis defined as $R_{\parallel}$. The axis perpendicular to the "seam" in the direction of the motion across the "seam" is labelled as $R_{\perp}$, and $\hat{n}_{\perp}$ is defined as a unit vector along $R_{\perp}$. If we further define the momentum component along $\hat{n}_{\perp}$ as $P_{\perp}$ and that along $R_{\parallel}$ with a unit vector $\hat{n}_{\parallel}$ as $P_{\parallel}$, the separable assumption, namely

$$E = \frac{P_{\perp}^2}{2M_{\perp}} + \frac{P_{\parallel}^2}{2M_{\parallel}} + E_i, \quad (3.39)$$

holds to a high accuracy. In Eq. (3.39), $M_{\perp}$ and $M_{\parallel}$ are defined by $M_{\perp}^{-1} = \hat{n}_{\perp} \cdot \underline{M}^{-1} \cdot \hat{n}_{\perp}$ and $M_{\parallel}^{-1} = \hat{n}_{\parallel} \cdot \underline{M}^{-1} \cdot \hat{n}_{\parallel}$, with the inverse mass tensor $\underline{M}^{-1}$ given as

$$\underline{M}^{-1} = \begin{pmatrix} \mu^{-1} & 0 \\ 0 & m^{-1} \end{pmatrix},$$

where μ is the reduced mass between H and FH and m is the reduced mass of FH. To find the local transition probability, p_t, we integrate $P_{\perp}$, as defined in Eq. (3.39), around the branch point and take the imaginary part of the action integral.

At the "seam" the trajectory may either propagate continuously on the upper electronic-field surface or switch onto the lower surface. The local probability of staying on the surface is $(1-p_t)$ and that of switching surfaces is p_t. If a trajectory does switch surfaces, adjustment of momenta is made to conserve the total energy in the decoupling approximation, where the change in momenta goes to the perpendicular component.

After transition to the lower electronic-field surface the trajectory may propagate either to the product valley, in which case Reaction (3.38) takes place, or back to the reactant side, where the inelastic process (at collision energy of 0.049 eV)

$$F(^2P_{3/2}) + H_2(v=0) + \hbar\omega(1.06 \ \mu m) \rightarrow F^*(^2P_{1/2}) + H_2(v')$$

$$(3.40)$$

occurs. This process involves net absorption of a photon, in contrast to Reaction (3.38) which involves no net photon absorption. The transition to $F^*(^2P_{1/2}) + H_2(v=0)$ or $F(^2P_{3/2}) + H_2(v=1)$ in the absence of the laser is impossible since the spin-orbit splitting of fluorine atom is 0.05 eV and the vibrational quantum of H_2 is 0.55 eV. In the field-free case the quantum state v=2 of HF is slightly more populated than the state v=3, and the "inversion" ratio is 0.75. The elastic scattering probability is 0.27. By shining an intense laser on the dynamic process we see that the reaction probability changes from 0.63 to 0.61 and 0.74 for field intensities of 100 GW/cm^2 and 10 TW/cm^2, respectively (63). The inversion ratio increases from 0.79 to 1.64 as the field intensity increases, and both of these numbers are larger than that in the field-free case. It is also observed that the laser field can induce transitions to $F^*(^2P_{1/2}) + H_2(v')$ where the vibrational quantum v' is either 0 or 1. Probabilities of these processes are small (below 10^{-1}) but increase as the field intensity increases. The process of producing v'=1 is especially interesting, because there are two kinds of excitations involved at the same time: F atom is electronically excited and H_2 molecule vibrationally excited.

C. *Bound-Continuum Processes*

By bound-continuum processes in general we mean those in which one or more discrete states of a system can interact with the continuous spectrum. Perhaps the most well-studied example is the damping of discrete atomic states coupled to the continuum of the vacuum electromagnetic field (69). This coupling

leads, of course, to the finite lifetimes of excited atomic
states and the corresponding natural linewidths in their
emission spectrum. Another example, linking the realms of
atomic and solid state physics, is that of surface molecules
and chemisorption (70,71). (The reader is referred to
Section III.D of this article for a discussion of laser/sur-
face-catalyzed processes.) The surface molecule is formed as
a result of the discrete states of the adsorbate atom (adatom)
mixing with a continuum, the conduction band states of the
solid surface. Analogous to the case of radiation coupling in
the first example, these discrete adatom states are also
damped, and the broadening can be described by an adatom density of states which depends directly on the strength of the
bound-continuum interaction (71).

Bound-continuum processes are of course not restricted to
the interaction between atomic states and the continuum. In
molecular physics, especially in molecular collision physics,
however, the bound-continuum problem differs from that in
atomic physics in one important respect: not only do the
electronic but also the various nuclear degrees of freedom -
rotational, vibrational and translational - couple with the
continuum states differently, and the complication of heavy-
particle dynamics will have to be considered. At present, this
complication is viewed adversely or as a blessing in disguise
depending on the viewpoints of those treating the problem,
since it holds out the promise of the possibility to exploit
the good services of classical mechanics (43)(see the discus-
tion in Section II.C.2).

In this section we will be discussing a number of bound-
continuum molecular rate processes in which heavy particle
dynamics is involved. Further, these will be considered in
the presence of intense laser radiation. The processes of
collision-induced spontaneous emission and photodissociation
involve the interaction of the molecular system with the con-
tinuums of multimode electromagnetic radiation and dissociative levels respectively, whereas that of field-influenced
(both FA and FM) collisional ionization occurs under the
coupling of bound states to the continuum of free electron
states. Before proceeding to consider these processes indivi-
dually, we will examine briefly several treatments applicable
to the general problem of bound-continuum molecular rate processes.

The problem can be dealt with on the basis of either
classical, semiclassical (43,72-74) or quantum mechanical
theories (75-88). In the classical theory one works directly
with transition probabilities, which are usually computed in
terms of bound-state widths obtainable as products of Landau-
Zener transition probabilities (72-73). In the semiclassical

theories, however, classical quantities such as action integrals are used but combined according to the principle of superposition of probability amplitudes (72). These amplitudes can be calculated either in a Landau-Zener fashion (72) or by making use of the complex branch-point methodology introduced in Section II.C.2 (43,74). The quantum mechanical theories fall mainly into three classes. The first (72,75-78) makes use of a local complex adiabatic potential to account for transitions into the continuum, whereas the second (79-82) is a coupled-channels treatment based on discretization of the continuum. The third approach derives from the Laplace Transformation method to treat nonstationary problems (83,84). This last quantum mechanical approach has not enjoyed as widespread an application as the others mentioned but it seems to be the most theoretically appealing in that the inherent complexities of the bound-continuum problem are incorporated very naturally into the theoretical framework. No limiting assumptions need to be made concerning the bound-continuum system, such as Franck-Condon considerations or two-state approximations, as are often required in other treatments. However, its general usage in the field of molecular rate processes is still restricted by formal and practical difficulties. In view of the fact that its application to molecular collisions is much less well documented than the other treatments, we will present a brief outline of its main features here. Individual examples illustrating the semiclassical and quantum mechanical coupled-channels treatments will be discussed later in this section.

For simplicity we consider the problem of one discrete state $|\phi>$ embedded in a continuum of states $|\varepsilon>$. The discrete state is described by a potential surface $V_d(R)$ where R stands for some internuclear degree of freedom and ε is some continuously varying energy parameter. The complete set of basis states for the system is then $\{|\phi>,|\varepsilon>\}$, with the orthogonal condition $<\phi|\varepsilon> = 0$. One then attempts to solve the Schroedinger equation

$$\left(-\frac{1}{2\mu}\frac{\partial^2}{\partial R^2} + H\right)|\psi> = E_0|\psi>, \qquad \hbar = 1 \qquad (3.41)$$

where μ is the reduced mass and E_0 the total energy of the system. H is the total Hamiltonian less the kinetic energy corresponding to R and has the following properties (in what follows $< >$ indicates integration over electronic coordinates only):

$$<\phi|H|\phi> = V_d(R) \qquad (3.42)$$

$$<\varepsilon'|H|\varepsilon> = \varepsilon\delta(\varepsilon-\varepsilon'). \qquad (3.43)$$

Also, the quantity

$$<\phi|H|\varepsilon> \equiv V_{d\varepsilon}(R) \tag{3.44}$$

gives the strength of the bound-continuum interaction. If the continuum of states $|\varepsilon>$ could be discretized into $|\varepsilon_1>$, $|\varepsilon_2>$ .... etc., H would be represented in the matrix form

$$
\underset{\sim}{H} = \begin{pmatrix} V_d & V_{d\varepsilon_1} & V_{d\varepsilon_2} \cdots \\ V^*_{d\varepsilon_1} & \varepsilon_1 & \bigcirc \\ V^*_{d\varepsilon_2} & \bigcirc & \varepsilon_2 \\ \vdots & & & \ddots \end{pmatrix} \;\; . \tag{3.45}
$$

Since in general discretization may not be possible, it will be convenient to partition H into two portions, one having a continuous and the other a discrete spectrum:

$$H = H_0 + |\phi> V_d <\phi| . \tag{3.46}$$

It can be seen that

$$<\phi|H_0|\phi> = 0 \tag{3.47}$$

$$<\varepsilon|H_0|\varepsilon'> = \varepsilon\delta(\varepsilon-\varepsilon') \tag{3.48}$$

and $<\phi|H_0|\varepsilon> = V_{d\varepsilon}$. $\tag{3.49}$

With H partitioned in this manner, it is then possible to write the formal solution for $|\psi>$ as the Laplace transform of the states $F(k)G(k)|\phi>$:

$$|\psi> = \int_C dk\, e^{-ikR} F(k)G(k)|\phi> . \tag{3.50}$$

In Eq. (3.50) the momentum k is related to an energy variable E(k) by

$$2\mu E(k) = 2\mu E_0 - k^2 ; \tag{3.51}$$

G(k) is the Green's operator for the partial Hamiltonian H_0

$$G(k) = [E(k) - H_0]^{-1} ; \tag{3.52}$$

and C is a suitable contour in the complex k-plane specified below. $F(k)$ is determined by substituting Eq. (3.50) in Eq. (3.41) and is given by

$$F(k) = <\phi|G|\phi>^{-1}\exp\{\frac{iR}{V_d}\int^k dk'<\phi|G(k')|\phi>^{-1}\} \qquad (3.53)$$

provided C is any contour such that

$$\frac{i}{R}e^{-ikR}F(k)<\phi|G|\phi>|\phi>|_C = 0 \quad . \qquad (3.54)$$

$|\psi>$ can then be written

$$|\psi> = N\int_C dk\ \frac{G(k)|\phi>}{<\phi|G(k)|\phi>}\ \exp\{\frac{iR}{V_d}\int^k dk'\ \frac{1}{<\phi|G(k')|\phi>} - ikR\} \qquad (3.55)$$

where N is a normalization constant. Using the completeness relation

$$I = |\phi><\phi| + \int_0^\infty d\varepsilon\ \rho(\varepsilon)|\varepsilon><\varepsilon| \qquad (3.56)$$

where I is the identity operator and $\rho(\varepsilon)$ is the density of states, it can be deduced that

$$<\phi|G(k)|\phi> = \left[E(k) - \int_0^\infty d\varepsilon\ \frac{|V_{d\varepsilon}|^2}{E(k)-\varepsilon}\right]^{-1} \qquad (3.57)$$

and

$$<\varepsilon|G(k)|\phi> = \frac{V_{d\varepsilon}^*<\phi|G|\phi>}{\rho(\varepsilon)[E(k)-\varepsilon]} \quad . \qquad (3.58)$$

The physical quantities of interest are $<\phi|\psi>$ and $<\varepsilon|\psi>$ as R→∞, as these represent probability amplitudes at asymptotic conditions for the system to remain in the discrete state $|\phi>$ or leak into a continuous channel characterized by ε.

The formalism just described has only been used successfully to treat collision problems where the potential surface $V_d(R)$ is linear in R and where straight line trajectories for heavy particles are assumed (84).

Finally there is yet another quantum mechanical approach which has not been applied to any molecular bound-continuum problems but which may prove to be of potential practical importance. This is the Green's function approach to treat quasi-free scattering processes within the impulse approximation (85), and has been applied with success to A(p,2p)B scattering problems in nuclear reactions for the study of the hole structure of atomic nuclei (86,87). To see how the formalism may be applicable to a particular kind of bound-continuum processes, namely, collisional ionization, we consider the Penning ionization situation

$$A* + B \to A + B^+ + e^-. \tag{3.59}$$

(A* means an excited electronic state of the atom A.) Within the impulse approximation, a virtual electron is considered to be produced by the direct process

$$B \longrightarrow B^+ + e^-(\text{virtual}) \tag{3.60a}$$

which then interacts with the projectile atom A*

$$A* + e^-(\text{virtual}) \longrightarrow A + e^-. \tag{3.60b}$$

In this picture the A* + B collision is mediated by a single electron. The process in diagrammatic form is represented in Figure 14.

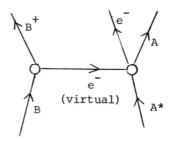

FIGURE 14. *Diagrammatic representation of the collisional ionization process* $A* + B \to A + B^+ + e^-$ *within the impulse approximation.*

Thus the dynamics involving a di-atom collision is simplified to one of electron-atom scattering, and the total scattering amplitude for the process can be expressed as a product of an electron-atom inelastic scattering amplitude and a form factor of the ionized atom. Admittedly this impulse approximation has limitations and is expected to be good only for very energetic collisions or in cases where the atom to be ionized has only a single electron or a loosely bound valence electron. However, this approach allows one to calculate differential cross sections (with respect to emitted electronic energy) in a straightforward fashion and, as in the Laplace transformation approach, deals with bound-continuum interactions in a rigorous manner. Recently an improved impulse approximation has been proposed for the problem of collisional ionization and its relation to other approximations examined (88). It is expected that this approach will play an increasingly important role in molecular bound-continuum problems.

We will now proceed to discuss the individual bound-continuum processes. Collision-induced spontaneous emission and photodissociation will be discussed within the semiclassical framework whereas collisional ionization will be dealt with quantum mechanically.

1. Collision-Induced Spontaneous Emission. The problem of collision-induced emission in the presence of intense laser radiation can be of importance in several contexts. From the theoretical viewpoint, the interference between field-free nonadiabatic and multimode radiative couplings (the possibility of emission at a continuous range of frequencies corresponding to particular molecular configurations) deserves special attention. From the practical perspective this problem is relevant to the investigation of how premature emission (within collision regions) of a radiatively excited molecular collision system, which is to operate as a chemical laser in asymptotic regions, can affect the performance of the laser.

We consider a process of the type

$$A + B + \hbar\omega \rightarrow A + B + \hbar\omega' \qquad (3.61)$$

where ω is the frequency of the incident laser radiation and ω' that of the emitted radiation. To describe the interaction of the collision system with the incident laser we can make use of the electronic-field surfaces introduced in Section II. For a two-state system their generation is illustrated in Figures 15 and 16. In these figures W_1 and W_2 denote the field-free adiabatic surfaces and E_i the electronic-field-surfaces. R represents an internuclear degree of freedom. The avoided crossing near R_0 in Figure 16 is a result of radiative coupling and its dimensions depend on the strength of the coupling. It is near this region that photon absorption and radiative excitation from W_1 and W_2 is most likely to take place.

The electronic-field surfaces E_i can be regarded as forming a spectrum for spontaneous emission if radiative coupling with the vacuum field is also considered. This interaction involves

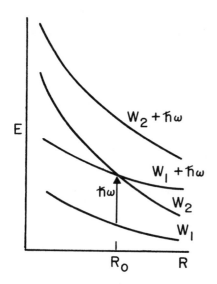 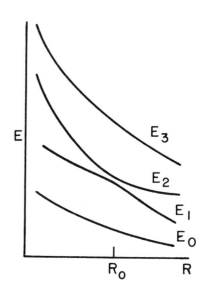

FIGURE 15 (left). Schematic drawing of the potential surfaces for the molecular collision-plus-field system in the diabatic representation. R_0 is the configuration at which the field is in resonance with the surfaces W_1 and W_2.

FIGURE 16 (right). Schematic drawing of the electronic-field surfaces E_0, E_1, E_2 and E_3. The avoided crossing between E_1 and E_2 is caused by the radiative coupling d_{12} [cf. Eq. (2.87)].

in principle an infinity of field modes, since a continuum of emission frequencies is possible corresponding to the different nuclear configurations at which emission can take place. After emission of a photon, it is assumed that the system will be de-excited back to the ground state W_1.

The semiclassical picture allows us to think of the photon of frequency ω' as being emitted near the configuration R_ε such that

$$\hbar\omega' = E_1(R_\varepsilon) - E_0(R_\varepsilon). \tag{3.62}$$

Emission-electronic-field surfaces $E_+^{(e)}$ describing the dynamics of this event can then be constructed analogously to the electronic-field surfaces. These are illustrated in Figures 17 and 18. It should be noticed that the dimensions of the

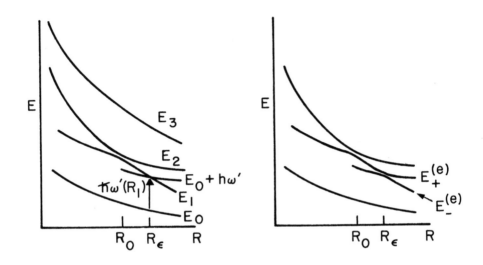

FIGURE 17 (left). Schematic drawing of the diabatic surfaces for the description of single-mode emission at R_ϵ.

FIGURE 18 (right). Schematic drawing of emission - electronic-field surfaces for single-mode emission at R_ϵ. The avoided crossing is caused by emission radiative coupling

emission avoided crossing are much smaller than those of the absorption one since the former is generated through coupling with the vacuum field. The dynamical picture then emerges that if the system is still propagating on the surface $E_+^{(e)}$ at asymptotic regions, it will have emitted a photon of frequency ω' and become de-excited to the ground state W_1.

To know the probability of emission at the particular configuration R_ϵ, however, we have to determine specifically the trajectories which would avoid emission before and which would lead to emission when that configuration is reached. These

semiclassical trajectories accounting for pre-emission loss can
be better visualized if we first consider the case of discretized allowable emission configurations. A system propagating on the electronic-field surface E_1 will then encounter
a series of configurations at which it can emit. Let one of
these be $R_{\epsilon i}$. At this configuration the system is capable of
seeing emission coupling only at the frequency $\hbar\omega_i = E_i(R_{\epsilon i}) -
E_0(R_{\epsilon i})$; and couplings to other frequencies corresponding to
more distant configurations can be considered as yet not
'turned on'. Moreover, emission couplings to frequencies
corresponding to configurations prior to $R_{\epsilon i}$ are also assumed
to have negligible effect on the dynamics of the system at $R_{\epsilon i}$,
since it has 'survived' pre-emission loss at these configurations before reaching $R_{\epsilon i}$. Hence, at each possible emission
configuration, emission-electronic-field surfaces can be
generated independently of radiative couplings at other configurations. A schematic representation for the discrete case
is illustrated in Figure 19.

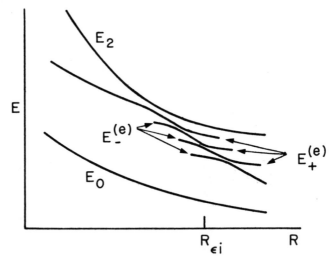

*FIGURE 19. Schematic drawing of emission electron-field
surfaces for discrete emission configurations. $R_{\epsilon i}$ is the
i^{th} emission site.*

When the emission avoided crossings become continuously
distributed along R, it is still physically meaningful to
treat each avoided crossing as generated by a localized
emission coupling. In this continuous limit the discrete
series of branch points will constitute a line of branch
points in the complex R plane. It will be expected to be

relatively close to the real axis compared to the location of
the absorption branch point because of the relative weakness
of the emission coupling. Whereas in the discrete case, a
trajectory, after rounding each emission branch point, may return to the real axis before going around the next, it will be
forced to deviate from the real axis throughout the configurations at which emission coupling is active and be propagated
along the line of branch points in the continuous case. The
imaginary action accumulated along this line will determine
the accumulated probability for making local transitions from
$E_+^{(e)}$ to $E_-^{(e)}$ continuously, that is, the accumulated probability
for not having emitted a photon while the molecular system
stays on E_1. At the actual emission site the trajectory
reverts back to the real axis and is propagated to asymptotic
regions on $E_-^{(e)}$.

The shape of the emission spectrum (as a function of the
emission configuration R_ϵ) is then found to be determined by
the shape factor

$$F_S(E,R_\epsilon) = \exp[-2\chi(E,R_\epsilon)]\{1 - \exp[-2\xi_\epsilon(E,R_\epsilon)]\} \quad (3.63)$$

where E is the total energy of the collision system. The
action factor χ accounts for pre-emission effects and ξ_ϵ determines the local emission probability. Since the emission
branch points tend to depart more from the real axis as R_ϵ
increases, $1-\exp(-2\xi_\epsilon)$ will increase and $\exp(-2\chi)$ decrease as
R_ϵ become large. Thus it is seen that pre-emission and
localized actual emission effects compete with each other for
the total emission probability at a particular configuration.
Results of model calculations indicate that this total probability depends critically only on the local emission probability at the actual emission configuration (43).

The present approach has two main drawbacks from the
theoretical point of view. First, from a set of electronic
surfaces generating a spectrum for spontaneous emission, only
a pair is selected to couple radiatively. In other words, a
two-state approximation is resorted to. Secondly, the emission
couplings at various configurations are treated locally to
produce a global emission loss surface. Even though it has
been demonstrated that both these approximations are justifi-
able (43), it will be desirable to further develop a formalism
which does not presume the Franck-Condon principle for the
conservation of nuclear kinetic energy during the emission
process so that the problem of off-resonance emission involving nuclear kinetic energy transfer to the radiation field can
be treated.

2. *Laser-Induced Collisional Ionization.* Collisional
ionization refers to processes where bound and continuum electronic states are coupled during atomic or molecular collisions. The result is a collisional means for a discrete
state to "leak" into an ionization continuum. Ionizing
collisions play an important role in gas-phase chemistry
since they represent a primary mechanism for depletion of
atomic or molecular species having electronic excitation
(stored in metastable electronic states), and for production
of ions. Much experimental and theoretical work on collisional
ionization has been carried out in recent years so that by now
it is a fairly well understood process (89).

Assisting or influencing collisional ionization by intense
laser radiation opens new vistas in studies of laser-induced
molecular rate processes. Such laser-induced collisional
ionization (involving colliding atoms or molecules A and B)
are characterized by

$$A(A^*) + B + \hbar\omega \rightarrow A + B^+ + e^-(\varepsilon) \qquad\qquad \text{PI} \qquad (3.64a)$$

$$\rightarrow AB^+ + e^-(\varepsilon) \qquad\qquad \text{AI} \qquad (3.64b)$$

where ω is the laser photon frequency, ε is the kinetic energy
of the emitted electron, and $A(A^*)$ indicates that species A may
or may not be in an excited state. Process (3.64a) is the
laser-field counterpart to field-free (FF) Penning ionization
(PI). Likewise, Process (3.64b) is the laser-field counterpart to FF associative ionization (AI). Processes (3.64) are
not of the sort involving photoionization before or after
collision. Rather, like their FF counterparts, ionization
proceeds primarily by virtue of collision.

Two recent sets of experiments depend significantly on the
occurrence of laser-induced ionizing collisons. In one (90,91)
laser light is used to produce two-photon ionization of small
amounts of Cs in Ar gas. In the other (92,93) ionization of
alkali vapors is observed when they are irradiated by a laser.
Additional more conclusive experimental evidence for the
occurrence of laser-induced ionizing collisions appeared very
recently (27). In this experiment PI and AI are observed in a
crossed-beam study of the $Li^*(2p)+Li^*(2p)+\hbar\omega$ laser-collision
system where $\hbar\omega$ is the photon energy of a laser tuned near the
$Li^*(2p)$ resonance line. The formation of associative ions
strongly indicates that a combined radiative-collisional process has been measured.

Ionizing collisions in the presence of intense laser fields
can be classified according to two categories. One category is
called *field-assisted* (FA) collisional ionization, and the
other *field-modified* (FM) collisional ionization. In FA cases

laser radiation "assists" ionization when photon absorption, occurring during collision, permits access to electronic continua that are energetically inaccessible under field-free conditions. In FM cases laser radiation "modifies" ionization when photon absorption, occurring again during collision, permits access to new regions of electronic continua that are already energetically accessible without the laser field, and to new continua that exist only by virtue of the matter-field interaction. The experimental work reported so far (27,90-93) deals with FA ionizing collisions. Experiments on FM cases are, however, being considered (94). Both FA and FM collisional ionization are characterized by two important features. One is that the continuum states are in energy resonance for photon absorption over large ranges of the internuclear separations during collision. The other is that radiative coupling between the discrete and continuum electronic states is typically very long-range. The resonance feature means that radiative-collisional transitions between bound and continuum states are favored over those between bound states, since the former can occur during practically the whole collision while the latter can occur only during that portion of the collision when the heavy particles are near a pseudo-crossing. This means that laser-promotion of bound-bound collisional transitions requires in general higher laser intensities than laser-promotion of bound-continuum collisional transitions. The recent progress in experimental observation of laser-induced collisional ionization is therefore not surprising.

In developing theoretical descriptions of laser-induced ionizing collisions it is natural to investigate first the possibility of adapting existing formalisms (72-82) for treating FF bound-continuum collisional transitions to include the presence of intense laser fields. The first theories (30,31, 95) put forward to account for radiative-collisional transitions are based on semiclassical treatment of the nuclear dynamics within a straight-line-path approximation, and include the laser field as a perturbation. At the University of Rochester we have developed a quantum-mechanical formalism for laser-induced ionizing collisions (81) that combines the discretization procedures used by Bellum and Micha (79,80) to treat nuclear dynamics for bound-continuum collisional transitions, and the so-called electronic-field representations (10,12,14,29) to treat the laser field in a non-perturbative manner. This leads to a coupled-channels description of Process (3.64) with transitions between discrete and "discretized" states mediated by radiative coupling.

In recent work (82,96) we have applied our quantum mechanical theory to the FM PI process

$$He^*(1s2s, {}^3S) + Ar + \hbar\omega \rightarrow He + Ar^+(3p^5, 2p) + e^-(\varepsilon)$$

$$(3.65)$$

specialized to the cases of $\hbar\omega$ corresponding to the 10.6 μm line of a CO_2 laser, and the 1.315 μm line of an iodine laser. We will briefly review this application of our theory.

We describe the space of bound and continuum internal matter-field states accessible during collision in terms of complete sets of bound (continuum) electronic states $\{\phi_{i(\varepsilon)}\}$ and photon number states $|N\rangle$, which together form an electronic-field representation for treating the collision dynamics. The Schroedinger equation for nuclei, electrons and laser field corresponds to a continuously infinite set of coupled equations when the total scattering wave function is expanded first in terms of this electronic-field representation and then in terms of angular momentum eigenfunctions (spherical harmonics) for the heavy-particle (hp) rotational motion. This system of coupled equations determines expansion coefficients $F_{i(\varepsilon),j}^{N,L}(R)$ which are radial wave functions (of the internuclear separation R) for the heavy particles moving in their L^{th} angular momentum state during collisions leading to $i(\varepsilon) \leftarrow j$ bound-bound (-continuum) transitions with N laser photons remaining afterwards.

To treat the continuously infinite set of coupled equations we incorporate into the formalism discretization procedures used for FF collisional ionization (79,80). We discretize by expanding hp wave functions $F_{\varepsilon,j}^{N,L}(R)$ (corresponding to bound-continuum transitions during collision) in a complete set of functions, $C_I(\varepsilon)$, of the continuous electron energy ε. As a result the continuously infinite set of coupled equations reduces to a discretely infinite set of coupled equations (see Reference 81).

We apply this discretization formalism to Process (3.65) at the same level of approximation used in its application to FF Penning ionization of Ar by $He^*(1s2s, {}^3S)$ (79,80). Namely, we approximate the basis functions $C_I(\varepsilon)$ by a set of step functions in ε, so that $C_I(\varepsilon) = \Delta\varepsilon^{-1/2}$ for ε in the Ith incremental continuum region centered at ε_I with size $\Delta\varepsilon$, and zero otherwise. Three additional assumptions are made: 1) Process (3.64) is treated in terms of a single discrete electronic state ϕ_d of (HeAr)* and a single electronic continuum of the (HeAr)$^+$ molecular ion; 2) matrix elements coupling two different continuum states ϕ_ε and $\phi_{\varepsilon'}$ are neglected; 3) each $I \leftarrow d$ collisional transition is described as occurring independently of collisional transitions from ϕ_d to ϕ_ε in other incremental continuum regions. These assumptions are discussed in detail in Reference 80. With them and a small enough choice of $\Delta\varepsilon$ we obtain a set of coupled equations for each incremental region along the ε scale. For the Ith $\Delta\varepsilon$ increment we have

$$[(\hbar^2/2m)\,d^2/dR^2 - L(L+1)/2mR^2 - N\hbar\omega - V_d(R) + E]F_{d,d}^{N,L}(R) =$$

$$V_{d,I}^C(R)F_{I,d}^{N,L}(R) + D_{d,I}^{N,N+1}(R)F_{I,d}^{N+1,L}(R) + D_{d,I}^{N,N-1}(R)F_{I,d}^{N-1,L}(R),$$

$$(3.66a)$$

$$[(\hbar^2/2m)\,d^2/dR^2 - L(L+1)/2mR^2 - N\hbar\omega - V_+(R) - \varepsilon_I + E]$$

$$\times F_{I,d}^{N,L}(R) = V_{I,d}^C(R)F_{d,d}^{N,L}(R) + D_{I,d}^{N,N+1}(R)F_{d,d}^{N+1,L}(R)$$

$$+ D_{I,d}^{N,N-1}(R)F_{d,d}^{N-1,L}(R),\qquad (3.66b)$$

where E is the total energy for electrons, nuclei and laser photons, and m is the reduced mass of the colliding atoms. $V_{I,d}^C$ and $D_{I,d}^{N,N+1}$ are the Coulomb and radiative interaction matrix elements between the discrete state d and "discretized" continuum regions I (see Reference 81). We treat the radiative coupling in the dipole approximation and neglect diagonal radiative coupling matrix elements, i.e., where $a=b$. In Eq. (3.66) $V_d(R)=\langle\phi_d|H_{e\ell}|\phi_d\rangle$ is the interaction potential of the discrete electronic state and $V_+(R)=\langle\phi_+|H_{e\ell}|\phi_+\rangle$ is the interaction potential of a state ϕ_+ of the molecular ion, where $H_{e\ell}$ is the usual electronic Hamiltonian. Eq. (3.66a) determines the hp motion in the reactant channel for a situation where N photons are in the laser field. We find $F_{d,d}^{N,L}$ (the hp wave function for elastic scattering with N photons remaining in the laser field) linked by Coulomb coupling to $F_{I,d}^{N,L}$ (the hp wave function for $I\leftarrow d$ transitions where N photons remain in the laser field), and by radiative coupling to $F_{I,d}^{N\pm1,L}$ (the hp wave function for $I\leftarrow d$ transition where $N\pm1$ photons remain in the laser field). By replacing N in Eq. (3.66b) by $N\pm1$, $N\pm2,\ldots,$ etc., we can in principle generate a complete set of coupled equations governing reactant-channel dynamics for which $I\leftarrow d$ transitions occur while all possible exchanges of photons one at a time with the laser field are accounted for during collision. Similarly, Eq. (3.66b) determines product-channel hp dynamics for N photons in the laser field, and an analogous complete set of coupled equations can be generated from it also.

In our calculations we restrict our attention to Process (3.64) where absorption of only a single laser photon occurs during collision, and no subsequent photon emission occurs. This is a very reasonable approximation, since during the time

$(10^{-13}$sec) of a typical collision two (or more) sequential
absorptions or absorption followed by emission are much less
likely than a single absorption. We incorporate this restriction by taking only the coupled equations generated from Eq.
(3.66) that link $F_{d,d}^{N(N-1),L}$ and $F_{I,d}^{N(N-1),L}$. That is, we couple
all hp wave functions corresponding to N (the initial number
of laser photons) and N-1 (the number of laser photons after
one has been absorbed. Furthermore, we neglect in this restricted set of coupled equations terms linking the N and N-1
hp wave functions to those for N+1 or N-2, corresponding to
photon emission or one additional photon absorption, respectively. The resulting simplified coupled equations can be
dynamically analyzed (see References 82 and 100) by assuming
ionization occurs in a Franck-Condon way, according to which
electrons are emitted with kinetic energies ε that to a good
approximation preserve the nuclear kinetic energies. This
dynamical analysis permits us to distinguish three continuum
regions important for FM collisional ionization. Namely, for
Process (3.65) electron emission governed primarily by Coulomb
coupling will occur in a narrow region near $\varepsilon_0 \approx 0.15$ E_h, just as
occurs under FF conditions; and electron emission governed
primarily by radiative coupling will occur in two other narrow
regions, one near $\varepsilon_+ \approx \varepsilon_0 + \hbar\omega$ and another near $\varepsilon_- \approx \varepsilon_0 - \hbar\omega$.

In Figures 20 and 21 we display the functions for $V_d(R)$ and
$V_+(R)$ from References 79 and 80 in a way illustrating the
three continuum regions near ε_+, ε_0 and ε_-. The ε_0 and ε_+
continuum regions are shown by Figure 20. Here $V_d(R)$ (corres-
ponding to a discrete electronic-field basis state comprised
of Φ_d and N laser photons) is degenerate near ε_0 with the
$V_+(R)+\varepsilon$ continuum (corresponding to the continuum electronic-
field basis states comprised of Φ_ε and N laser photons); and
$V_d(R)+\hbar\omega$ (corresponding to another discrete electronic-field
basis state comprised of Φ_d, one absorbed photon and N-1 laser
photons) is degenerate near ε_+ with the same $V_+(R)+\varepsilon$ continuum.
In Figure 21 the ε_0 and ε_- continuum regions are shown. Here
the same $V_d(R)$ is degenerate as in Figure 20 near ε_0 with the
$V_+(R)+\varepsilon$ continuum, and is also degenerate near ε_- with the
$V_++\hbar\omega+\varepsilon$ continuum (corresponding to other continuum electronic-
field basis states comprised of Φ_ε, one absorbed photon and
N-1 laser photons). We thus have a graphic picture in Figures
20 and 21 of the three important continuum regions. In addition to discrete electronic-field energies, we see as well
electronic-field continua corresponding to electronic-field
continuum basis states, and these continua share overlapping
regions.

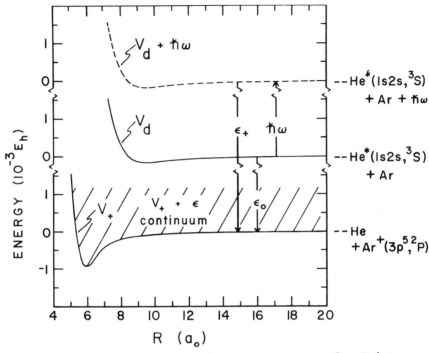

FIGURE 20. Relevant potential energy curves for He* + Ar + ℏω collisions.

A qualitative analysis of the R behavior of radiative coupling appears in Reference 81, and we suggested there that $D_{\varepsilon,d}(R)$ can be expected to be of much longer range than the Coulomb coupling. This conclusion is based on the different mechanisms responsible for the two couplings. Coulomb coupling depends on electron-electron interactions, and becomes appreciable only when regions of R are reached where overlap of the electronic charge clouds of the two colliding atoms also becomes appreciable. Radiative coupling, however, basically depends on the interaction of electronic transition dipole moments with the laser field, and these electronic transition dipole moments can exist at larger R where charge overlap between the atoms is still negligible.

The long range of $D_{\varepsilon,d}$ points to a possible complication in studying Process (3.65) both theoretically and experimentally. Namely, a clearcut investigation of Process (3.65) would be hindered if the radiative coupling persisted (with some non-negligible magnitude) out essentially to infinite separation of reactants and products. If this were the case, then, in addition to Process (3.65) which is a collisional means for

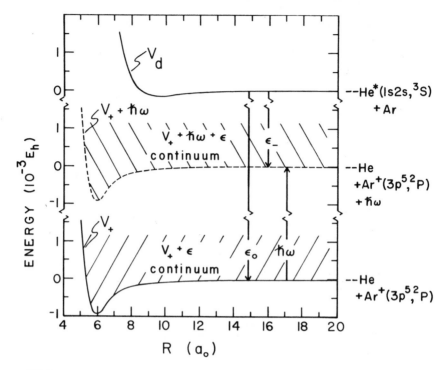

FIGURE 21. Relevant potential energy curves for He* + Ar + ℏω collisions. Here we see V_+ shifted up by ℏω, whereas V_d is shifted up by the ℏω in Figure 20.

letting the laser field couple bound and continuum electronic states), there would be competing non-collisional processes occurring by which direct interaction of the laser field with separated He*(1s2s,[3]S) and Ar atoms could produce He, Ar⁺ (3p⁵,[2]P) and free electrons without collision. We fortunately do not need to worry about such non-collisional processes competing with Process (3.65), because the dipole-coupling matrix elements for Process (3.65) vanish rigorously in the limit as R→∞, as we show in Reference 96.

We propose the following functional form to represent the radiative coupling:

$$D_{\varepsilon,d}(R) = (2/\pi)^{1/2}(2\varepsilon)^{-1/4}A_D\{1 + \exp[(R-R_0)/\Delta R]\}^{-1}.$$

$$(3.67)$$

As explained in Reference 96, we describe the ε dependence of the radiative coupling by the same emitted-electron density-of-states factor $(2/\pi)^{1/2}(2\varepsilon)^{-1/4}$ first introduced in the field-free work (79,80) for the Coulomb coupling. As can be seen, for each ε the radiative coupling of Eq. (3.67) exponentially approaches a saturated value of $(2/\pi)^{1/2}(2\varepsilon)^{-1/4}A_D$ for $R \ll R_0$ and exponentially approaches zero for $R \gg R_0$. So $D_{\varepsilon,d}(R)$ correctly goes to zero in the limit $R \to \infty$, as we discussed. The range of $D_{\varepsilon,d}(R)$ is determined by R_0, and the rate of the exponential approach to the saturated value for $R < R_0$, and to zero for $R > R_0$, is determined by ΔR. We choose R_0 to be $12a_0$, which is more than $2.5a_0$ beyond the potential minimum of V_d. Our choice for ΔR is $5a_0$, which represents a rather slow rate of exponential change in $D_{\varepsilon,d}(R)$. With these parameters Eq. (3.67) describes a fairly long-range radiative coupling that gradually changes with R. The magnitude of $D_{\varepsilon,d}$ is determined by A_D. In our calculations of Process (3.65) for a CO_2 laser we take $A_D=6.25 \times 10^{-5}$a.u., and for an iodine laser we take $A_D=15.678 \times 10^{-5}$a.u. As we discuss in Reference 96 these values of A_D correspond roughly to laser intensities of 1.6×10^8 W/cm^2 and 10^9 W/cm^2, respectively.

For a given initial hp collision energy E_i ($E_i=23.89 \times 10^{-4}E_h$ for the calculations reported here), the emitted-electron energy ε can be replaced as the continuum variable within any of the three continuum regions by the final (product-channel) collision energy E_f. The relationship between ε and E_f for each continuum region follows from conservation of total energy E (of heavy particles, electrons and laser field) asymptotically (i.e., as $R \to \infty$). For the ε_+ continuum region we find (see Reference 96),

$$E_f = E_i + \Delta E + \hbar\omega - \varepsilon. \qquad (3.68)$$

Here $\Delta E = V_d(\infty) - V_+(\infty)$ is the electronic difference between He*$(1s2s,^3S)$+Ar and He+Ar$^+$($3p^5,^2P$) ($\Delta E=0.149E_h$) (79,80). For the ε_- continuum region we find

$$E_f = E_i + \Delta E = \hbar\omega - \varepsilon, \qquad (3.69)$$

and for the ε_0 continuum region,

$$E_f = E_i + \Delta E - \varepsilon. \qquad (3.70)$$

In Figure 22 we display results of the FM Penning ionization cross sections $d\sigma/d\varepsilon$ calculated using our approach for Process (3.65) occurring with intense laser radiation of a CO_2 laser in one case, and of an iodine laser in another case. We plot $d\sigma/d\varepsilon$ against ε as described in References 79 and 80. A separate scale for each continuum region marks final collision energies (determined according to the preceding paragraph). The broad distribution in Figure 22 is the part of

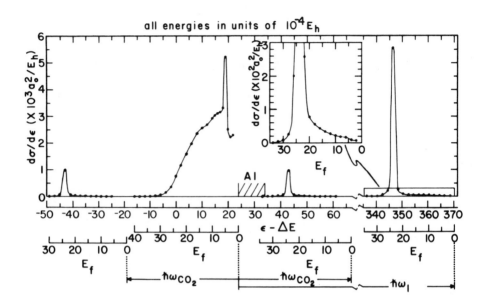

FIGURE 22.   Collisional ionization cross sections (per unit emitted-electron energy) for He* + Ar + ℏω collisions, where the CO_2 and iodine lasers are each used separately for ℏω.

the FM spectra in the ε_0 continuum region. The two small narrow distributions on either side of the broad distribution represent the calculations presented in our initial report (82) of these studies of FM ionizing collisions. These results describe Process (3.65) where a CO_2 laser of around 1.6×10^8 W/cm^2 is used. The large narrow distribution at the right of Figure 22 is the FM spectrum of the ε_+ region for the case of an iodine laser of intensity around 10^9 W/cm^2.

Whereas the laser-field intensity represented by the CO_2 laser calculations results in relatively weak FM contributions to the Penning ionization cross sections, the intensity represented by the iodine laser calculations is enough to bring the FM spectrum in the ε_+ region to a peak value exceeding that of the spectrum in the ε_0 region. The photon energy for the iodine laser is quite large compared to the scale of the ε axis in Figure 22, and a break in the axis has been made to conveniently include the iodine laser FM spectrum in the ε_+ region. In the insert we display on an expanded dσ/dε scale the portion of the iodine laser FM spectrum contained in the rectangular region as indicated. This expanded plot makes very evident the asymmetry of the FM spectrum with respect to its spectral peak.

The earlier field-free studies using discretization (79,80) and our present FM studies treat only Penning ionization, i.e., cases where $E_f>0$. This explains why the calculated values of $d\sigma/d\varepsilon$ for the broad distribution in Figure 22 do not extend to the right of $E_f=0$ (i.e., $\varepsilon-\Delta E\approx23\times10^{-4}E_h$). The portion of the ε_0 continuum region accessible for associative ionization (where $E_f<0$) can, however, be easily estimated. Referring to Figures 20 and 21 we see that V_+ has a well depth of less than $10\times10^{-4}E_h$. Therefore, E_f is restricted to values greater than roughly $-10\times10^{-4}E_h$ for any of the continuum regions of these calculations. The shaded area labeled AI in Figure 22 represents the associative ionization portion of the ε_0 spectral region. The figure shows that $\hbar\omega$ for the CO_2 laser is just within the bounds to keep the ε_+, ε_0 and ε_- continuum regions well separated. There is, of course, absolutely no question that the continuum regions for the iodine laser are well separated.

The calculated FM energy spectra show a comparatively enormous resonance cross section at $E_f\approx E_i$, whereas the spectrum in the ε_0 region sharply peaks at $E_f\approx5\times10^{-4}E_h < E_i$. The calculations thus confirm that FM collisional bound-continuum transitions (governed by radiative coupling) favor Penning ionization much more strongly than their field free counterparts governed by Coulomb coupling. Although our results depend on the radiative coupling, we feel our choice in Eq. (3.67) is physically reasonable enough to demonstrate the basic behavior of FM emitted-electron energy spectra.

We gain insight into the nature of the pronounced resonance peaks of the FM spectra from the behavior at fixed ε (or equivalently E_f) of the L-partial cross sections $d\sigma^{(L)}/d\varepsilon$ that contribute to $d\sigma/d\varepsilon$. Figure 23 shows plots of $d\sigma^{(L)}/d\varepsilon$ versus L from the iodine laser calculations for three neighboring $\Delta\varepsilon$ increments near the FM spectral peak and centered at E_f values of 22 (dashed line), 24 (solid line) and 26 (dotted line) $\times$ $10^{-4}E_h$, respectively. The first and third of these $\Delta\varepsilon$ increments fall on either side of the spectral peak, and the second is located at the spectral peak itself. For $L \lesssim 50$ the $d\sigma^{(L)}/d\varepsilon$ contributions for all three E_f values have similar behavior characterized by alternating maxima and minima. For $L \approx 50$, however, $d\sigma^{(L)}/d\varepsilon$ at the spectral peak shows a huge broad maximum that slowly decreases, while the $d\sigma^{(L)}/d\varepsilon$ on either side of the spectral peak quickly die out. Similar large cross-section contributions at the spectral peak that occur for $L \gtrsim 50$, and persist out beyond $L \approx 170$, also characterize the

CO_2-laser FM distribution (see Reference 82). They reflect
the radiative coupling's long-range behavior considering that
high L values correspond to large impact parameters and dis-
tances of closest approach. We can intuitively explain why at
high L large resonant $d\sigma^{(L)}/d\varepsilon$ result for the $\Delta\varepsilon$ increment at
the spectral peak, but negligible $d\sigma^{(L)}/d\varepsilon$ values result for
the other $\Delta\varepsilon$ increments. At the spectral peak where $E_f \simeq E_i$ we
find that the classical turning points for the effective po-
tentials, $V_d^{(L)}(R) = V_d(R) + L(L+1)/R^2$ and $V_+^{(L)}(R) = V_+(R) +$
$L(L+1)/R^2$, are nearly coincident for $L \gtrsim 50$, while on either
side of the spectral peak where $E_f \neq E_i$ we do not find this condition for the high-L classical turning points. We conclude,
therefore, that, when the reactant- and product-channel classical turning points coincide, the usual favorable conditions
for transition near a classical turning point are reinforced
enough to make the small but long-range radiative coupling
strengths effective in leading to such comparatively large

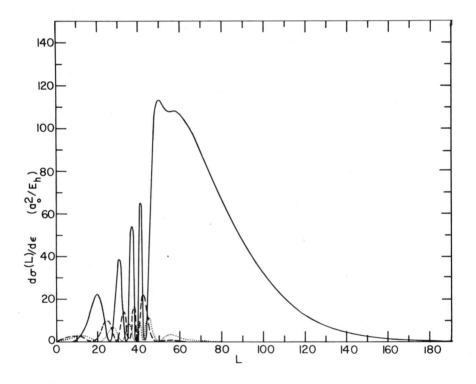

FIGURE 23. *L-partial contributions* $d\sigma^{(L)}/d\varepsilon$ *to the
iodine-laser field-modified emitted-electron energy spectrum
in the* ε_+ ($\approx \varepsilon_0 + \hbar\omega_I$) *continuum region.*

ionization cross-section contributions. Without this near coincidence of classical turning points at high L, the effectiveness of the radiative coupling's small but long-range strength is significantly reduced.

3. Photodissociation. Here we shall consider a molecule initially bound in the ground electronic state, and then later on a relatively intense laser field is switched on to excite the molecule to either an upper, repulsive electronic state or the vibrational continuum of the ground electronic state. The molecule then becomes unstable and flies apart into fragments. This kind of process involves a transition from a nuclear bound state to a nuclear continuum, and is therefore an example of bound-continuum processes. In Part a of this photodissociation subsection we shall first discuss unimolecular decay induced by single-photon absorption from an ultraviolet or visible laser field, and in Part b by multiphoton absorption from an infrared laser.

a. UV-visible. Utilizing the electronic-field representation we have developed a semiclassical theory of photodissociation (97). The advantages of this kind of treatment, except that of including intense field effects which we shall discuss later, are: i) the interaction between the molecule and the radiation field need not be weak and short as assumed in a perturbative type of treatment (98,99) and ii) one need not treat absorption and dissociation as two independent steps, but as a unified process. Due to the use of the electronic-field representation, a photodissociation problem actually becomes much like field-free (FF) unimolecular decay. There are several other treatments (most are quantum mechanical) which go beyond the perturbative theory, for example, References 14, 100-103. In the following paragraphs we shall briefly describe the theory we have developed and discuss some of its implications.

To begin with, we assume that the laser field is adiabatically switched on, as is usually the case when a molecular beam is crossed by a laser beam. From the initial distribution of FF states one can then find the initial distribution of the electronic-field states. One can propagate this distribution to a later time by using the time-evolution operator, or equivalently the Møller wave operator. For an experiment where final states of fragments are well characterized, one can then propagate a final state backward in time by a Møller wave operator and project the wave function onto the initial distribution. The expectation value thus obtained corresponds to the dissociation amplitude to that final state. This quantum description of photodissociation processes can be transformed

into a semiclassical theory by making classical-limit approximations to propagators and canonical transformations (104).
If we further define a critical surface as is often done in
transition state theories of chemical reactions, we can obtain
an expression for the dissociation rate into a specific final
state.

We shall use a one-dimensional example, as in a diatomic
molecule, to describe in words, instead of formulas, our semiclassical method as dictated by the expression we have derived.
The example we shall use is represented schematically in
Figure 24. In case a we show that a molecule, at t=-∞, is in

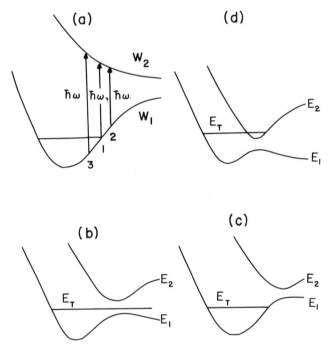

*FIGURE 24. One-dimensional example of direct photodissoci-
ation: (a) Two field-free adiabatic electronic surfaces W_1 and
W_2 are plotted as functions of the dissociation coordinate R.
$R_<$ and $R_>$ are the LHS and RHS turning points of a bound vibrational state on the lower surface W_1, respectively.   Arrows 1,
2, and 3 correspond, respectively, to the photon energies equal
to, smaller, and greater than the energy difference of W_2 and
W_1 at $R_>$. $R_< \leq R \leq R_>$ defines the internal region; (b) a pair
of electronic-field surfaces constructed from W_1 and W_2 with
the field frequency of case 1 shown in Figure 1(a); (c) same as
(b) with the frequency of case 2; (d) same as (b) with the frequency of case 3.*

a bound vibrational state of W_1. When a field is switched on, depending on the photon frequency we may be in one of the three cases of b, c or d of Figure 24. In case b the field frequency equals $\Delta W = W_2 - W_1$ at the outer turning point $R_>$. In cases c and d the laser frequency is less than and greater than ΔW, respectively. At t=0 the molecule can be at any position in the internal region, as defined by the region between the inner and outer turning points of W_1. Solving the problem semiclassically, one can start propagating trajectories from positions which are randomly selected in the internal region. The contribution from each trajectory is weighted by the absolute square of the vibrational wave function on W_1. On propagating outward, a trajectory may switch surfaces, be trapped by a well on E_2 or deflected by a potential barrier on E_1. Every event of these "incidents" (including not switching surfaces) contributes a damping factor to the dissociation probability to a certain final state.

Our method, which corresponds closely to physical intuition, gives a Golden Rule type expression for the dissociation rate in the weak-field limit. The Franck-Condon principle and angular distribution of dissociated fragments can also be easily visualized using the electronic-field representation. What is more important is that this theory predicts intense field effects which are amenable to experimental observation. One of these intense field effects is that the dissociation probability is in general exponentially dependent on the field intensity, in contrast to the linear dependence as predicted by a Golden Rule type expression. In an intense laser field the dissociation probability is an oscillatory function of laser frequency. The oscillatory behaviour is different from the Airy function behaviour derived from the Franck-Condon factor, but reduces to it in the weak-field limit.

b. IR. In its simplest form, IR photodissociation is a multiphoton process occurring on a single electronic potential energy surface. Through a series of photon absorptions the molecule is driven up the vibrational ladder until sufficient energy has been imparted to break the molecule apart. The excitation aspect of the overall process represents a familiar topic of study (105-107). However, transfer into the continuum, which is the defining step in photodissociation, is a less fully-examined topic. We therefore concentrate our attention upon it. Two goals suggest themselves: to model the dissociation process, and to see how dissociation acts to modify the dynamics of excitation (108).

Partitioning the wave function into discrete and continuum components permits us to focus on the discrete portion alone. The continuum part of the wave function does not appear in

explicit form; instead, we model its influence on the discrete
portion by a rigorously derived integral operator. We are left
with a nonlocal equation for the time dependence of the discrete level probability amplitudes. The nonlocality appears
as an integral over all past time. For preliminary purposes
we have simply replaced the integral by a local term. As our
research proceeds we expect to improve on this eyebrow-lifting
approximation.

The local equation has been used to treat the multiphoton
dissociation of two Morse oscillators. The first is chosen to
mimic $HCl(^1\Sigma^+)$, with well parameters drawn from Herzberg (109)
and the dipole moment function adapted from Kaiser's data (47).
As expected, this oscillator does not dissociate. The second
oscillator has a shallower, narrower well, so that it supports
only eleven bound states compared to twenty-four in the first
example. With the same dipole function as before we see dissociation when the field quantum is resonant with a low-lying
transition and the field intensity is large ($\approx 10^{14}$ W/cm^2).

In future applications we expect to apply a more accurate
form of the method to cases in which dissociation is more
likely to occur. To do so we shall need to incorporate
additional nuclear degrees of freedom in the system dynamics.
It seems evident that a quasicontinuous region in the upper
discrete level spectrum is necessary if we are to see significant probability for molecular dissociation in fields of
reasonable intensity (say, 10^9-10^{12} W/cm^2).

D. Heterogeneous Catalysis: Laser/Surface-Catalyzed Processes

Preceding sections of this article have dealt with the
influence of intense laser radiation on gas-phase molecular
rate processes. In this section we turn our attention to the
study of laser-induced rate processes occurring at solid
surfaces. Surface chemistry plays a dominant role in the field
of catalysis, and the combination of heterogeneous catalysis
with laser chemistry is therefore of great potential importance.

In order for a heterogeneous chemical reaction to proceed
at a solid surface, a number of fundamental rate processes must
occur (110):

1. Diffusion of reactants to the surface,
2. Adsorption of the reactants on the surface,
3. Reaction on the surface (including migration of reactants on the surface),
4. Desorption of products, and
5. Diffusion of products from the surface.

The diffusion steps, 1 and 5, are rarely rate determining for
gaseous/surface heterogeneous reactions, and we shall not

consider the effects of lasers on these processes in the discussion below. We should note, however, that the effects of lasers on these two steps might be of some importance in the case of reactions taking place at a liquid/surface interface.

In contrast to the vast number of theoretical investigations concerning the influence of laser radiation on gasphase chemical rate processes (29,105,106,111,112) comparatively little work has been devoted to an understanding of laser-induced chemistry at solid surfaces. However, experimental investigations of steps 2 (adsorption) (113), 3 (surface reaction) (114) and 4 (desorption) (115,116) have been undertaken by several groups. These studies have shown that laser radiation may substantially affect the rates of desorption of atoms from surfaces, as well as selectively promote chemical reactions. Further, they have presented evidence that these effects are not due to simple thermal heating on the surface. The work on desorption outlined in References 115 and 116 is of particular interest as these effects are observed at surprisingly low laser power, 1-10 W/cm^2, as compared to the power of 10^6 W/cm^2, or more, characteristic of References 105,106,111 and 112.

1. Desorption. We have initiated the development of theoretical models capable of describing laser-induced surface processes. We will begin our discussion of these models with a review of our quantum-theoretical treatment of laser-stimulated desorption (LSDE) (117). The model we will outline here is closely related to models which have been used in the past to study thermal (field-free) desorption from surfaces (118) as well as laser-stimulated dissociation of gas-phase molecules (119). The primary distinction from previous work on gas-phase molecules lies in the inclusion of surface-lattice dynamics. We will present some analytic results which should be of aid in the interpretation of experimental data on LSDE.

We will consider only motion of the adsorbed atom (adatom), A, perpendicular to the surface and follow Reference 118 in approximating the adatom-surface interaction by a truncated anharmonic potential having N+1 bound levels ($0 \leqslant n \leqslant N$) which are assumed to be the same as those of the corresponding untruncated oscillator. An adatom is assumed to have evaporated once it reaches level N+1. Then, we can write the Hamiltonian for an adatom bound at the surface, and coupled to both the acoustic modes of the surface lattice, as well as to an external coherent electromagnetic field in the form

$$H = \hbar(\omega_A - \varepsilon^* a^\dagger a) a^\dagger a + \sum_j \hbar\omega_j b_j^\dagger b_j + \hbar\sum_\ell (K_\ell b_\ell a^\dagger + K_\ell^* b_\ell^\dagger a)$$
$$+ h[V(t) a^\dagger + V^*(t) a] \tag{3.71}$$

where ε^* is the anharmonic correction as in work by Narducci (120).

The first two terms represent the unperturbed elastic energies of the adatom mode and surface-lattice modes, respectively. That is, $a^\dagger$(a) creates (annihilates) an optical phonon of frequency $\omega_A = (k/M_A)^{1/2}$ where k is the elastic force constant and M_A the mass of the adatom. Similarly, $b_j^\dagger$(b_j) creates (annihilates) an acoustic phonon of frequency ω_j. The third term represents the perturbation coupling A to the lattice, where K_ℓ may be given in terms of lattice and adatom mode variables in the form

$$\hbar K_\ell = \hbar K_\ell^* = -\hbar k e_{\ell,z} (4M_A M_S \omega_A \omega_\ell N)^{-1/2} . \tag{3.72}$$

Here, $e_{\ell,z}$ stands for the z component of the polarization vector of lattice mode ℓ, M_S is the mass of a surface atom and N is the number of acoustic-phonon modes. The last contribution to the Hamiltonian gives the coupling of the adatom vibrational mode to the external electromagnetic field. For example, for a sinusoidal driving field of the form $E\cos(\omega_L t)$

$$V(t) = \frac{\mu E}{2\hbar} e^{-i\omega_L t} = \frac{\omega_R}{2} e^{-i\omega_L t} , \tag{3.73}$$

where μ stands for the electric-dipole transition moment and ω_R for the Rabi frequency. Note that we have taken the rotating-wave approximation in both adatom-lattice and adatom-field interaction terms.

We will first consider the case in which the adatom-surface potential is assumed to be harmonic. That is, we set $\varepsilon^*=0$ in Eq. (3.71). The Hamiltonian is now that of a driven, damped harmonic oscillator, with the adatom vibrational mode playing the role of the oscillator, damped by the lattice modes and driven by the coherent electromagnetic field. Assuming that the surface-lattice, or "reservoir", variables obey Markoffian statistics, we may apply standard techniques of quantum-statistical mechanics (121) to find the thermal average (over lattice modes) of the mean number of quanta in the adatom vibrational mode at time t:

$$\langle n(t) \rangle = \langle a^\dagger(t) a(t) \rangle$$

$$= \langle n(t=0) \rangle \exp(-\gamma t) + \bar{n}[1-\exp(-\gamma t)]$$

$$+ i\langle a(t=0) \rangle \int_0^t dt' \, V^*(t') \exp\{\gamma(t'-t)-(i\omega_A+\gamma/2)t'\}$$

$$-i<a^{\dagger}(t=0)> \int_0^t dt'V(t')\exp\{\gamma(t'-t)+(i\omega_A-\gamma/2)t'\}$$

$$+\int_0^t dt'\ \exp(\gamma(t'-t))\times[V(t')g^*(t')+V^*(t')g(t')].$$

(3.74)

Here, $g(t) = \int_0^t dt'\ V(t')\exp\{[i\omega_A+\gamma/2](t'-t)\}$, (3.75)

$$\bar{n} = \frac{1}{\exp(\hbar\omega_A/kT)-1}$$

(3.76)

$$\gamma = 2\pi g(\omega_A)\ |K(\omega_A)|^2 ,$$

(3.77)

where $g(\omega_A)$ stands for the density of lattice modes, and both g and the coupling constant K are evaluated at the frequency of the adatom oscillator. Also, ω_A now stands for the observed (renormalized) vibrational mode frequency, and T for the surface temperature.

The first two terms in Eq. (3.74) describe the approach to thermal equilibrium in the absence of the electromagnetic field. If we assume that the oscillator occupies the ground state at t=0, and that the surface temperature is low enough that thermal desorption effects may be ignored (that is, $\bar{n} \ll N+1$), and, finally, that V(t) is of the form given by Eq. (3.73), then Eq. (3.74) takes the simple form

$$<n(t)> = \frac{(\omega_R/2)^2}{(\omega_A-\omega_L)^2+(\gamma/2)^2}$$

$$\times\ \{1+\exp(-\gamma t)-2\exp(-\gamma t/2)\cos(\omega_L-\omega_A)t\}.$$

(3.78)

We may note that if the damping is set equal to zero, we recover the usual result for the undamped oscillator (119). The steady state value of <n> is just

$$<n> = \frac{(\omega_R/2)^2}{(\omega_A-\omega_L)^2+(\gamma/2)^2}$$

(3.79)

or at resonance,

$$\langle n \rangle = (\omega_R/\gamma)^2 , \tag{3.80}$$

determined by the Rabi frequency and damping constant γ alone.
For laser-stimulated desorption to occur, we must have

$$(\omega_R/\gamma)^2 \geqslant N+1 . \tag{3.81}$$

If we assume that surface-lattice properties are identical
with those of the bulk-lattice and adopt a Debye model (122)
for the frequency spectrum $g(\omega)$, then

$$(\omega_R/\gamma)^2 = (2/9\pi e_{\ell,z}^2)^2 (M_S/M_A)^2 \frac{\omega_R^2 \omega_D^6}{\omega_A^8} \tag{3.82}$$

where ω_D stands for the Debye frequency. Thus, setting
$e_{\ell,z} = 1$, $M_A \simeq M_S$, $\omega_A = .5\times10^{13}$ Hz, $\mu = 1$ (in atomic mass
units), the laser power at 10 W/cm^2 and requiring the excita-
tion of the oscillator, N+1, to be of order 10, we obtain an
order of magnitude estimate for ω_D lying between 10^{14} and
10^{15} Hz (ω_D is on the order of , for typical solids, 10^{14} Hz),
to account for the observations of Djidjoev et al. The
sensitive dependence of the width γ on surface-lattice
dynamics indicates the desirability of incorporating a more
realistic treatment of these dynamics in future developments
of this model.

Many of the approximations made in the model outlined
above are common to the models of References 118 and 119, and
we refer the reader to these sources for a discussion of their
range of validity. Two points warrant further discussion.
First, we have assumed that the value of the electric field
amplitude at a given site on the surface is known. In general,
this value will be site dependent and require the solution of
a complicated boundary value problem for its determination
(123). Second, we have ignored the coupling of the electric
field to the lattice vibrations, either directly or indirectly
(through such mechanisms as photon-conduction electron,
conduction electron-phonon coupling). For some surfaces, such
as metals, which possess a strong absorption characteristic in
the infrared, these effects may not be negligible.

So far we have treated the adatom-surface interaction in
the harmonic approximation. Now we will consider the anharmonic effects. For a Morse potential the anharmonicity ε^*
is related to the natural frequency of the adatom vibration

ω_A, and the binding energy D by $\epsilon^* = \hbar\omega_A^2/4D$. When $\epsilon^* \neq 0$, instead of the solution of Eq. (3.74), the average excitation <n> can be obtained by solving the more general set of equations

$$d<a>/dt = -iV(t) - i[i\omega_e(t)+\gamma/2]<a>, \qquad (3.83)$$

$$d<n>/dt = -iV(t)(<a>-<a^\dagger>) - \gamma(<n>-\bar{n}). \qquad (3.84)$$

Here < > denotes the reservoir thermal average under the Markoff approximation (121). In deriving Eqs. (3.83) and (3.84) we have used the factorization approximation (120). The reservoir-induced damping constant γ and the thermal average reservoir quantum number $\bar{n}$ are given by the same forms as in Eqs. (3.76) and (3.77), except that the constant frequency ω_A is replaced by the time-dependent effective frequency $\omega_e(t)$ defined by

$$\omega_e(t) = \omega_A - 2\epsilon^*(<n(t)> + 1/2). \qquad (3.85)$$

If the anharmonicity shift, $2\epsilon^*<n>$, is small the coupled Eqs. (3.83) and (3.84) may be solved by means of an iterative procedure whose details are presented elsewhere (124). Some numerical results are presented in Figure 25. In our calculations, we have again taken the low temperature limit (setting $\bar{n} = 0$) and have assumed that the initial state of the adatom vibrational mode is the ground state.

Figure 25 shows the average excitation <n> (or the average number of photons absorbed by the adatom) as a function of the laser power for several values of the damping factor γ and the anharmonicity ϵ^*. For low excitations, <n> is linearly proportional to the laser power I as in the case of the harmonic potential ($\epsilon^* = 0$). For high excitations, the absorption curves are bent and are saturated to the relation $<n> \propto I^{1/3}$. The saturation results from detuning from resonance. That is, as the local vibrational mode becomes highly excited, the effective frequency (due to anharmonicity) becomes strongly red-shifted away from resonance with the laser frequency. The dashed lines in Figure 25 show that the average excitation is inversely proportional to the square of the damping factor γ, which is characterized by surface-lattice dynamics. When the coupling is large, the damping factor γ is also large and a higher laser power is required for the desorption process than that for the weak-coupling case in which the adatom has a large lifetime.

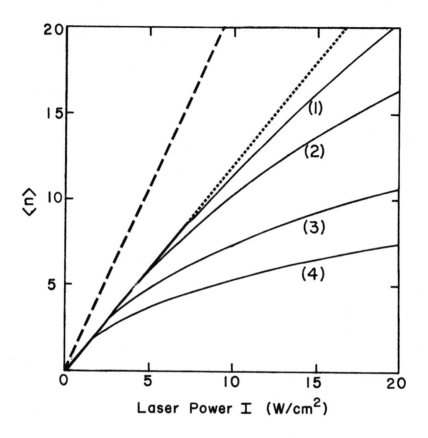

FIGURE 25. Laser power dependence of the average excitation ⟨n⟩ of the adatom. Dashed lines show the linear dependence of ⟨n⟩ on I when ε* = 0 for γ = 1.5×10⁸ sec⁻¹ (----), and γ = 2.0×10⁸ sec⁻¹ (·····). Solid lines show the nonlinear effects of ⟨n⟩ for γ = 2.0×10⁸ sec⁻¹ and (1) ε* = 10⁶ sec⁻¹, (2) ε* = 2×10⁶ sec⁻¹, (3) ε* = 5×10⁶ sec⁻¹, and (4) ε* = 10⁷ sec⁻¹.

Potential areas of application of LSDE include:
1. Selective enhancement of desorption-limited chemical reactions,
2. Removal of undesirable impurities or "poisons" from surfaces (see Reference 125 for an example of a "brute force", that is, thermal and non-resonant, application of LSDE),
3. Study of surface-lattice dynamics (which LSDE is strongly dependent upon through the damping constant γ), and

4. Study of the adatom-surface bond.

2. Migration. A logical extension of our desorption model allows us to study the effects of laser radiation on the migration of adatoms over the surface. We simply add terms to the Hamiltonian of Eq. (3.71) which allow for the transport of the adatom among the various sites on the surface. That is, our new Hamiltonian has the form,

$$K = H + T \tag{3.86}$$

where

$$T = E_o \sum_n c_n^\dagger c_n + \sum_{n \neq m} \phi_{nm} c_n^\dagger c_m (a^\dagger + a)$$

$$+ \hbar \sum_{n,\ell} \omega_\ell c_n^\dagger c_n (X_\ell^n b_\ell + X_\ell^{n*} b_\ell^\dagger). \tag{3.87}$$

Here, the operators $c_n^\dagger$, c_n create or annihilate an adatom at the lattice site n on the surface, and they can be written in the form $c_n^\dagger c_m = |n><m|$, where $|m>$ is the site, or "Wannier," representation of the wavefunction of the motion of the adatom parallel to the surface in the periodic potential of the lattice. The second term is the "key" term, representing vibration-induced inter-site transitions. The basic idea is that excitation of the local vibrational mode of the adatom via the coupling to the electromagnetic field leads to an enhancement of the rate of such intersite "hopping". The third term is an addition to the energy E_O of the adsorbed atom due to the fact that the energy is changed by lattice vibrations. The coupling constants ϕ_{nm} and X_ℓ^n are introduced in a phenomenological fashion here, but could in principle be calculated if a potential of interaction between adatom and lattice atoms is chosen and if one solves the Schroedinger equation for the adatom located at an equilibrium distance from the surface and moving parallel to the surface with the lattice atoms fixed in equilibrium positions (see Reference 125 for a more detailed discussion of this and other points relating to the field-free model).

A set of coupled, quantum-mechanical Langevin equations for the operators $c_n^\dagger c_n$ (giving information on migration) and $a^\dagger a$ (describing desorption) may be derived starting from the microscopic model given by the Hamiltonian K. These equations of motion are currently under investigation in our laboratory.

Laser-stimulated migration of adatoms would:
1. Facilitate studies of diffusion mechanisms, by reducing

complications due to heating of the substrate,
    2. Allow for enhancement of surface diffusion limited reactions, by increasing the rates of encounter of reactants,
    3. Allow for preferential enhancement of the mobilities of selected species in a multicomponent environment,
    4. Allow the study of the potential of interaction between adatom and lattice atoms.

    3. New Directions. We conclude this section by briefly mentioning several lines of research undertaken very recently in our laboratory.

    First, the single-frequency laser stimulated surface desorption process is being extended to a multiple-frequency selective desorption process in which a set of very low intensity lasers at frequencies $\omega_k = \omega + k(\omega_A - \omega)/N$, k=1,2...N are used. The rate of energy absorption of the adatom stimulated by a set of N-frequency lasers will be N^2 times those of the single-frequency stimulation case. Second, the problem of atom-surface scattering in the presence of intense laser radiation is also being attacked by using both quantal and classical models. In our initial model, the laser is assumed to excite one of the local vibrational modes of the surface-lattice. In the third line of research we attempt to deal with the process in which adsorbed atoms are ionized under the concerted actions of projectile atoms (molecules) and laser radiation incident on the surface. This problem is a natural extension of our work on gas-phase FA and FM collision-induced ionization, and will be treated using the Green's function approach to quasi-free scattering processes within the impulse approximation.

REFERENCES

1. Macomber, J.D., "The Dynamics of Spectroscopic Transitions" (Wiley, New York, 1976). An excellent condensed history of the study of transient phenomena is given in Chapter 1.

2. Loudon, R., "The Quantum Theory of Light" (Clarendon, Oxford, 1973).

3. Louisell, W. H., "Quantum Statistical Properties of Radiation" (Wiley, New York, 1973).

4. Sargent, M., III, Scully, M. O., and Lamb, W. E., "Laser Physics" (Addison-Wesley, Reading, Massachusetts, 1974).

5. Allen, L., and Eberly, J. H., "Optical Resonance and Two-Level Atoms" (Wiley, New York, 1975).
6. Bialynicki-Birula, I., and Bialynicka-Birula, Z., *Phys. Rev. A 14*, 1101 (1976).
7. Glauber, R. J., *Phys. Rev. 131*, 2766 (1963).
8. Miller, W. H., private communication.
9. Zimmerman, I. H., Yuan, J. M., and George, T. F., *J. Chem. Phys. 66*, 2638 (1977).
10. (a) Yuan, J. M., George, T. F., and McLafferty, F. J., *Chem. Phys. Lett. 40*, 163 (1976); (b) Yuan, J. M., Laing, J. R., and George, T. F., *J. Chem. Phys. 66*, 1107 (1977).
11. Gudzenko, L. I., and Yakovlenko, S. I., *Sov. Phys. JETP 35*, 877 (1972); Perel'man, N. F., and Kovaskii, V. A., *Sov. Phys. JETP 36*, 436 (1973); Mikhailov, A. A., *Opt. Spectr. 34*, 581 (1973); Lisitsa,V. S., and Yakovlenko, S. I., *Sov. Phys. JETP 39*, 759 (1974); Varfolomeev, A. A., *Sov. Phys. JETP 39*, 985 (1974); Vitlina, R. Z., Chaplik, A. V., and Entin, M. V., *Sov. Phys. JETP 40*, 829 (1975); Gudzenko, L. I., and Yakovlenko, S. I., *Sov. Phys. Tech. Phys. 20*, 150 (1975); Vetchinkin, S. I., Bakhrakh, V. L., and Umanskii, I. M., *Opt. Spectrosc. 40*, 28 (1976).
12. Kroll, N. M., and Watson, K. M., *Phys. Rev. A 8*, 804 (1973); ibid., *A 13*, 1018 (1976).
13. Gersten, J. I., and Mittleman, M. H., *J. Phys. B 9*, 383 (1976).
14. Lau, A. M. F., *Phys. Rev. A 13*, 139 (1976); ibid., *A 16*, 1535 (1977); Lau, A. M. F., and Rhodes, C. K., *Phys. Rev. A 15*, 1570 (1977); ibid., *A 16*, 2392 (1977).
15. Berman, P. R., *Appl. Phys. 6*, 283 (1975).
16. Copeland, D. A., and Tang, C. L., *J. Chem. Phys. 65*, 3161 (1976).
17. Geltman, S., *J. Phys. B 9*, L559 (1976).
18. Jortner, J. and Ben-Reuven, A., *Chem. Phys. Lett. 41*, 401 (1976).
19. Milonni, P. W., *J. Chem. Phys. 66*, 3715 (1977).
20. Knight, P. L., *Phys. Lett. 61A*, 22 (1977).
21. Payne, M. G., Choi, C. W. and Nayfeh, M. H., in "Abstracts from the International Conference on Multiphoton Processes" (University of Rochester, 1977), pp. 190-191.
22. Bowden, C. M., Stettler, J. D., and Witriol, N. M., *J. Phys. B 10*, 1789 (1977).
23. Light, J. and Szöke, A., *Phys. Rev. A*, in press.
24. Falcone, R. W., Green, W. R., White, J. C. Young, J. F., and Harris, S. E., *Phys. Rev. A 15*, 1333 (1977).
25. Collins, C. B., Johnson, B. W., Mirza, M. Y., Popescu, D., and Popescu, I., *Phys. Rev. A 10*, 813 (1974).

26. Weingartshofer, A., Holmes, J. K., Caudle, G. and Clarke, E. M., *Phys. Rev. Lett. 39,* 269 (1977).

27. Hellfeld, A. V., Caddick, J., and Weiner, J., *Phys. Rev. Lett. 40,* 1369 (1978).

28. Fedorov, M. V., Kudrevatova, O. V., Makarov, V. P., and Samokhin, A. A., *Opt. Commun. 13,* 299 (1975).

29. George, T. F., Zimmerman, I. H., Yuan, J. M., Laing, J. R., and DeVries, P. L., *Acc. Chem. Res. 10,* 449 (1977).

30. Geltman, S., *J. Phys. B 10,* 3057 (1977).

31. Nayfeh, M. H., *Phys. Rev. A 16,* 927 (1977).

32. Cody, R. J., Sabety-Dzvonik, M. J. and Jackson, W. M., *J. Chem. Phys. 66,* 2145 (1977).

33. Callender, R. H., Gersten, J. I., Leigh, R. W., and Yang, J. L., *Phys. Rev. A 14,* 1672 (1976).

34. Chin, S. L., *Phys. Lett. 61A,* 311 (1977).

35. Moore, C. B., ed., "Chemical and Biochemical Applications of Lasers," Vol. III (Academic Press, New York, 1977).

36. Miller, W. H., and George, T. F., *J. Chem. Phys. 56,* 5637 (1972).

37. Power, E. A., and Zienau, S., *Phil. Trans. Roy. Soc. A 251,* 427 (1959).

38. Bender, C. F., and Davidson, E. R., *J. Chem. Phys. 49,* 4989 (1968).

39. Zimmerman, I. H., and George, T. F., *Chem. Phys. 7,* 323 (1975).

40. Zimmerman, I. H., and George, T. F., *J. Chem. Phys. 63,* 2109 (1975).

41. Baer, M., *Chem. Phys. 15,* 49 (1976).

42. Laing, J. R., George, T. F., Zimmerman, I. H., and Lin, Y.-W., *J. Chem. Phys. 63,* 842 (1975).

43. Lam, K. S., Zimmerman, I. H., Yuan, J. M., Laing, J. R., and George, T. F., *Chem. Phys. 26,* 455 (1977).

44. Feyman, R. P., and Hibbs, A. R., "Quantum Mechanics and Path Integrals" (McGraw-Hill, New York, 1965).

45. Pechukas, P., *Phys. Rev. 181,* 174 (1969).

46. Zimmerman, I.H. and George, T.F., unpublished.

47. Kaiser, E. W., *J. Chem. Phys. 53,* 1686 (1970).

48. DeVries, P. L. and George, T. F., *Mol. Phys. 36,* 151 (1978).

49. DeVries, P. L., Mahlab, M. S., and George, T. F., *Phys. Rev. A 17,* 546 (1978).

50. Dunning, T. H., Jr., and Hay, P. J., *J. Chem. Phys. 69,* 134 (1978) and private communication with Dunning, T. H., Jr.

51. DeVries, P. L., and George, T. F., *Phys. Rev. A 18,* October (1978).

52. DeVries, P. L., and George, T. F., *Mol. Phys.,* in press.

53. Marx, B. R., Simons, J., and Allen, L., *J. Phys. B 11*, L273 (1978); McClean, W. A., and Swain, S., *J. Phys. B 11*, 1717 (1978); Agustini, P., Georges, A. T., Wheatley, S. E., Lambroupoulos, P., and Levenson, M. D., *J. Phys. B 11*, 1733 (1978).

54. Mollow, B. R., *Phys. Rev. 175*, 1555 (1968); Lecompte, C., Mainfray, G., Manus, C. and Sanchez, F., *Phys. Rev. Lett. 32*, 265 (1974); ibid., *Phys. Rev. A 11*, 1009 (1975); Arslanbekov, T. U., *Sov. J. Quantum Electron. 6*, 117 (1976); Smirnova, T. N., and Tikhonov, E. A., *Sov. J. Quantum Electron. 7*, 621 (1977).

55. Lompre, L. A., Mainfray, G., Manus, C., Repoux, S., and Thebault, J., *Phys. Rev. Lett. 36*, 949 (1976); Lompre, L. A., Mainfray, G., Manus, C., and Thebault, J., *Phys. Rev. A 15*, 1604 (1977).

56. Lee, H. W., DeVries, P. L., and George, T. F., *J. Chem. Phys. 69*, Sept. 15 (1978).

57. Lee, H. W., DeVries, P. L., Zimmerman, I. H., George, T. F., *Mol. Phys.*, in press.

58. Lee, H. W., and George, T. F., *J. Phys. Chem.*, submitted.

59. Parker, J. H., and Pimentel, G. C., *J. Chem. Phys. 51*, 91 (1969).

60. Muckerman, J. T., *J. Chem. Phys. 57*, 3388 (1972).

61. Schatz, G. C., Bowman, J. M., and Kuppermann, A., *J. Chem. Phys. 63*, 674 (1975); Redmon, M. J., and Wyatt, R. E., *Int. J. Quant. Chem. S 9*, 403 (1975); Connor, J. N. L, Jakubetz, W., and Manz, J., *Mol. Phys. 29*, 347 (1975).

62. Komornicki, A., Morokuma, K., and George, T. F., *J. Chem. Phys. 67*, 5012 (1977).

63. Yuan, J. M., and George, T. F., *J. Chem. Phys.*, submitted.

64. Bendazzoli, G. L., Raimondi, M., Garetz, B. A., George, T. F., and Morokuma, K., *Theoret. Chim. Acta. (Berl.) 44*, 341 (1977).

65. Komornicki, A., George, T. F., and Morokuma, K., *J. Chem. Phys., 65*, 48 (1976).

66. Jaffe, R. L., Morokuma, K., and George, T. F., *J. Chem. Phys. 63*, 3417 (1975).

67. Tully, J. C., and Preston, R. K., *J. Chem. Phys. 55*, 562 (1971).

68. Stine, J. R., and Muckerman, J. T., *J. Chem. Phys. 65*, 3975 (1976).

69. Heitler, W., "The Quantum Theory of Radiation," 3rd ed. (Oxford University, London, 1970), § 16.

70. Newns, D. M., *Phys. Rev. 178*, 1123 (1969).

71. Gadzuk, J. W., *Surface Sci. 43*, 44 (1974).

72. Miller, W. H., *J. Chem. Phys. 52*, 3563 (1970).

73. Miller, W. H., and Morgner, H., *J. Chem. Phys.* 67, 4923 (1977).

74. Lam, K. S., Bellum, J. C., and George, T. F., *Chem. Phys.*, in press.

75. Nakamura, H., *J. Phys. Soc. Japan* 26, 1473 (1969).

76. Hickman, A. P., and Morgner, H., *J. Phys. B 9*, 1765 (1976).

77. Hickman, A. P., Issacson, A. D., and Miller, W. H., *J. Chem. Phys.* 66, 1483 (1977); ibid., 66, 1492 (1977).

78. Bieniek, R. J., *Phys. Rev. A*, in press.

79. Bellum, J. C., and Micha, D. A., *Chem. Phys.* 20, 121 (1977).

80. Bellum, J. C., and Micha, D. A., *Phys. Rev. A*, in press.

81. Bellum, J. C., and George, T. F., *J. Chem. Phys.* 68, 134 (1978).

82. Bellum, J. C., Lam, K. S., and George, T. F., *J. Chem. Phys.* 69, 1781 (1978). .

83. Demkov, Y. N., *Sov. Phys. Doklady* 11, 138 (1966).

84. Demkov, Y. N., and Osherov, V. I., *Sov. Phys. JETP 26*, 916 (1968).

85. Boffi, S., in "Nuclear Optical Model Potential," ed. by Boffi, S., and Passatore, G., (Springer-Verlag, New York, (1976), pp. 44-67.

86. Gross, D. H. E., and Lipperheide, R., *Nuclear Phys. A 150*, 449 (1970).

87. Wille, V., and Lipperheide, R., *Nuclear Phys. A 189*, 113 (1972).

88. Nakamura, H., Shirai, T., and Nakai, Y., *Phys. Rev. A 17*, 1892 (1978).

89. For reviews of collisional ionization in the absence of laser radiation, see: (a) Muschlitz, E. E., Jr., *Adv. Chem. Phys. 10*, 171 (1966); (b) Berry, R. S., in "Molecular Beams and Reaction Kinetics," ed. by Ch. Schlier (Academic Press, New York, 1970), p. 193 ff.; (c) Rundel, R. D., and Stebbings, R. F., in "Case Studies in Atomic Collision Physics," Vol. 2, ed. by McDaniel, E. W., and McDowell, M. R. C., (North-Holland, Amsterdam, 1972), p. 549 ff.; (d) Niehaus, A., *Ber. Bunsenges. Phys. Chem.* 77, 632 (1973).

90. Hurst, G. S., Nayfeh, M. H., and Young, J. P., *Appl. Phys. Lett. 30*, 229 (1977); ibid., *Phys. Rev. A 15*, 2283 (1977).

91. Nayfeh, M. H., Hurst, G. S., Payne, M. G., and Young, J. P., *Phys. Rev. Lett. 39*, 604 (1977).

92. Lucatorto, T. B., and McIlrath, T. J., *Phys. Rev. Lett. 37*, 428 (1976).

93. McIlrath, T. J., and Lucatorto, T. B., *Phys. Rev. Lett. 38*, 1390 (1977).

94. Winn, J. S., private communication.

95. Nayfeh, M. H. and Payne, M. G., *Phys. Rev. A 17*, 1695 (1978).
96. Bellum, J. C., and George, T. F., *Phys. Rev. A*, submitted.
97. Yuan, J. M., and George, T. F., *J. Chem. Phys. 68*, 3040 (1978).
98. Dunn, G. H., *Phys. Rev. 172*, 1 (1965).
99. Holdy, K. E., Klotz, L. C., and Wilson, K. R., *J. Chem. Phys. 52*, 4588 (1970).
100. Shapiro, M., *Isr. J. Chem. 11*, 691 (1973); ibid., *J. Chem. Phys. 56*, 2582 (1972).
101. Band, Y. B., and Freed, K. F., *J. Chem. Phys. 63*, 3382, 4479 (1975).
102. Heller, E., *J. Chem. Phys. 68*, 2066, 3891 (1978).
103. Zimmerman, I. H., Yuan, J. M., and George, T. F., *Mol. Phys.*, in press.
104. Miller, W. H., *Adv. Chem. Phys. 25*, 69 (1974).
105. "Abstracts from the International Conference on Multiphoton Processes" (University of Rochester, 1977).
106. Ambartzumian, R. V., and Letokhov, V. S., in Reference 35, p. 167 ff.
107. Mukamel, S. and Jortner, J., *Chem. Phys. Lett. 40*, 150 (1976); Black, J. G., Yablonovitch, E., Bloembergen, N., and Mukamel, S., *Phys. Rev. Lett. 38*, 1131 (1977); Cantrell, C. D., and Galbraith, H. W., *Opt. Comm. 18*, 513 (1976).
108. Zimmerman, I. H., Druger, S. D., and George, T. F., *Chem. Phys. Lett.*, submitted.
109. Herzberg, G., "Spectra of Diatomic Molecules" (Van Nostrand Reinhold, 1950) p. 534.
110. Castellan, G. W., "Physical Chemistry" (Addison-Wesley, Reading, Massachusetts, 1964), p. 649.
111. Stone, J., Goodman, M. F., and Dows, D. A., *Chem. Phys. Lett. 44*, 411 (1976).
112. Bloembergen, N. and Yablonovitch, E., *Physics Today 31*, 23 (1978).
113. Kochelashvilli, K. S., Karlov, N. V., Orlov, A. N., Petrov, R. P., Petrov, Yu. N., and Prokhorov, A. M., *JETP Lett. 21*, 302 (1975).
114. Khmelev, A. V., Apollonov, V. V., Borman, V. D., Nikolaev, B. I., Sazykin, A. A., Troyan, V. I., Firsov, K. N., and Frolov, B. A., *Sov. J. Quantum Electron. 7*, 1302 (1977).
115. Djidjoev, M. S., Kohkhlov, R. V., Kiselev, A. V., Lygin, V. I., Namiot, V. A., Osipov, A. I., Panchenko, V. I., and Provotorov, B. I., in "Tunable Lasers and Applications," ed. by Mooradian, A., Jaeger, T., and Stokseth, P., (Springer-Verlag, Berlin, 1976), p. 100 ff.

116. Karlov, N. V., Petrov, R. P., Petrov, Yu. N., and Prokhorov, A. M., *JETP Lett. 24,* 258 (1976).

117. Slutsky, M. S., and George, T. F., *Chem. Phys. Lett. 57,* 474 (1978).

118. Garrison, B. J., Diestler, D. J., and Adelman, S. A., *J. Chem. Phys. 67,* 4317 (1977).

119. Aldridge, J. P., Birely, J. H., Cantrell, C. D., and Cartwright, D. C., in "Physics of Quantum Electronics," Vol. 4, ed. by Jacobs, S. F., Sargent, M., III, Scully, M. O., and Walker, C. T., (Addison-Wesley, Reading, Massachusetts, 1976) p. 57 ff.

120. Narducci, L. M., Mitra, S. S., Shatos, R. A., and Coulter, C. A., *Phys. Rev. A 16,* 247 (1977).

121. See Reference 3, pp. 344-47, 366-67.

122. Reif, F., "Fundamentals of Statistical and Thermal Physics," (McGraw-Hill, New York, 1965), pp. 411-18.

123. Metiu, H., *J. Chem. Phys.,* submitted.

124. Lin, J. T., Slutsky, M. S., and George, T. F., manuscript in preparation.

125. Schwirzke, F., Oren, Lena, Talmadge, S., and Taylor, R. J., *Phys. Rev. Lett. 40,* 1181 (1978).

126. Kitahara, K., Metiu, H., Ross, J., and Sibley, Robert, *J. Chem. Phys. 65,* 2871 (1976).

MULTIPHOTON DISSOCIATION OF GAS PHASE IONS USING
LOW INTENSITY CW LASER RADIATION[1]

R. L. Woodin[2]
D. S. Bomse[3]
J. L. Beauchamp

Division of Chemistry and Chemical Engineering
California Institute of Technology
Pasadena, California

I. INTRODUCTION

Multiphoton excitation of molecules by infrared radiation
offers a form of molecular activation which has captured the
imagination of a sizable audience of chemists and physicists.
Extensive research in laboratories around the world has been
spurred on by the promise of chemically and isotopically
selective infrared laser induced processes . Indeed, work has
progressed so rapidly in this field that although an extensive
review of infrared multiphoton photochemistry was presented in
Volume III of this series (1), key developments during the
intervening year warrant a new look at the subject. In particular, it has been demonstrated in our laboratory that
multiphoton excitation can be achieved readily using low power
laser sources (several watts) (2-5). Thus photodissociation
experiments need not be limited to excitation by megawatt,
pulsed lasers.
Lee and co-workers have clearly demonstrated that molecules
in the collision free environment of supersonic molecular
beams readily undergo multiphoton absorption and decomposition

[1]*This work was supported by the United States Department
of Energy.*
[2]*Present address: Exxon Research and Engineering Company,
Corporate Research Laboratories, Linden, New Jersey.*
[3]*NSF Pre-Doctoral Fellow, 1976-1979.*

unaided by bimolecular interactions (6-8). From Yablonovitch's observation (9) that the level of multiphoton excitation depends on energy fluence rather than power one infers multiphoton excitation can be effected using low power sources if the target molecules are irradiated under collisionless conditions for times approaching several seconds. An obvious requirement for the success of such an experiment is that net absorption rates must exceed spontaneous emission rates. The techniques of ion cyclotron resonance spectroscopy (ICR) (10-18) are uniquely suited for studying slow multiphoton dissociation of gas phase ions. In ICR experiments ions may be trapped and irradiated under nearly collisionless conditions for times up to several seconds. The ability to detect and mass analyze reactants and products at any instant of time during the trapping and irradiation cycle provides directly information regarding reaction pathways, isotopic selectivity and dissociation rates. Furthermore, buffer gases can be added to allow up to 10^3 collisions during the storage period, thus probing the effect of bimolecular interactions on multiphoton dissociation.

In this chapter we briefly review the experimental methodology and chemical applications of ion cyclotron resonance spectroscopy. With this introduction, recent studies of multiphoton dissociation of ions using low power cw infrared laser radiation are reviewed in considerable detail.

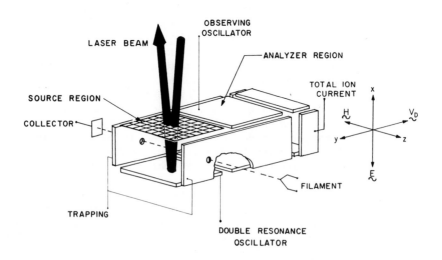

FIGURE 1. Cutaway view of cyclotron resonance cell. The electron beam is collinear with the magnetic field. Laser beam positioning is indicated for low intensity IR laser photodissociation studies (described in Section IIC).

II. ION CYCLOTRON RESONANCE SPECTROSCOPY

Ion trapping techniques permit gas phase ions to be confined in a spatially well defined region for extended and accurately controlled periods of time (11). As a result, these techniques have applications covering a wide range of phenomena associated with ion chemistry (10). Ion cyclotron resonance studies have, in the past, been concerned mainly with ion-molecule reaction processes. Recently, however, ICR techniques have found use in gas phase ion photochemical studies (12-18).

A. *Experimental Methodology*

In ICR experiments, ions formed by electron impact ionization undergo cyclotron motion in a strong magnetic field with frequency ω, given in cgs units by equation (1),

$$\omega = \frac{qH}{mc} \tag{1}$$

where q is the charge of the ion, H the magnetic field strength, m the ion mass and c the speed of light. A sketch of the ICR cell used in our laboratory is shown in Figure 1. Cyclotron motion contains the ions in the x-y plane while appropriate potentials on the trapping plates restrict ion motion in the z direction. Positive or negative ions are selected by utilizing positive or negative trapping voltages, respectively. Ions are trapped in the source region, Figure 2(a), by removing the source drift voltages and applying a negative bias to the analyzer drift plates. After a suitable delay drift potentials are switched to create an electric field between drift plates, Figure 2(b). The ions drift into the analyzer region, where a radio-frequency oscillator is connected to a drift plate. If the oscillator frequency coincides with the cyclotron frequency of an ion, an absorption of power by the ions occurs which can be readily observed with a marginal oscillator detector. Power absorption results in an increasing radius of ion cyclotron orbit as depicted in Figure 2(b). A typical experimental timing sequence is shown in Figure 3. The trapping sequence is initiated by a 10-100 msec ionizing electron beam pulse. After a specified trapping time the ions are drifted into the analyzer region where they absorb power, producing a transient in the marginal oscillator signal which is monitored by a boxcar integrator. Kinetic information is obtained by observing a particular ion while slowly varying the trapping time. Over a pressure range of 10^{-8}-10^{-5} Torr and trapping times of 1-10 sec ions undergo up

Potentials Applied in Trapping Mode
(Side View of ICR Cell)

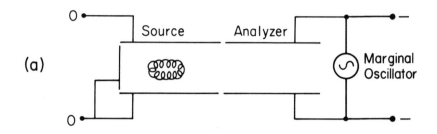

(a)

Potentials Applied in Detect Mode

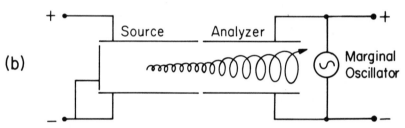

(b)

FIGURE 2. *Potentials applied to the trapped ion cell in
(a) trapping and (b) detect modes.*

to 1000 collisions with neutral molecules. At these pressures
and with typically less than 10^5 ions in the cell, neutral
concentrations are greatly in excess of ion concentrations and
pseudo-first order kinetics describe the ion-molecule
reactions.

ICR photochemical studies using low intensity uv-vis radiation have been performed using high pressure Xe or Hg-Xe arc
lamps in conjunction with a monochromator or band pass
filters (12-18). High resolution photolyses (19,20) and some
of the two photon experiments utilize ion lasers and tunable
dye lasers (13,17,21). In the vis-uv studies the photon beam
is directed into the source region along the y-axis of Figure
1, care being taken not to have the light strike any part of
the cell. In order to perform the IR multiphoton dissociation
experiments to be discussed in Sections IV and V the configuration of Figure 1 is used. The IR beam from a line tunable
Apollo Model 550A cw CO_2 laser is brought into the cell
through a 92% transmittance mesh which replaces one of the
source plates. The beam is reflected back from the highly
polished lower plate and collected outside the apparatus by a
graphite beam stop. Laser intensities in the cell can be

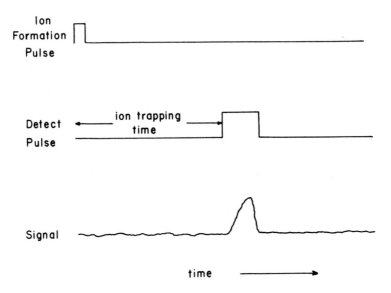

FIGURE 3. Timing sequence for trapped ion ICR experiments
(11). Ions can be irradiated during any portion of the
trapping period.

varied from 1-100 W/cm^2. A calibrated beam splitter diverts
a fraction of the infrared beam to a pyroelectric detector
allowing continuous power measurement. A fast mechanical
shutter (rise time approximately 5 msec) admits the laser beam
into the cell for any portion of the timing sequence shown in
Figure 3. Ions can be irradiated on alternate trapping cycles
and corresponding ion intensities (laser off and laser on) are
monitored by a two channel boxcar integrator. These signals
are then processed in a straightforward fashion to yield
photodissociation rate constants, even in the presence of ion
loss due to diffusion and reaction.

To date all ion photochemical studies utilizing ICR techniques have been performed at ambient temperature. Pressure
measurements are carried out with ionization gauges calibrated
against a capacitance manometer at higher pressure. Sample
mixtures are prepared using parallel multiple inlet manifolds.
Other experimental details may be found in individual
references.

B. Chemical Applications of Ion Cyclotron Resonance Spectroscopy

Over the past few years a tremendous variety of organic and inorganic chemistry has been explored with ICR techniques. Acid-base reactions (22-24), organic displacement and elimination reactions (25,26), metal clustering reactions (27), and transition metal ligand displacement reactions (27,28) are only a few of the areas which have been fruitfully explored. The attachment of alkali metal ions to organic molecules serves to illustrate ICR experimental techniques.

Application of ICR spectroscopy to the investigation of fast Li^+ transfer processes has been especially successful (29,30). A trapped ion spectrum of Li^+ clustering with 3-pentanone is shown in Figure 4 (31). The trapped ion source is filled with Li^+ generated by thermionic emission from a glass bead filament (29,30). In the presence of 3-pentanone reactions (2) and (3) are readily observed (Figure 4). While

$$Li^+ + (C_2H_5)_2CO \xrightarrow{k_1} (C_2H_5)_2COLi^+ \qquad (2)$$

$$(C_2H_5)_2COLi^+ + (C_2H_5)_2CO \xrightarrow{k_2} [(C_2H_5)_2CO]_2Li^+ \qquad (3)$$

the data in Figure 4 imply the reaction sequence (2) and (3), positive identification of reaction pathways is afforded by ion cyclotron double resonance (10). A double resonance radio frequency oscillator, which is connected to the source region of the ICR cell (Figure 1) may be tuned to the cyclotron frequency of one ion while the detection oscillator observes a different ion. The amplitude of the double resonance oscillator is adjusted so the ions are ejected from the cell in a time short compared to the time between collisions. Ejection of a reactant ion leads to a concomitant diminution of observed product ion intensity, thus confirming ion-molecule reaction pathways. For the example in Figure 4, ejection of Li^+ results in the disappearance of signal for $(C_2H_5)_2COLi^+$ and $[(C_2H_5)_2CO]_2Li^+$. Furthermore, ejection of $(C_2H_5)_2COLi^+$ causes the $[(C_2H_5)_2CO]_2Li^+$ signal to be reduced to zero, thus confirming reactions (2) and (3). Double resonance experiments thereby provide a simple, direct method for sorting out ion-molecule reaction sequences.

Of particular interest in the experiment of Figure 4 is the observation that the clustering reactions (2) and (3) are bimolecular, even at very low pressures (less than 10^{-7} Torr) (31). The reaction exothermicity (estimated to be 47 kcal/mol in process (2)) is initially present as vibrational

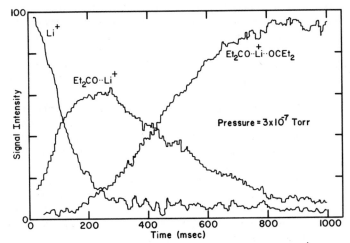

FIGURE 4. *Typical trapped ion spectrim of Li^+ clustering with $(C_2H_5)_2CO$ (31). $(C_2H_5)_2CO$ pressure is 3 x 10^{-7} Torr. Li^+ ions are generated in the first 10 msec by thermionic emission from a glass bead filament.*

excitation of the adduct. In the absence of exothermic reaction channels the fate of excited species formed by a process such as (2) is generally thought to be governed by collisional stabilization (process (4)) or unimolecular decomposition to reactants (process (5)). However, under conditions

$$[(C_2H_5)_2COLi^+]^* + M \longrightarrow (C_2H_5)_2COLi^+ + M \qquad (4)$$

$$[(C_2H_5)_2COLi^+]^* \longrightarrow Li^+ + (C_2H_5)_2CO \qquad (5)$$

where collisions are infrequent (for example, Figure 4) an alternative mechanism for deactivation of vibrationally excited species may be infrared radiative stabilization (process (6)) (32). Observation of processes which may

$$[(C_2H_5)_2COLi^+]^* \longrightarrow (C_2H_5)_2COLi^+ + h\nu \qquad (6)$$

involve radiative deactivation are especially pertinent to postulated mechanisms of molecular aggregation in interstellar space, where particle densities are so low as to preclude collisional stabilization of excited species (33).

In ICR experiments ion densities are too small to permit conventional absorption spectroscopy. Therefore, photochemically generated excited states of ions must be observed

indirectly.   For example, monitoring photodissociation
processes as a function of wavelength yields photoexcitation
spectra in cases where photon energies exceed thermodynamic
thresholds for ion decomposition. Band intensities in photoexcitation spectra are a product of the gas phase extinction
coefficient and the quantum yield for decomposition (14).
Since the latter is zero below the energy threshold for dissociation, only states of higher energy can be detected. In
many cases photodissociation band shapes are identical to
absorption band shapes for the species in condensed phase,
indicating quantum yields for dissociation which are constant
and very likely near unity in the gas phase (14).   Excited
states below threshold can be observed if some change in
reactivity follows absorption (34).   Often, comparisons of
photodissociation spectra with photoelectron spectra (PES) of
the corresponding neutrals are extremely useful in identifying
excited electronic states of ions (13,17). Usually only a few
of the states revealed in photoelectron spectra are accessible
by electric dipole selection rules from the ground state of
the ion. Photodissociation spectra are thus useful to confirm excited state assignments.

Ion photodissociation studies provide information relating
to the structure of gas phase ions. Dunbar has used high
resolution photodissociation spectroscopy to show that *cis* and
trans isomers of 1,3,5-hexatriene radical cations interconvert
on a time scale of seconds (19). Recently, Dunbar used the
same technique to distinguish among the possible $C_3H_5Cl^+$
isomers formed by ion-molecule reactions (20). Several groups
have used photodissociation techniques to measure the relative
populations of $C_7H_7^+$ isomers (35-37).

Acid-base properties of molecules in excited states have
been studied using ICR photochemical techniques (15,16,18).
The Li^+ affinity of a base B (defined as the enthalpy change
in a reaction such as (2)) is particularly sensitive to the
charge distribution within the base since Li^+ binding has been
shown to be primarily electrostatic (38). Therefore, photodissociation of BLi^+ complexes provides insight into changes in
electron density which accompany electronic excitation. For
example, Figure 5 shows the photodissociation spectrum of Li^+
complexed to *p*-methoxybenzaldehyde, obtained by monitoring the
decomposition, process (7), and compares it to the gas phase

$$CH_3O-\!\!\left\langle\bigcirc\right\rangle\!\!-C\!\!\underset{H}{\overset{O\cdot\cdot Li^+}{\lessgtr}} \;+\; h\nu \;\rightarrow\; CH_3O-\!\!\left\langle\bigcirc\right\rangle\!\!-C\!\!\underset{H}{\overset{O}{\lessgtr}} \;+\; Li^+ \qquad (7)$$

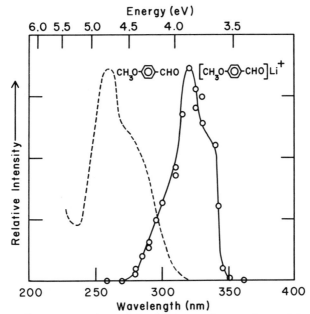

FIGURE 5. *Comparison of the gas-phase absorption spectrum of paramethoxybenzaldehyde with the photodissociation spectrum of the corresponding Li*$^+$ *complex (18).*

absorption spectrum of the free base. The red shift of the photodissociation band relative to the neutral absorption band results from an increased dipole moment and hence higher basicity of the excited relative to the ground electronic state. Full details of the spectroscopic and thermodynamic considerations involved in this analysis are given elsewhere (18).

Photodetachment of negative ions, generalized in equation (8), has also been investigated using ICR methods. High

$$A^- + h\nu \rightarrow A + e^- \tag{8}$$

resolution studies permit accurate measurement of electron affinities (39-47) and probe excited states of negative ions (40-41).

Photodissociation spectra of gas phase ions in comparison with comparable spectra measured in solution provide insight into solvation effects. While only a few examples of ion photodissociation have been discussed in this article a thorough review of the subject has been published recently (48).

III. STUDIES OF MULTIPHOTON DISSOCIATION PROCESSES USING
      VISIBLE RADIATION

In our laboratory, attempts to probe the lower excited
states of benzene radical cation, $C_6H_6^+$, unexpectedly led to
discovery of photodissociation at photon energies below
thermodynamic threshold (13). The dependence of the observed
dissociation rate on laser power and neutral gas pressure plus
comparison with photoelectron spectra (PES) indicate a
sequential, two-photon absorption mechanism. Similar results
have since been observed in several other systems (13,17,21),
reactions (9), (10), and (11). Figure 6 compares the PES of

$$C_6H_6^+ \xrightarrow{h\nu} C_6H_5^+ + H \qquad \Delta H = 3.9 \ eV \qquad\qquad (9)$$

$$C_6H_5CN^+ \xrightarrow{h\nu} C_6H_4^+ + HCN \qquad \Delta H = 3.2 \ eV \qquad\qquad (10)$$

$$C_6H_5Br^+ \xrightarrow{h\nu} C_6H_5^+ + Br \qquad \Delta H = 3.5 \ eV \qquad\qquad (11)$$

cyanobenzene reported by Rabalais and Colton (49) to the
photodissociation spectrum for the direct (single photon)
process and to the excitation function for the two photon
process in $C_6H_5CN^+$. The latter was obtained using argon ion
laser lines and dye laser output to cover the spectral range
from 4545-6050 Å (2.05-2.73 eV). The energy axis of the
photoelectron spectrum is adjusted such that the first
adiabatic ionization potential of cyanobenzene is zero on the
photodissociation scale. This allows a direct comparison
between peak locations in the photodissociation spectrum and
excited state energy levels observed by PES. Since the photodissociation experiments provide no information regarding the
symmetry of excited states, comparison with PES is useful,
particularly for systems such as benzene, where PES band
assignments have been made. In $C_6H_5CN^+$ the two photon process
(circles in Figure 6) coincides with a sharp PES band whereas
there is no overlap between the single photon process (squares
in Figure 6) and the photoelectron spectrum. The lack of
overlap suggests strongly that the observed transition is
$\pi - \pi^*$ in nature, populating a state whose detection is forbidden by PES selection rules. Similarly, in the case of
$C_6H_6^+$, although a complete two photon excitation function was
not obtained, comparison of the partial function with the PES
of benzene shows an energy match with the first excited $^2A_{2u}$
state of the ion (13).

For reactions (9)-(11) the two photon process is quenched
rapidly as pressure is increased. Thus, a sequential
excitation is thought to occur, illustrated for benzene by the

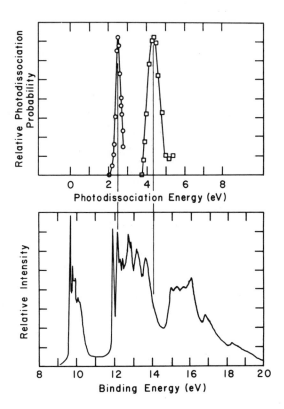

FIGURE 6. *Comparison of the photoelectron spectrum of cyanobenzene (49) to the photodissociation spectrum for the two photon process, obtained using laser light (○), and the excitation function for the single photon process at 100 A resolution (□). The energy axis of the photoelectron spectrum is adjusted such that the first adiabatic ionization potential of cyanobenzene is zero on the photodissociation energy scale. The photodissociation probabilities for the one and two photon processes are not directly comparable.*

following mechanism

$$C_6H_6^+ \underset{k_3[C_6H_6]}{\overset{\Phi\sigma_1}{\rightleftharpoons}} [C_6H_6^+]^* \overset{\Phi\sigma_2}{\longrightarrow} C_6H_5^+ + H \qquad (12)$$

where Φ is the photon flux, σ_1 and σ_2 are cross sections for the absorption of the first and second photon, respectively, k_3 is the rate constant for collisional deactivation and $[C_6H_6]$ is the number density of benzene neutrals. The excited species, $[C_6H_6^+]^*$, lifetime must approach 1 sec (the time between ion-molecule collisions) which implies internal conversion from the $^2A_{2u}$ state to a high vibrational level of the ground electronic state ($^2E_{1g}$) competes favorably with fluorescence. These investigations complement studies of fluorescence from polyatomic molecular ions in detailing excited state photochemical and photophysical processes (50).

Similar kinetic treatments of equation (12) have been proposed by Orlowski, Freiser and Beauchamp (17), and by Dunbar and Fu (21). Dunbar reports σ_1 and σ_2 for $C_6H_5Br^+$ are 4.4×10^{-18} cm^2 and 7.9×10^{-18} cm^2, respectively. Both groups obtain collisional deactivation rate constants which indicate nearly unit efficiency for relaxation per collision. Dunbar determined $k_3 = (1.1 \pm 0.3) \times 10^{-9}$ cm^3 molecule sec^{-1} for $[C_6H_5Br^+]^*$. Reported values for k_3 are $(8.0 \pm 0.5) \times 10^{-10}$ cm^3 molecule^{-1} sec^{-1} and $(1.1 \pm 0.1) \times 10^{-9}$ cm^3 molecule^{-1} sec^{-1} for $[C_6H_6^+]^*$ and $[C_6H_5CN^+]^*$, respectively. These results are in excellent agreement with the measured rate constants for symmetric charge transfer of $(8.0 \pm 1.0) \times 10^{-10}$ cm^3 molecule^{-1} sec^{-1} in benzene and $(1.0 \pm 0.1) \times 10^{-9}$ cm^3 molecule^{-1} sec^{-1} in cyanobenzene. Symmetric charge transfer thus appears to be extremely efficient in deactivating vibrationally excited molecules containing several eV of internal energy.

IV. MULTIPHOTON DISSOCIATION OF IONS WITH LOW INTENSITY CW INFRARED LASER RADIATION

While the studies briefly described above have provided interesting insights into ion photochemical processes using visible and uv radiation, the most exciting recent development in this expanding field of research is the observation of ion multiphoton dissociation processes using relatively low power cw infrared lasers. The scope of low intensity IR laser multiphoton dissociation processes studied to date is discussed in Section IV.A. Extensive work has been carried out on ions derived from diethyl ether, $(C_2H_5)_2O$, and perfluoropropylene, C_3F_6 (Section IV.B) with the most detailed information concerning photodissociation of $[(C_2H_5)_2O]_2H^+$ and $C_3F_6^+$ (Sections IV.C and IV.D).

A. Summary of Observed Photodissociation Processes

ICR techniques described in Section II have been used successfully to observe infrared multiphoton dissociation of ions at cw CO_2 laser intensities of 1-100 W/cm^2 (2-5). Table I lists those photodissociation reactions which have been observed. In each case the decomposition pathway has been

TABLE I. *Observed Low Intensity IR Multiphoton Dissociation Reactions*

Reactant	Products	ΔH (kcal/mole)	n[a]
[(C$_2$H$_5$)$_2$O]$_2$H$^+$[b,c]	(C$_2$H$_5$)$_2$OH$^+$ + (C$_2$H$_5$)$_2$O	31	12
(C$_2$H$_5$)$_2$OH$^+$[b,c]	C$_2$H$_5$OH$_2^+$ + C$_2$H$_4$	27	10
(C$_2$D$_5$)$_2$OD$^+$[b,c]	C$_2$D$_5$OD$_2^+$ + C$_2$D$_4$	27	10
(C$_2$D$_5$)(C$_2$H$_5$)OH$^+$[b,c]	C$_2$D$_5$OH$_2^+$ + C$_2$H$_4$	27	10
CH$_3$CHOC$_2$H$_5^+$[b,c]	CH$_3$CHOH$^+$ + C$_2$H$_4$	34	13
C$_2$H$_5$OH$_2^+$[b,c]	CH$_3$CHOH$^+$ + H$_2$	21	8
(CH$_2$CH$_2$CH$_2$CH$_2$O)$_2$H$^+$[b,c]	CH$_2$CH$_2$CH$_2$CH$_2$OH$^+$ + CH$_2$CH$_2$CH$_2$CH$_2$O	31	12
C$_3$F$_6^+$[d,e]	C$_2$F$_4^+$ + CF$_2$	56[f]	20
C$_6$H$_{12}^+$[d]	C$_5$H$_9^+$ + CH$_3$	30	11
CpRh(CO)$_2$H$^+$[b,g]	CpRhH$^+$ + 2CO	--[h]	--[h]

[a] Calculated for 944 cm^{-1} except for C$_3$F$_6^+$ which is given for 1047 cm^{-1} (maximum of photo-dissociation probability curve).
[b] Ion-molecule reaction product.
[c] Ref. (2)–(4).
[d] Formed directly by electron impact ionization.
[e] Ref. (3)–(5).
[f] Ref. (34).
[g] Cp = (η^5–C$_5$H$_5$).
[h] Value not known.

confirmed by direct observation of the product ion and by
double resonance techniques. The identity of the neutral
product is inferred from mass balance considerations.   All
observed reactions are identical to the lowest energy thermal
decomposition pathway, where known. Enthalpy changes are
calculated from ion heats of formation. Activation energies
are unknown and may be in excess of the reaction endo-
thermicity. The minimum number of absorbed photons needed to
reach thermodynamic threshold for dissociation, n, represents
the enthalpy change divided by the photon energy. For all
ions studied the observed decomposition pathway is wavelength invariant.

In almost every case (the exceptions are discussed in
Section IV.D) semi-log plots of fractional dissociation yield
as a function of irradiation time are linear, and the slope
defines a photodissociation rate. Fractional dissociation
yield of a given ion is the ratio of ion intensity during
laser irradiation to ion intensity without irradiation at
identical trapping times.

1. *Diethyl Ether System.* Major ions in the electron impact mass spectrum of diethyl ether at long trapping times and
low pressures are $(C_2H_5)_2OH^+$ and $CH_3CHOC_2H_5^+$ (51). At higher
diethyl ether pressures (>5 x 10^{-7} Torr) and long trapping
times (>500 msec) appreciable amounts of proton bound dimer
are formed by process (13). Over the pressure range of these

$$(C_2H_5)_2OH^+ + (C_2H_5)_2O \rightarrow [(C_2H_5)_2O]_2H^+ \tag{13}$$

ICR experiments dimer formation is bimolecular with rate constant, k = (1.9 ± 0.2) x 10^{-11} cm^3 molecule^{-1} sec^{-1}. A
typical experiment monitoring $[(C_2H_5)_2O]_2H^+$ is shown in Figure
7.   Ions are produced by a 10 msec electron beam pulse and
stored for up to 2 sec. At 1 sec of trapping time the
remaining $(C_2H_5)_2OH^+$ is rapidly ejected by double resonance,
thus preventing further formation of the dimer. This is
evidenced (Figure 7, upper trace) by the constant abundance of
$[(C_2H_5)_2O]_2H^+$ after 1 sec. The laser, tuned to 944 cm^{-1}, is
gated on at 1 sec of trapping time coincident with ejection of
$(C_2H_5)_2OH^+$ and effects an exponential decay of the dimer
(Figure 7, lower trace) with a power of 14 W/cm^2. At this
laser power no appreciable photodissociation of $(C_2H_5)_2OH^+$ is
observed, and the increase in abundance of this species
exactly matches the decrease in abundance of the proton bound
dimer verifying that there is only one active decay channel,
equation (14).

$$[(C_2H_5)_2O]_2H^+ + nh\nu \rightarrow (C_2H_5)_2OH^+ + (C_2H_5)_2O \tag{14}$$

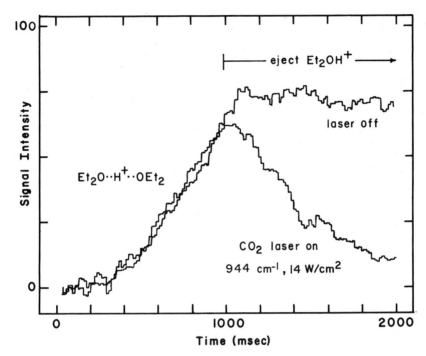

FIGURE 7. Ion intensity versus trapping time for a typical multiphoton dissociation experiment (2). At a diethyl ether pressure of 5.5 x 10⁻⁷ Torr ions are formed by a 10 msec 70 eV electron beam pulse. The upper trace is the proton bound dimer signal with the laser off. Ejection of $(C_2H_5)_2OH^+$ beginning at 1 second trapping time halts further dimer formation. CW irradiation by the infrared laser coincident with ejection of $(C_2H_5)_2OH^+$ (lower trace) results in photodissociation of the dimer. At this pressure the time between collisions is approximately 50 msec.

At low pressures where proton bound dimer formation is not significant multiphoton dissociation of the protonated ether, $(C_2H_5)_2OH^+$, can be studied. The laser induced process and postulated four-center intermediate (52) are shown in equation (15). The other major ion present at long times,

$$(C_2H_5)_2OH^+ + nh\nu \longrightarrow \left[\begin{array}{c} H^+ \\ C_2H_5O \!-\! CH_2 \\ \vdots \quad \vdots \\ H \!-\! CH_2 \end{array} \right]^* \longrightarrow$$

$$C_2H_5OH_2^+ + CH_2CH_2 \qquad (15)$$

$CH_3CHOC_2H_5^+$ also undergoes infrared multiphoton dissociation, equation (16).

$$CH_3CHOC_2H_5^+ + nh\nu \longrightarrow CH_3CHOH^+ + CH_2CH_2 \qquad (16)$$

Ions derived from electron impact ionization of $(C_2D_5)_2O$ undergo the identical reactions as do the corresponding unlabelled species. However, with the partially deuterated ether, $C_2H_5OC_2D_5$ an interesting isotope effect is observed in the decomposition of the protonated molecular ion. Chemical ionization of $C_2D_5OC_2H_5$ at low (12 eV) electron energies using cyclohexane as protonating agent allows for selective forma-tion of $(C_2H_5)(C_2D_5)OH^+$ with only trace amounts of $(C_2H_5)(C_2D_5)OD^+$. By analogy with equation (15) there are two possible product ions from the decomposition of $(C_2H_5)(C_2D_5)OH^+$, equation (17). Yet during laser irradiation,

$$
\begin{matrix}
H \\
C_2H_5OC_2D_5 \\
+
\end{matrix}
+ nh\nu
\left[
\begin{array}{l}
\longrightarrow C_2D_5OH_2^+ + CH_2CH_2 \qquad (17a) \\
\\
\longrightarrow C_2H_5OHD^{+\cdot} + CD_2CD_2 \qquad (17b)
\end{array}
\right.
$$

$C_2D_5OH_2^+$ is the only product detected, equation (17a). Thus hydrogen transfer in the four center intermediate is more facile than deuterium transfer. Consideration of ion detec-tion limits in this experiment provides a lower limit for the combined primary and secondary isotope effects (defined as the ratio of rates of product ion formation) as ≥ 6. It is assumed that the observed specificity arises from energetics of decomposition and is in no way attributed to selective laser pumping of only one half of the ion. In comparison, when $(C_2D_5)(C_2H_5)OH^+$ is formed by highly exothermic proton transfer such that the protonated ether internal energy greatly exceeds the threshold for decomposition in accordance with equation (17), the observed isotope effect is ~ 2 (53). These results imply that multiphoton dissociation occurs at an energy only slightly in excess of thermodynamic threshold. Large primary isotope effects have also been reported for metastable ion decompositions at threshold energies (54).

   2. *Perfluoropropylene System.* Electron impact ionization (14–70 eV) of perfluoropropylene, $CF_3CF=CF_2$, produces only four major ions: $C_3F_6^+$, $C_3F_5^+$, $C_2F_4^+$, and CF_3^+ (12). Fluoride abstraction by CF_3^+ occurs, reaction (18), leaving three

$$CF_3^+ + C_3F_6 \longrightarrow C_3F_5^+ + CF_4 \qquad (18)$$

ions at long trapping times. Dissociation of the parent cation $C_3F_6^+$, reaction (19), is the only observed infrared laser

$$C_3F_6^+ + nh\nu \rightarrow C_2F_4^+ + CF_2 \qquad \Delta H = 56 \; kcal/mole \qquad (19)$$

induced process (55). The product ion is both stable to laser irradiation and chemically unreactive.

$C_3F_5^+$ is totally unaffected by laser irradiation despite the availability of a decay channel, reaction (20), which has an energy requirement comparable to that of the observed photo-

$$C_3F_5^+ + nh\nu \not\longrightarrow CF_3^+ + C_2F_2 \qquad \Delta H \cong 53 \; kcal/mole \qquad (20)$$

dissociation reaction (55). The inertness of $C_3F_5^+$ toward the laser field proves that those photodissociation reactions which are observed result from absorption by the ions and are not due to some general non-specific heating of the ICR cell contents.

B. *Variation of Photodissociation Probabilities with Laser Wavelength*

The wavelength dependence of infrared multiphoton dissociation of four different ions has been studied. Analysis of these photodissociation spectra is not as straightforward as the vis-uv photodissociation spectra discussed in Sections II and III because the infrared studies have been limited to the CO_2 laser wavelength range, 925-1090 cm^{-1}. However, comparisons with gas phase spectra of pertinent neutrals and consideration of changes in bonding due to ionization allows a reasonable interpretation of the features present in the infrared photodissociation spectra. The analysis rests on the assumption that multiphoton dissociation spectra somewhat mimick the small signal absorption spectrum (refer to discussion in Section V.A) as is observed in pulsed laser photolyses of neutrals (1,56). Thus multiphoton dissociation represents one of the few techniques available for obtaining spectral information on gas phase ions.

1. Perfluoropropylene. Figure 8 shows the variation of $C_3F_6^+$ photodissociation probability, P_D, with laser wavelength. P_D is defined (in this case) as the fraction of ions decomposed during 2 sec of irradiation at 34 W/cm^2 in 4.8×10^{-7} Torr C_3F_6. The two sets of data are for electron impact ionization at 70 and 20 eV. Also shown in Figure 8 is the gas phase perfluoropropylene absorption spectrum over the same wavelength range. The infrared absorption band of the

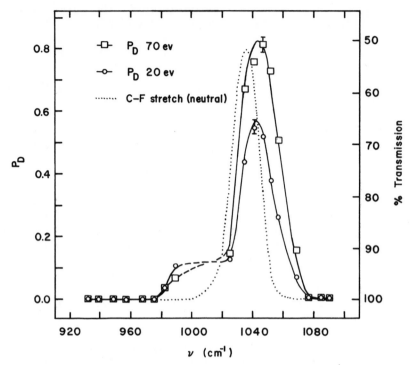

FIGURE 8. Photodissociation spectrum of $C_3F_6^+$ over the
CO_2 laser spectral range (3). Left ordinate is fraction of
$C_3F_6^+$ dissociated after 2 seconds of irradiation at 34 W/cm^2.
The two solid curves are for ionization energies of 70 eV (□)
and 20 eV (○). Perfluoropropylene pressure is 4.8 x 10^{-7}
Torr. Dotted line is infrared absorption spectrum of perfluoropropylene at 0.8 Torr in a 10 cm length cell.

neutral molecule (λ_{max} = 1037 cm^{-1}) has been identified as a
C—F stretch of A' symmetry (57). Comparison with other
fluorinated species suggests that this vibrational mode
involves only the CF_3 group. Since the lowest energy ionization process in C_3F_6 entails removing an electron from the
C—C double bond, this particular vibrational frequency is
unperturbed in going from the neutral to the cation. Therefore the photodissociation spectrum and the neutral absorption
band are nearly superimposed. Although the infrared absorption spectrum of the neutral shows a combination band at 978
cm^{-1}, its intensity relative to the major peak at 1037 cm^{-1} is
considerably smaller than the feature at 985 cm^{-1} in the $C_3F_6^+$
photodissociation spectrum. The small peak in the photodissociation spectrum occurs at an energy which is too low to
attribute it to a v = 1 → v = 2 transition. Thus, it is

tentatively assigned as a combination band. Attempts to probe in detail the characteristics of $C_3F_6^+$ photodissociation at 985 cm^{-1} were thwarted by a lack of laser intensity in that spectral region.

Energy deposition into internal degrees of freedom of ions formed by electron impact ionization increases with increasing electron energy. Hence the population of vibrationally excited $C_3F_6^+$ will be greater when formed with 70 eV electrons than with 20 eV electrons. The increase in photodissociation yield at 70 eV compared to 20 eV (Figure 8) is attributed to the presence of "hot" ions formed at 70 eV.

    2. *Diethyl Ether*. The wavelength dependences for multiphoton dissociation of $(C_2H_5)_2OH^+$ and $[(C_2H_5)_2O]_2H^+$ are shown in Figure 9a and data for $(C_2D_5)_2OD^+$ are shown in Figure 9b. Also shown are the gas phase absorption spectra of the corresponding neutrals over the range of CO_2 laser wavelengths

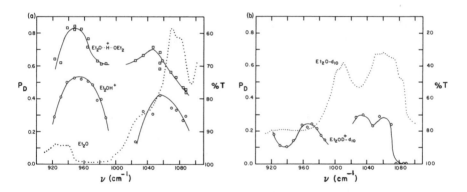

FIGURE 9. (a) Photodissociation spectra of $(C_2H_5)_2OH^+$ (○) and $[(C_2H_5)_2O]_2H^+$ (□) over the CO_2 laser spectral range. For $(C_2H_5)_2OH^+$, P_D is the fraction of ions dissociated after 1.9 seconds of irradiation at 48 W/cm^2; $(C_2H_5)_2O$ pressure is 8.8 x 10^{-8} Torr. P_D for $[(C_2H_5)_2O]_2H^+$ is defined as the fraction of ions dissociated after 2.0 seconds of irradiation at 10 W/cm^2; $(C_2H_5)_2O$ pressure is 4.7 x 10^{-7} Torr. Ionization energy for both experiments is 14 eV. Dotted line is the infrared absorption spectrum of diethyl ether at 20 Torr in a 10 cm length cell. (b) Photodissociation spectrum of $(C_2D_5)_2OD^+$ over the CO_2 laser spectral range. Experimental conditions are the same as for photodissociation of $(C_2H_5)_2OH^+$ in (a). Dotted line is the infrared absorption spectrum of $(C_2D_5)_2O$ at 16 Torr in a 10 cm length cell.

(note the change of scale in the axes for percent transmission). For both $(C_2H_5)_2OH^+$ and $(C_2D_5)_2OD^+$ experimental conditions were nearly identical; therefore differences in P_D values for the monomer species are a direct measure of differences in cross sections for multiphoton dissociation at each wavelength. No such direct comparison regarding dissociation cross sections can be made between the protonated ether and the proton bound dimer owing to differences in laser power and ether pressure for the two experiments.

Protonation of the ether molecule introduces three new degrees of freedom whose vibrational frequencies are all expected to lie outside the 925-1090 cm^{-1} region. The remaining vibrational bands should differ from those of the parent ether primarily as a result of perturbations to force constants introduced by a positive charge at oxygen and, secondly, from the effect of increased mass. Observed bands in the infrared spectrum of diethyl ether from 900-1100 cm^{-1} have been assigned to combinations of C–C stretches, C–O stretches and methylene wags (58). Upon protonation, rehybridization at oxygen to add p-character to the newly formed O–H bond results in increased s-character in the C–O bonds. The resulting blue shift in C–O stretching frequencies appears to be reflected in the photodissociation spectrum of $(C_2H_5)_2OH^+$. Specifically, the P_D maxima at 955 cm^{-1} and 1048 cm^{-1} correlate with the peak at 930 cm^{-1} and the shoulder at 1040 cm^{-1}, respectively. Addition of a second ether molecule to form the proton bound dimer distributes the charge between both ether moities with the result that vibrational frequencies of the dimer are more like those of the neutral ether. This is verified in Figure 9a. The absorption spectrum of $(C_2D_5)_2O$ (Figure 9b) shows maxima at 1010 cm^{-1} (assigned to C–O stretch) and 1060 cm^{-1} (assigned as methylene bends). In the $(C_2D_5)_2OD^+$ photodissociation spectrum the former is blue shifted to 1035 cm^{-1} and no frequency change is observed for the methylene bending mode. This is in accord with arguments presented for $(C_2H_5)_2O$. Below 980 cm^{-1} the $(C_2D_5)_2O$ absorption spectrum exhibits a broad absorption which extends to 840 cm^{-1}. The photodissociation spectrum in the same region cannot be correlated with specific $(C_2D_5)_2O$ vibrational modes due to the lack of distinguishable features in the absorption spectrum.

C. Effects of Collisions on Multiphoton Dissociation Probabilities

The role of collisions in multiphoton absorption and dissociation processes remains in dispute and must be better understood before large-scale reaction systems can be designed and implemented. A majority of authors have concluded that collisions decrease the efficiency of multiphoton dissociation (59-62) and reduce isotopic selectivity (63,64). However, reported measurements of transmitted pulse energy indicate that collisions can enhance multiphoton absorption (1,60,63, 65). Quigley (63) observes that collisional enhancement of multiphoton absorption cross sections is inversely proportional to laser pulse energy suggesting that the ICR experiments may be particularly sensitive to variations in pressure.

1. Proton-Bound Dimer of Diethyl Ether. Collisional effects both prior to and during irradiation have been studied in decompositions of $[(C_2H_5)_2O]_2H^+$. Since proton-bound dimer formation (equation (14)) is bimolecular and exothermic, question arose as to whether or not $[(C_2H_5)_2O]_2H^+$ is in a highly vibrationally excited state prior to irradiation. An experiment similar to the one depicted in Figure 7 was carried out, modified such that the laser was delayed following onset of $(C_2H_5)_2OH^+$ ejection for up to 900 msec. At various diethyl ether pressures and buffer gas (SF$_6$ and i-C$_4$H$_{10}$) pressures *no change* in $[(C_2H_5)_2O]_2H^+$ photodissociation rate is observed with increasing laser delay. Invariance of dissociation rates with laser delay (at constant pressure) implies that $[(C_2H_5)_2O]_2H^+$ is vibrationally relaxed prior to irradiation.

The data in Figure 10 indicate that $[(C_2H_5)_2O]_2H^+$ multiphoton dissociation rates decrease with increasing pressure at constant laser intensity. In addition to diethyl ether, results obtained by adding either of two buffer gases, SF$_6$ and i-C$_4$H$_{10}$, to a small amount of diethyl ether are included. To allow for a direct comparison of deactivation efficiencies of the three gases, dissociation rates are plotted as a function of ion-molecule collision frequency (66). Both SF$_6$ and i-C$_4$H$_{10}$ appear to be more effective than diethyl ether at quenching dissociation.

2. Perfluoropropylene. In sharp contrast to the collisional effects demonstrated in Figure 10 for $[(C_2H_5)_2O]_2H^+$, $C_3F_6^+$ multiphoton dissociation rates increase with increasing pressure up to $\sim 5 \times 10^{-6}$ Torr as shown in Figure 11. These results are obtained for C$_3$F$_6$, Ar, N$_2$ and SF$_6$ as buffer gases. In order to minimize possible hot ion effects (as seen in Figure 8) and to avoid ionization of buffer gases other than C$_3$F$_6$, an electron impact ionization energy of 14 eV was used. Only C$_3$F$_6$ (dashed line in Figure 11) shows a marked tendency to quench dissociation, beginning at pressures above 1.5 $\times$ 10^{-6} Torr. This is attributed to deactivation by symmetric charge transfer as was discussed in Section III for two photon

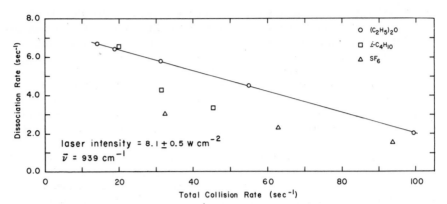

FIGURE 10. $[(C_2H_5)_2O]_2H^+$ multiphoton dissociation rate as a function of added buffer gases: SF_6 ($\triangle$) and $i\text{-}C_4H_{10}$ ($\square$). Dissociation rate is plotted as a function of total collision rate $((C_2H_5)_2O$ plus buffer gas) to allow direct comparison of collision efficiencies. SF_6 or $i\text{-}C_4H_{10}$ are added to 3.7×10^{-7} Torr of diethyl ether. Ionization energy is 14 eV.

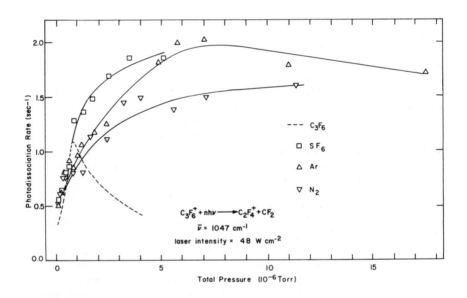

FIGURE 11. $C_3F_6^+$ multiphoton dissociation rate as a function of added buffer gases: Ar($\triangle$), N_2($\triangledown$) and SF_6($\square$) (3). The dashed line indicates $C_3F_6^+$ multiphoton dissociation rate with C_3F_6 as the added buffer gas. Ar, N_2, or SF_6 are added to 1.2×10^{-7} Torr C_3F_6. The abscissa indicates total pressure.

processes. In Figure 11 photodissociation rates are plotted against buffer gas pressure rather than collision frequency because for all four gases used rate constants for collision with $C_3F_6^+$ are nearly identical (66).

D. *Variation of Multiphoton Dissociation Probabilities with Laser Intensity*

Variation of multiphoton dissociation probabilities with laser intensity probes the mechanism of the excitation process and offers a means of comparison with megawatt pulsed laser experimental results. In studies of SF_6 at high laser powers it is observed that the dissociation probability per pulse is proportional to $(I_{las})^n$ where I_{las} is the laser intensity. Experimentally derived values (1) of n vary from 1.5 to 14. The use of focused laser beams leads to large uncertainties in measurements of power density and irradiated volumes, possibly accounting for the lack of agreement in reported values of n. The effects of laser power and pressure are intimately related since, at constant pressure, increased laser power reduces the time scale for excitation. Hence, fewer collisions are experienced during excitation. Therefore the ideal experiment for determining the variation of photodissociation yield with I_{las} would utilize an unfocused laser beam and should allow extrapolation to a low pressure limit. Such conditions are readily obtained in the ICR studies of slow multiphoton dissociation.

1. Proton Bound Dimer of Diethyl Ether. Careful examination of multiphoton dissociation of $[(C_2H_5)_2O]_2H^+$ reveals an induction period prior to the onset of decomposition. Induction period is defined as the time delay between opening the shutter and the first observable photodissociation. The variation of induction period with laser irradiation is illustrated in Figure 12 for $[(C_2H_5)_2O]_2H^+$ at constant ether pressure. When corrected for a shutter rise time of 5 msec, the data in Figure 12 closely fit equation (21) (solid line in

$$(Induction\ Period)\ \times\ (I_{las}) = 0.3\ J/cm^2 \qquad (21)$$

Figure 12), indicating an energy fluence threshold of 0.3 J/cm^2. Observation of an energy fluence threshold is in agreement with pulsed laser multiphoton dissociation experiments on SF_6 (9).

Assuming decomposition is rapid compared to excitation in the ICR experiments (Section V.C), the induction period represents the time required to pump ions to energies

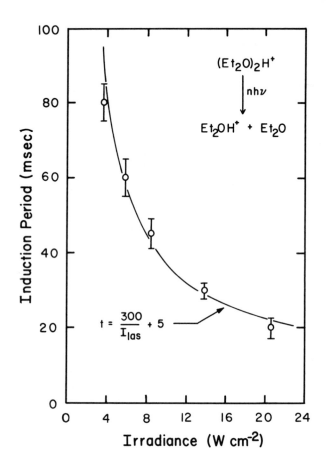

FIGURE 12. Induction period for multiphoton dissociation
of [(C$_2$H$_5$)$_2$O]$_2$H$^+$ as a function of laser intensity. Induction
period is defined as the time between shutter opening and
initial measurable photodissociation. Diethyl ether pressure
is 5.8 x 10^{-7} Torr, ionization energy is 14 eV and laser wavelength is 939 cm^{-1}. The timing sequence is the same as in
Figure 7. A close fit to the data is obtained by the equation
given above (solid line) which takes into account the 5 msec
shutter opening time. Induction periods (t) in msec and laser
intensities (I$_{las}$) in J cm^{-2} s^{-1} give an energy fluence
threshold of 300 mJ/cm^2.

sufficient for dissociation. This argument predicts that
photodissociation rates are pressure independent at pressures
such that the time between collisions is long compared to the
observed induction period. This criterion defines the low

pressure limit. Extrapolation of data such as in Figure 11 to the low pressure limit appropriate to the laser intensity yields the collision free photodissociation rate. A plot of the logarithm of collision free dissociation rate as a function of ln (I_{las}) (Figure 13) is linear with slope 0.84 ± 0.22. Therefore, within experimental error, multiphoton dissociation of $[(C_2H_5)_2O]_2H^+$ is first order in laser intensity. Over the range of laser intensities (4-20 W/cm^2), which permit accurate measurement of dissociation rates, no saturation effects are observed. A cross section for multiphoton excitation can be defined as the ratio of the low pressure photodissociation rate to photon flux. The calculated cross section for multiphoton dissociation of $[(C_2H_5)_2O]_2H^+$ is $(2.0 \pm 0.5) \times 10^{-20}$ cm^2.

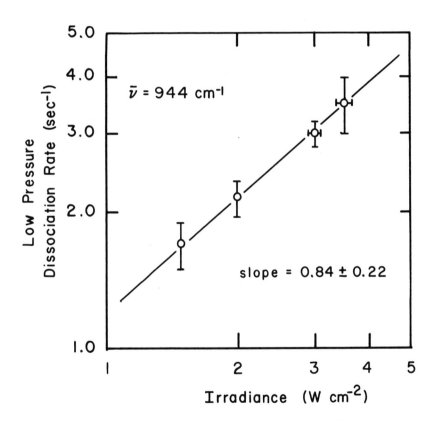

FIGURE 13. *Log-log plot indicating the dependence of the photodissociation rate $[(C_2H_5)_2O]_2H^+$ on the first power of laser intensity. Extrapolation to obtain low pressure rates uses induction period data from Figure 12, as discussed in Section IV.D.*

2. *Perfluoropropylene*. Since collisions in $C_3F_6^+$ are seen (Section IV.C) to enhance multiphoton infrared dissociation, extrapolation of these data to a low pressure limit is not straightforward. $C_3F_6^+$ photodissociation has been studied as a function of laser power at several pressures. For C_3F_6 pressures $\gtrsim 1 \times 10^{-6}$ Torr and C_3F_6 plus buffer gas (N_2) mixtures at total pressures $\gtrsim 1 \times 10^{-6}$ Torr photodissociation rates increase linearly with increasing irradiance up to the highest attainable laser intensity, 100 W/cm^2. At C_3F_6 pressures $< 1 \times 10^{-6}$ Torr the $C_3F_6^+$ signal does not exhibit a simple exponential decay with time at higher laser powers. The observed dissociation process can be described by the decay of two separate ion populations with different rate constants. While a complete quantitative analysis is not currently available, the mechanistic implications of this result are discussed in Section V.

V. MECHANISM OF INFRARED MULTIPHOTON DISSOCIATION

While it is tempting to apply current theories (67-79) for megawatt pulsed laser multiphoton dissociation to the low intensity IR photolyses of gas phase ions, there are inherent differences between the two types of experiments which necessitate modification of the existing developed theories. In particular, the time scale of the ICR experiments requires consideration of spontaneous emission as a viable deactivation mechanism. Equally important, the low laser intensities obviate power broadening (74-76) as a mechanism for overcoming anharmonicities in hot band absorptions. In addition any mechanistic considerations must allow for collisional enhancement of dissociation rates (as observed in $C_3F_6^+$).

A. Excitation Processes

Several authors have pointed out that at high levels of vibrational excitation the spectrum of transitions is nearly continuous (71,80,81). This realm of molecular excitation is generally known as the quasi-continuum of vibrational states. Theories of excitation through the quasi-continuum treat the process as a sequence of incoherent single photon events. Photoacoustic experiments indicate that for SF_6 absorption cross sections in the quasi-continuum decrease exponentially with increasing internal excitation (82). Model calculations (8), using cross sections obtained from the photoacoustic results, agree well with pulsed laser multiphoton dissociation

yields. Applying equation (21) to megawatt powers and short
pulse durations predicts induction periods commensurate with
the model calculations mentioned above. Thus absorption
through the quasi-continuum is expected to be similar for both
high power pulse laser excitation and low power cw experiments. The only difference may arise from deactivation due to
spontaneous emission in the low intensity photolyses. Estimates of spontaneous emission rates are 1-100 sec^{-1} (32).

The significant difference between megawatt multiphoton
activation and the ICR experiments described in Section IV
involves the mechanism(s) of initial excitation to the quasi-
continuum. The ease of populating the quasi-continuum will
depend on the number of photons required to reach this region.
Molecules possessing many degrees of freedom will have a significant amount of internal energy at ambient temperature.
The combination of many vibrational modes and appreciable
thermal energy content serves to locate such molecules very
nearly in the quasi-continuum prior to laser excitation. It
is thus natural to consider large molecule and small molecule limits. Although not all vibrational frequencies of
$[(C_2H_5)_2O]_2H^+$ are known, approximating the frequencies with
those reported (58) for $(C_2H_5)_2O$ permits an estimate of the
density of vibrational states at given internal excitations of
$[(C_2H_5)_2O]_2H^+$. Using the well-known Whitten-Rabinovitch
approximation (83), the densities of states at energies corresponding to absorption of one and two IR photons (1000 cm^{-1})
are 120 states/cm^{-1} and 9 x 10^4 states/cm^{-1}, respectively.
The cw CO_2 laser bandwidth of ~ 0.001 cm^{-1} implies that for
$[(C_2H_5)_2O]_2H^+$, following resonant absorption of one IR photon,
a near continuum of states (where the level separations and
the laser bandwidth are comparable) is available for successive absorptions. Furthermore, at ambient temperatures the
thermal energy content (3-4 kcal/mole) in excess of the zero
point energy for $[(C_2H_5)_2O]_2H^+$ is comparable to excitation by
a single photon. This puts the proton bound dimer of diethyl
ether in the large molecule category, where the density of
vibrational states is large at low energies and multiphoton
excitation is expected to be facile.

Density of state calculations for $C_3F_6^+$ using normal mode
frequencies measured (57) for C_3F_6 are 5 states/cm^{-1} and 213
states/cm^{-1} at energies corresponding to absorption of one and
two IR photons, respectively. The quasi-continuum of vibrational states in $C_3F_6^+$ is not reached until an ion has
absorbed approximately 4 photons; therefore $C_3F_6^+$ is classified
as a "small" molecule. The mechanism by which these four
photons are absorbed is significant in determining the $C_3F_6^+$
multiphoton dissociation rate. Sequential excitations within
one vibrational mode are not possible because the highest

available laser intensities, corresponding to a Rabi frequency
5×10^{-3} cm^{-1} for a 1.0 D transition moment (84), cannot overcome vibrational anharmonicities. Among available theoretical
treatments an excitation mechanism based on the formalism
of a single resonant absorption appears best suited to these
experiments (77,78). A two-level system assumed to be in
resonance with the laser field, is pumped and then depopulated
by intramolecular vibrational energy transfer. When the
resonant mode returns to the ground state it is free to absorb
another photon. The process is repeated until the quasi-
continuum is reached. The requirement of a single resonant
vibrational mode is consistent with the sharp frequency
dependence observed in the $C_3F_6^+$ photodissociation spectrum.
The slow step in the excitation process is intramolecular V-V
transfer at low excitation levels, since at all laser powers
used resonant transitions from the ground state are saturated
(84). Therefore, this simplified model predicts photodissociation rates to be independent of photon flux because
intramolecular V-V transfer is a nonradiative process. Even
though photolysis of $C_3F_6^+$ shows a change in behavior at low
pressures and high laser power, it does not exhibit the predicted saturation effect, indicating that modifications to the
theory are needed to describe low intensity infrared multiphoton dissociation processes.

B. Effects of Collisions on Multiphoton Excitation

For species at the large molecule limit, excitation occurs
exclusively via sequential photon absorptions through the
quasi-continuum and collisions act only to depopulate excited
vibrational states. This is observed in the decrease in
$[(C_2H_5)_2O]_2H^+$ dissociation rates with increasing pressure,
Figure 10.
Collisions have three possible effects on the excitation
of small molecules. First, and most obvious, is simple
deactivation as described in the above paragraph. Second,
collisions can enhance the rate of intramolecular V-V transfer,
thus increasing the rate at which the molecule is pumped to
the vibrational quasi-continuum. The third effect results
from the narrowness of the laser linewidth. At a given laser
frequency only a few rotational states lead to a resonant
transition from the ground vibrational level. These states
are depleted by photodissociation and collisional repopulation
of crucial rotational levels may be involved in the rate-
limiting process.
Both collision-induced intramolecular V-V transfer (85,86)
and collisional repopulation of depleted rotational states
would account for an increase in photodissociation rate with

increasing pressure (63). To date no experimental results
conclusively distinguish between either mechanism.

The two population decay observed in $C_3F_6^+$ decomposition
at low pressure and high laser intensity (Section IV.D)
results from either the presence of vibrationally excited ions
prior to irradiation, or varying rates for collisional redistribution of rotational states. In the first case, the
fraction of $C_3F_6^+$ with appreciable vibrational excitation
decomposes rapidly because it has already passed the intramolecular V-V transfer bottleneck. At the lowest pressure
used (8×10^{-8} - 2×10^{-7} Torr) there are not sufficient
numbers of collisions to deactivate vibrationally "hot" $C_3F_6^+$
formed by electron impact ionization (Section IV.B, Figure 8).
The alternative explanation for the two population decay is
derived from observations by Polanyi and Woodall (87,88) who
reported that probabilities for collisional induced rotational
transitions vary inversely with the change in rotational
energy. During the course of infrared laser photolysis the
involvement of species originally in rotational states
energetically far removed from the resonant state leads to a
slowing of repopulation rates, and hence a change in the
observed dissociation rate.

The obvious complexities in low intensity, infrared
multiphoton dissociation preclude further detailing of the
excitation process. It is reasonable to expect that modeling
calculations similar to those performed for megawatt pulsed
laser photolyses of SF_6 will provide new insight into the
nature of the phenomena.

C. *Unimolecular Dissociation of Activated Molecules*

Observation of a large isotope effect in the photodissociation of $(C_2H_5)(C_2D_5)OH^+$ (Section IV.A) as well as the
presence of only a single decomposition channel in all
reactions studied (Table I) establishes that dissociation
occurs before the molecule can absorb an appreciable amount of
energy above threshold. The relatively slow time scale for
low intensity IR laser excitation implies that standard
statistical treatments of unimolecular reactions (i.e., RRKM
theory (83)) can be utilized to describe the decomposition
step.

VI. SUMMARY AND PROGNOSIS

Two important conclusions which can be drawn from the ICR
experiments discussed in this chapter are (1) low power cw
CO_2 laser radiation effects infrared multiphoton dissociation

and (2) for species at the small molecule limit collisions
can enhance multiphoton dissociation rates. Ion populations
can be completely destroyed on a time scale of 1 sec with
energy fluences comparable to those used in megawatt infrared
pulsed laser investigations. Observed energy fluence thresholds are also comparable for both types of experiments. The
combined advantages of ICR techniques and spatial and
temporal uniformity of cw irradiation facilitate studies of the
effects of laser wavelength and power as well as collisions
on dissociation probabilities.

Low intensity multiphoton dissociation of ions provides a
convenient method for obtaining hitherto unknown vibrational
spectra of gas phase ions. This will allow differentiation
among structural isomers of ions. The availability of CO,
hydrogen halide, and optically pumped FIR lasers implies
multiphoton dissociation can be studied at a wide range of
infrared wavelengths.

REFERENCES

1. Ambartzumian, R. V., and Letokhov, V. S., *in* "Chemical
    and Biochemical Applications of Lasers" (C. Bradley
    Moore, ed.), Vol. III. Academic Press, New York,
    (1977).
2. Woodin, R. L., Bomse, D. S., and Beauchamp, J. L.,
    J. Am. Chem. Soc. 100, 3248 (1978).
3. Bomse, D. S., Woodin, R. L., and Beauchamp, J. L., *in*
    "Advances in Laser Chemistry" (A. H. Zewail, ed.),
    Springer Series in Chemical Physics, Springer, Berlin,
    Heidelberg, New York, (1978).
4. Woodin, R. L., Bomse, D. S., and Beauchamp, J. L.,
    J. Am. Chem. Soc., to be submitted.
5. Bomse, D. S., Woodin, R. L., and Beauchamp, J. L., *Chem.
    Phys. Lett.*, to be submitted.
6. Coggiola, M. J., Schulz, P. A., Lee, Y. T., and Shen,
    Y. R., *Phys. Rev. Lett. 38*, 17 (1977).
7. Grant, E. R., Schulz, P. A., Sudbo, Aa. S., Coggiola,
    M. J., Shen, Y. R., and Lee, Y. T., Proceedings Inter-
    national Conference on "Multiphoton Processes"
    Rochester, New York, (1977), to be published.
8. Grant, E. R., Schulz, P. A., Sudbo, Aa. S., Shen, Y. R.,
    and Lee, Y. T., *Phys. Rev. Lett. 40*, 115 (1978).
9. Kolodner, P., Winterfeld, C. W., and Yablonovitch, E.,
    Opt. Commun. 20, 119 (1977).
10. Lehman, T. A., and Bursey, M. M., "Ion Cyclotron
    Resonance Spectrometry", Wiley-Interscience, New York,
    (1976); Beauchamp, J. L., *Ann. Rev. Phys. Chem. 22*,

    527 (1971).
11. McMahon, T. B., and Beauchamp, J. L., *Rev. Sci. Inst.*
    43, 509 (1972).
12. For a full discussion of the ion-molecule chemistry of
    perfluoropropylene see Freiser, B. S., and Beauchamp,
    J. L., *J. Am. Chem. Soc. 96*, 6260 (1974).
13. Freiser, B. S., and Beauchamp, J. L., *Chem. Phys. Lett.*
    35, 35 (1975).
14. Freiser, B. S., and Beauchamp, J. L., *J. Am. Chem. Soc.*
    98, 3136 (1976).
15. Freiser, B. S., and Beauchamp, J. L., *J. Am. Chem. Soc.*
    98, 265 (1976).
16. Freiser, B. S., Staley, R. H., and Beauchamp, J. L.,
    Chem. Phys. Lett. 39, 49 (1976).
17. Orlowski, T. E., Freiser, B. S., and Beauchamp, J. L.,
    Chem. Phys. 16, 439 (1976).
18. Freiser, B. S., and Beauchamp, J. L., *J. Am. Chem. Soc.*
    99, 3214 (1977).
19. Dunbar, R. C., and Teng, H. Ho-I., *J. Am. Chem. Soc.*
    100, 2279 (1978).
20. Orth, R. G., and Dunbar, R. C., *J. Am. Chem. Soc. 100*,
    5949 (1978).
21. Dunbar, R. C., and Fu, E. W., *J. Phys. Chem. 81*, 1531
    (1977).
22. Wolf, J. F., Staley, R. H., Koppel, I., Taagepera, M.,
    McIver, Jr., R. T., Beauchamp, J. L., and Taft, R. W.,
    J. Am. Chem. Soc. 99, 5417 (1977).
23. Staley, R. H., and Beauchamp, J. L., *J. Am. Chem. Soc.*
    97, 5920 (1975).
24. Murphy, M. K., and Beauchamp, J. L., *J. Am. Chem. Soc.*
    98, 1433 (1976).
25. Beauchamp, J. L., Holtz, D., Woodgate, S. D., and Patt,
    S. L., *J. Am. Chem. Soc. 94*, 2798 (1972).
26. Ridge, D. P., and Beauchamp, J. L., *J. Am. Chem. Soc.*
    96, 637 (1974); Ridge, D. P., and Beauchamp, J. L.,
    J. Am. Chem. Soc., 96, 3595 (1974).
27. Foster, Michael S., and Beauchamp, J. L., *J. Am. Chem.*
    Soc. 97, 4808 (1975).
28. Corderman, R. R., and Beauchamp, J. L., *J. Am. Chem. Soc.*
    98, 3998 (1976).
29. Staley, R. H., and Beauchamp, J. L., *J. Am. Chem. Soc.*
    97, 5920 (1975).
30. Woodin, R. L., and Beauchamp, J. L., *J. Am. Chem. Soc.*
    100, 501 (1978).
31. Woodin, R. L., and Beauchamp, J. L., *Chem. Phys.*, to be
    published.
32. Dunbar, R. C., *Spectrochim. Acta 31A*, 797 (1975).
33. Herbst, E., *Astrophys. J. 205*, 94 (1976); Herbst, E., and
    Klemperer, W., *Astrophys. J. 185*, 505 (1973).

34. Kramer, J. M., and Dunbar, R. C., *J. Am. Chem. Soc. 94*, 4346 (1972).

35. Dunbar, R. C., *J. Am. Chem. Soc. 95*, 472 (1973).

36. Dunbar, R. C., *J. Am. Chem. Soc. 97*, 1382 (1975).

37. McCreary, D. A., and Freiser, B. S., *J. Am. Chem. Soc. 100*, 2902 (1978).

38. Woodin, R. L., Houle, F. A., and Goddard III, W. A., *Chem. Phys. 14*, 461 (1976).

39. Reed, K. J., and Brauman, J. I., *J. Am. Chem. Soc. 97*, 1625 (1975).

40. Richardson, J. H., Stephenson, L. M., and Brauman, J. I., *J. Am. Chem. Soc. 97*, 1161 (1975).

41. Richardson, J. H., Stephenson, L. M., and Brauman, J. I., *J. Chem. Phys. 62*, 1580 (1975).

42. Richardson, J. H., Stephenson, L. M., and Brauman, J. I., *Chem. Phys. Lett. 30*, 17 (1975).

43. Richardson, J. H., Stephenson, L. M., and Brauman, J. I., *Chem. Phys. Lett. 25*, 321 (1974).

44. Richardson, J. H., Stephenson, L. M., and Brauman, J. I., *Chem. Phys. Lett. 25*, 318 (1974).

45. Richardson, J. H., Stephenson, L. M., and Brauman, J. I., *J. Chem. Phys. 59*, 5068 (1973).

46. Smyth, K. C., and Brauman, J. I., *J. Chem. Phys. 56*, 4620 (1972).

47. Symth, K. C., and Brauman, J. I., *J. Chem. Phys. 56*, 1132 (1972).

48. Richardson, J. H., *Appl. Spect. Rev. 12*, 159 (1976).

49. Rabalais, J. W., and Colton, R. J., *J. Electron. Spect. 1*, 83 (1972).

50. Allen, M., and Maier, J. P., *Chem. Phys. Lett. 34*, 442 (1975); Maier, J. P., private communication.

51. Beauchamp, J. L., Ph.D. Thesis, Harvard University, 1967.

52. Tsang, C. W., and Harrison, A. G., *Org. Mass Spectrom. 3*, 647 (1970).

53. Holtz, D., and Beauchamp, J. L., unpublished results.

54. Williams, Dudley, and Hvistendahl, George, *J. Am. Chem. Soc. 96*, 6753 (1974).

55. Berman, D. W., Bomse, D. S., and Beauchamp, J. L., unpublished photoionization results.

56. Hartford, Jr., A., *Chem. Phys. Lett. 53*, 503 (1978).

57. Nielsen, J. R., Claassen, H. H., and Smith, D. C., *J. Chem. Phys. 20*, 1916 (1952).

58. Wieser, H., Laidlaw, W. G., Krueger, P. J., and Fuhrer, H., *Spectrochim. Acta, 24A*, 1055 (1968).

59. Ambartzumian, R. V., Gorokhov, Yu. A., Letokhov, V. S., Makarov, G. N., and Puretzkii, A. A., *Zh. Eksp. Teor. Fiz. 71*, 440 (1976) *[Sov. Phys. JETP (1976)]*.

60. Grunwald, E., Olszyna, K. J., Dever, D. F., and Knishkowy, B., *J. Am. Chem. Soc. 99*, 6515 (1977).

61. Lyman, J. L., Rockwood, S. D., and Freund, S. M., *J. Chem. Phys. 67*, 4545 (1977).

62. Bado, P., and van den Bers, H., *J. Chem. Phys. 68*, 4188 (1978).

63. Quigley, G. P., *in* "Advances in Laser Chemistry" (A. H. Zewail, ed.), Springer Series in Chemical Physics, Springer, Berlin, Heidelberg, New York, (1978) and references contained therein.

64. Bittenson, S., and Houston, P. L., *J. Chem. Phys. 67*, 4819 (1977).

65. Ham. D. O., and Rothschild, M., *Optics Lett. 1*, 28 (1977).

66. Bass, L., Su, T., Chesnavich, W. J., and Bowers, M. T., *Chem. Phys. Lett. 34*, 119 (1975).

67. Cantrell, C. D., and Galbraith, H. W., *Opt. Commun. 18*, 573 (1976).

68. Cantrell, C. D., and Galbraith, H. W., *Opt. Commun. 21*, 374 (1977).

69. Cantrell, C. D., Freund, S. M., and Lyman. J. L., *in* "Laser Handbook", Vol. III, North Holland Publishing, Amsterdam, to be published.

70. Cantrell, C. D., *in* "Laser Spectroscopy", (J. L. Hall and J. L. Carlsten, eds.), Vol. III. Springer Series in Optical Sciences, Springer, Berlin, Heidelberg, New York, (1977).

71. Bloembergen, N., *Opt. Commun. 15*, 416 (1975).

72. Yablonovitch, E., *Opt. Lett. 1*, 87 (1977).

73. Shultz, M. J., and Yablonovitch, E., *J. Chem. Phys. 68*, 3007 (1978).

74. Goodman, M. F., Stone, J., and Dows, D. A., *J. Chem. Phys. 65*, 5052 (1976).

75, Mukamel, S., and Jortner, J., *Chem. Phys. Lett. 40*, 150 (1976).

76. Mukamel, S., and Jortner, J., *J. Chem. Phys. 65*, 5204 (1976).

77. Tamir, M., and Levine, R. D., *Chem. Phys. Lett. 46*, 208 (1977).

78. Hodgkinson, D. P., and Briggs, J. S., *Chem. Phys. Lett. 43*, 451 (1976).

79. Quack. M., *J. Chem. Phys. 69*, 1294 (1978).

80. Isenor, N. R., Merchant, V., Hallsworth, R. S., and Richardson, M. S., *Can. J. Phys. 51*, 1281 (1973).

81. Akulin, V. M., Alimprev, S. S., Karlov, N. V., and Shelepin, L. A., *Zh. Eksp. Teor. Fiz. 69*, 836 (1975) [*Sov. Phys. JETP 42*, 427 (1975)].

82. Black, J. G., Yablonovitch, E., Bloembergen, N., and Mukamel, S., *Phys. Rev. Lett. 38*, 1131 (1977).

83. Robinson, P. J., and Holbrook, K. A., "Unimolecular Reactions", Wiley-Interscience, New York, (1972).

84. Allen. L., and Eberly, J. H., "Optical Resonance and
     Two-Level Atoms", Wiley-Interscience, New York, (1975).
85. Frankel, Jr., D. S., and Manuccia, T. J., *Chem. Phys.
     Lett.* *54*, 451 (1978).
86. Knudtson, J. T., and Flynn, G. W., *J. Chem. Phys. 58*,
     1467 (1973).
87. Polanyi, J. C., and Woodall, K. B., *J. Chem. Phys. 56*,
     1563 (1972).
88. Ding, A. M. G., and Polanyi, J. C., *Chem. Phys. 10*, 39
     (1975).

PHOTOCHEMICAL FIXATION OF THE NUCLEIC ACID
DOUBLE HELIX UTILIZING PSORALENS

John E. Hearst

Department of Chemistry
University of California
Berkeley, California

I. INTRODUCTION

The photochemistry of the psoralens with nucleic acids has
a varied utility in elucidating nucleic acid structure in
solution and in intact cells and virus particles. Some of the
more promising experiments involve rapid kinetics with 15 ns
laser pulses or slower intercalation kinetics utilizing intense CW lasers. This introduction provides a brief but
complete description of the background nucleic acid chemistry
and psoralen photochemistry required for an understanding of
the laser experiments described in Sections II and III.

Watson and Crick hypothesized the DNA double helix in 1953
(1). In this structure the purine and pyrimidine bases, which
are attached to the 1' position of deoxyribose in the deoxy-
ribose-phosphate chain, hold two antiparallel sugar-phosphate
chains in the double-helical structure by complementary
hydrogen bonding. The double-helical hydrogen-bonded structure is a property of most naturally occurring deoxyribonucleic acid and is also, in a modified form in which the
bases are not perpendicular to the helix axis, a property of
much of the secondary structure in single-stranded ribonucleic
acid molecules (2). RNA is most often found in nature as a
single-stranded molecule.

During DNA replication the strands of double helix are
separated, each serving as template for a newly synthesized
progeny strand (3). For this reason alone, experimental
methods of probing the kinetics of such fundamental processes
as helix unwinding are of great importance to an understanding
of cellular dynamics. There are, however, other compelling

reasons for developing a photochemical probe for helix structure. Two of the functions of RNA are: structural, as in the
ribosome, and sequence coding, as in messenger RNA. In both
of these cases local secondary or helical structure is an
essential feature for the proper functioning of RNA molecules.
Psoralen photochemistry is a uniquely powerful method for
mapping such structural features in both RNA and single-
stranded DNA.

A. DNA Duplex Melting

In all nucleic acids the helix-coil transition has a
positive enthalpy change so DNA unwinds and the strands separate as the temperature increases. Strand separation is a
highly cooperative phenomenon occurring in a temperature
interval of less than 6°C. For this reason the helix-coil
transition is often likened to a phase transition of a one
dimensional lattice and is called "melting" (4). The "melting
temperature" is defined as the temperature at the midpoint of
the transition. There is an increase in absorbance at 260 nm
upon loss of the stacking interactions between bases in the
double-helical structures, so the "melting" of DNA is readily
followed by absorbance as well as by changes in hydrodynamic
properties. Many solution properties effect the stability of
the nucleic acid helix. For example, lowering ionic strength
lowers the melting temperature because charge repulsion
between the negative phosphate groups is greater at low ionic
strengths (5). High pH also destabilizes the helix because
two of the bases are titrated at high pH disrupting the
hydrogen bonding in the helix. The pH at which the helix-coil
transition occurs is ionic strength and temperature dependent,
but typically lies between pH 11 and pH 13.
If the strands have not completely separated, returning
the DNA sample to helix stabilizing conditions such as lower
temperature or pH results in rapid reformation of the helical
structure (renaturation) (6). If the strands have completely
separated, initiation of renaturation is a slow event and
renaturation of the DNA requires both specific chemical conditions and long times (7,8). After strand separation, recovery
of the helical structure can be blocked by reaction with
formaldehyde, which forms Schiff's bases with the amino groups
of the bases, and prevents reformation of hydrogen bonds.
Finally, many organic solvents also cause strand separation.
The most commonly used solvent in electron microscopy is a
mixture of formamide and water in which melting temperatures
are often reduced from typical values of greater than 90°C at
physiological salt concentrations to values of 30°C to 40°C
(9).

The melting temperature of DNA is dependent upon base composition; DNAs with a high concentration of adenine-thymine base pairs have lower melting temperatures than DNAs of high guanine-cytosine composition (5). From studies on model compounds the enthalpy of formation of an A-U base pair in helical RNA is approximately 6.9 kcal/mole base pair while that of a G-C base pair is 13.8 kcal/mole base pair (10). These figures are approximate because nearest neighbor effects are substantial, but in general the conclusion can be drawn that the A-T rich regions of helical DNA are less stable than the G-C rich regions and that the initiation of DNA melting occurs in A-T rich regions.

The sequence of bases in DNA contains the message of life; it is therefore not surprising that microheterogeneity in stability occurs along its length and that the temperature range over which the melt occurs is dominated by this microheterogeneity in situations where the DNA itself is homogeneous (i.e., every molecule of DNA is identical). The intermediate states occurring within the melting temperature range have been studied optically (4), by electron microscopy (11) and by density gradient sedimentation equilibrium (12).

Preparation of samples of DNA for observation in the electron microscope most often requires spreading the DNA in a protein (cytochrome C) monolayer on the surface of a trough of water or buffer (9). Maintaining ideal spreading conditions as well as conditions which will partially melt the DNA is difficult, but such experiments have been successful (11). Open loops in the A-T rich regions of duplex DNA can be observed and mapped to specific sites to a resolution of 30 to 50 base pairs.

Partially melted molecules can also be observed in alkaline CsCl gradients, because the buoyant density of double-stranded helical DNA in alkaline CsCl is typically 1.69 g/cc while that of single-stranded DNA is 1.76 g/cc (12).

B. *Secondary Structure in Single-Stranded Nucleic Acids*

Both single-stranded RNA and DNA form helical structures by folding back, forming hairpins and hairpin loop structures. The best characterized of such structures occur in the transfer RNAs, which are relatively small molecules (75 nucleotides) X-ray crystallography has revealed the complete tertiary structures of several of the tRNAs (13,14).

Hairpins and looped hairpins have been observed by our laboratory using electron microscopy in DNA and RNA (15,16). Our ability to reproducibly observe and map these figures has been greatly improved by photoreacting the nucleic acids with

psoralens. This stabilizes helical regions by covalent cross-
linkage, enabling the use of more favorable chemical conditions for electron microscopy because helix stabilizing
conditions need not be retained during the spreading process.

We have used this photofixation technique to solve many
structural questions relating to single stranded nucleic acid
molecules. An example is presented in Fig. 1 where the helix-
coil transition of the single-stranded DNA of the virus fd has
been followed in the electron microscope by photocrosslinkage
of secondary structure as a function of ionic strength (17).
The regions of secondary structure in many cases are regions
of important biological function such as the replication
origin of DNA synthesis or regions which initiate RNA tran-
scription or RNA processing. When this same experiment is

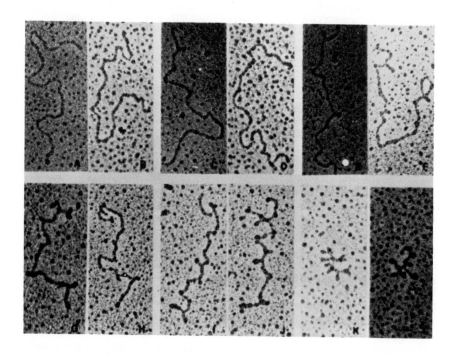

FIGURE 1. *Secondary structures of single-stranded (±)*
HindII-RFI DNA. fd RFI DNA was cut by the restriction enzyme
HindII, denatured, neutralized, and crosslinked at 15° in
different NaCl concentrations. (A) 0 mM, (B) 10 mM, (C,D) 20
mM, (E,F) 25 mM, (G,H) 30 mM, (I,J) 40 mM, (K,L) 100 mM.
Magnification × 72,300.

repeated on the intact virus particle, a specific orientation of the circular fd DNA in the filamentous virus particle can be demonstrated (18).

In a similar manner many structural loops in the 16S ribosomal RNA of *E. coli* have been mapped for the first time. At the present time there are no other methods for determining these structures.

C. *Psoralen Photochemistry*

The photoactivity of the psoralens as skin sensitizers dates back to ancient Egyptian and Assyrian writings (older than 1400 BC) in which extracts of the plant *Ammi majus* which are now known to contain psoralens were reported to induce sun tanning and to provide cure from skin disease when topically applied and exposed to the sun. The compounds today are used with long wavelength ultraviolet light for relief from serious cases of psoriasis.

The psoralens shown in Fig. 2 are all furocoumarins which intercalate between base pairs in the nucleic acid double helix and photoreact when excited with 320 nm to 380 nm light. Compounds I and V are natural products; compounds II, III and IV are synthetic psoralens developed in my laboratory for enhanced solubility in water and enhanced binding to DNA (19).

Upon photoexcitation either the 4'-5' double bond or the 3-4 double bond of the psoralen reacts with a 5-6 double bond of an adjacent pyrimidine in the DNA apparently forming a cyclobutane bridge. Some fraction of the resulting monoadducts (approximately 50%) are capable of forming a second cyclobutane bridge to another adjacent pyrimidine of the opposite nucleic acid chain by absorption of a second photon in the same wavelength region, thus linking the two strands together covalently (20,21).

Because the efficiency of photoreaction with helix is two orders of magnitude greater than with single separated strands, the photoreaction provides a remarkable assay both *in vitro* and *in vivo* for the existence of helix. Quantum yields are of the order of 1% for both steps in this cross-linking reaction, and at saturation *in vitro*, DNA can be reacted to the extent of 1 psoralen for every 2.5 base pairs.

In addition to the above examples of the utilization of this photochemistry to reveal nucleic acid structure in fd viral DNA and ribosomal RNA, similar studies have been completed on double-strand SV40 viral DNA and on nuclear DNA of higher cells revealing a 200 base pair structural repeat associated with chromatin structure (22,23). In these last cases it was observed that the basic proteins (histones) interacting with the DNA *in vivo* protect that DNA from psoralen

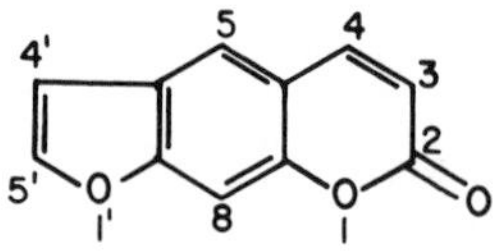

I. Psoralen

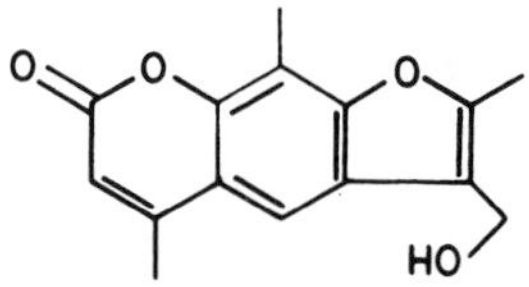

II. 4' Hydroxymethyl
4,5',8 trimethylpsoralen

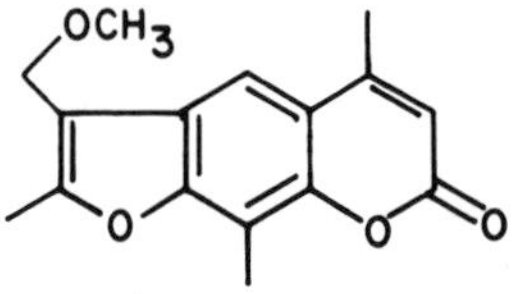

III. 4' Methoxymethyl
4,5',8 trimethyl-
psoralen

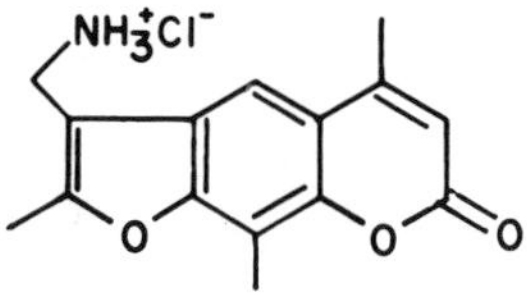

IV. 4' Aminomethyl
4,5',8 trimethylpsoralen
hydrochloride

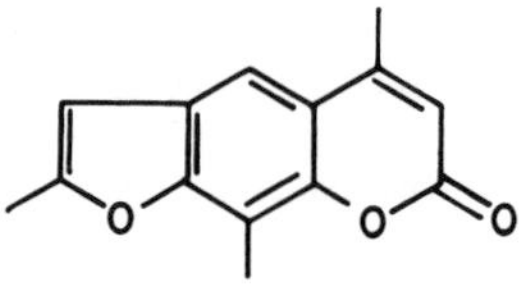

V. 4,5'8 Trimethylpsoralen

FIGURE 2. *Structures of psoralens.*

photocrosslinkage. Since these histones are found in structural units called nucleosomes which contain 8 histones and
approximately 140 base pairs of DNA it is only the inter-
nucleosomal DNA which can be crosslinked. When such DNA is
isolated and spread for electron microscopy under denaturing
conditions, repeated loops of 160 to 200 base pair are found
in the DNA, arising from the nucleosome protection and the
periodicity of the pattern (24,25).

Thus, there is much to be learned from psoralen photochemistry relating to structural questions not easily
approached in any other way. Despite these major successes
in elucidating static nucleic acid structure, it is my belief
that the major utility of psoralen photochemistry will be in
studying dynamic processes *in vivo*. Such experiments require
knowledge of the rapid kinetics of psoralen photochemistry
with nucleic acids and the utilization of lasers.

II.   RAPID PULSE KINETICS OF PSORALEN PHOTOADDITION TO DNA AND
OF PSORALEN CROSSLINK FORMATION

Very early in our studies of psoralen photochemistry the
need for a method of controlling monoaddition versus crosslink
formation became essential. This need arises from the desire
to do two types of experiments. First, psoralen constitutes
an ideal photolabel for helical structure. A major objective
is to isolate and purify ribosomal RNA and form monoadducts of
psoralen to it. (I will use ribosomal RNA in this example
although the approach is generally applicable to any structures containing nucleic acids.) The ribosome can then be re-
constituted by careful addition of the appropriate proteins,
and after reformation of the biologically active ribosome,
crosslinks formed by further irradiation with 320 to 380 nm
light. In this way helical regions in the ribosomal RNA in
the ribosome can be mapped. This procedure is necessary
because the ribosomal proteins apparently protect the RNA from
direct reaction with psoralens in the intact ribosome.

Another motive for controlled formation of monoadduct
arises from the fact that the genetic consequences of monoadduct formation and crosslink formation are expected to be
very different. A study of the carcinogenicity of monoadduct
versus crosslink requires reliable control of this reaction.
Such a study will constitute an important model of the molecular mechanisms of mutagenesis and carcinogenesis.

There are two methods other than fast pulse experiments
in which monoaddition can be favored, but neither provide as
convenient or as reliable a method of forming only monoadduct:
1) Since crosslink formation is a two photon event, low light
doses always favor monoadduct, although degree of reaction
must remain low as well. 2) Very long wavelength UV (380-390)
also favors monoadduct formation since a slight blue shift is
expected in the absorption spectrum of the monoadduct relative
to the unreacted psoralen. Irradiation in the absorption
tail therefore favors monoaddition although experimentally
this technique has enhanced monoaddition by only a factor of
four after long-term irradiation (26).

We have chosen the rapid-pulse technique which bases its
effectiveness upon the assumption that the photochemistry,
which is believed to involve a triplet excited state, is slow
relative to 15 ns light pulses (27).

A. Experimental

We have utilized three laser systems for controlled monoaddition of psoralen to DNA. A frequency doubled (347 nm), Q-
switched ruby laser having a pulse width of 15 ns and delivering 50 mJ per pulse was used for the experiments described in
this manuscript (27). The third harmonic (354.7 nm) of a
neodymium-YAG laser, deliverying 16 mJ per pulse in 15 ns
pulses has also been used. A focused nitrogen laser which
delivers 10 ns pulses at 337.1 nm and 3 mJ per pulse is the
third laser system we have worked with. In each case controlled monoaddition occurs. In the case of the neodymium-YAG
laser, when highly focused, we have observed some single pulse
crosslink formation which has been related to the very high
peak intensities delivered by this system. The mechanism for
this crosslink formation has not been established.

A typical cell for irradiation is cylindrical (3 mm to 8
mm diameter × 10 to 20 mm pathlength) with quartz end windows.
DNA concentrations range from 8 µg/ml to 50 µg/ml during
irradiation. Psoralen concentrations between 1 psoralen/2
base pairs to 1 psoralen/8 base pairs of DNA are typical conditions. 4'Aminomethyl-4'5',8 trimethylpsoralen (AMT) is the
only psoralen investigated in pulsed experiments to date (19).
It has been chosen because of its strong intercalation binding
constant to DNA which enhances the probability of photoreaction with DNA and because of its high water solubility
which provides experimental flexibility.

The AMT is synthesized containing a tritium label so
photoaddition to DNA can be followed by measuring bound radio-
activity/µg of DNA (19). The assay for crosslink formation
involves the denaturation of the helical DNA at pH 13 and room
temperature followed by rapid neutralization to pH 7.
Molecules which are not crosslinked do not renature under
these conditions while crosslinked molecules do. The ratio of
double-stranded to single-stranded molecules is measured by
density gradient sedimentation equilibrium in the analytical
ultracentrifuge.

In a typical laser pulse experiment approximately 2 AMT
molecules are added to 1000 base pairs of DNA by one pulse.
Assuming the photoefficiency of the second photon reaction
which leads to crosslink is approximately the same as that of
monoaddition, the probability of crosslink formation with two
pulses is $(.002)^2 = 4 \times 10^{-6}$ per base pair. In order to form
50% crosslinked molecules with two laser pulses, a very high
molecular weight DNA must be used. Based on this calculation
the DNA molecules should be at least 2.5×10^5 base pairs long
to insure an average of one crosslink per molecule. Such a
DNA molecule would have a molecular weight of 165×10^6
daltons. There are not many convenient sources of homogeneous

DNA of such high molecular weight. The DNA used in our experiments was isolated from an *E. coli* bacterial virus called T-4 and has a molecular weight of 113×10^6 daltons. Single-strand nicks in the DNA compromise the assay and were therefore avoided to the degree which was experimentally possible.

B. Results

The experimental design is presented in Fig. 3. After the first pulse of light, the excess psoralen which did not react with the DNA was extracted leaving behind only the drug added to the DNA by one pulse. From the radioactivity bound, 1.7 to 2.8 AMT molecules/1000 base pairs of T-4 DNA were bound by the first pulse. This sample was then exposed to additional pulses of light and the distribution of single- and double-stranded DNA in each sample determined by CsCl banding. Fig. 4 shows the results of this experiment. Briefly, the conclusions are as follows: 1) One pulse produces no detectable crosslinked molecules. 2) Two pulses result in 67% of the molecules having at least one crosslink. 3) Assuming that the distribution of crosslinks in the DNA population of molecules is Poisson, the average number of crosslinks formed by two pulses was 1.1 crosslinks/molecule. 4) Additional pulses create more and more crosslinked molecules until at 6 pulses nearly no single-stranded molecules remain. That which does remain appears to be predominantly nicked DNA. 5) Such a result cannot be explained on the basis of the lower dose of light received by the one pulse sample versus the two pulse sample; for on this basis if two pulses form 1.1 crosslinks/molecule, one pulse should form 0.27 crosslinks/molecule since crosslink formation is a two photon process. This amount of crosslinking would be easily detectable by our assay in the one pulse sample. The results were totally negative.

C. Discussion

We conclude that a 15 ns pulse of light, despite being energetic enough, is too brief to allow for the two photon process which leads to crosslink formation. In order to understand this result Brian Johnston, Andrew Kung, C. Bradley Moore and I have just completed a more sophisticated version of the above experiment in which crosslink formation was measured as a function of the time delay between the first 15 ns pulse and a second 15 ns pulse with time delays from 100 ns to 10 s. It has been concluded that there is a first order

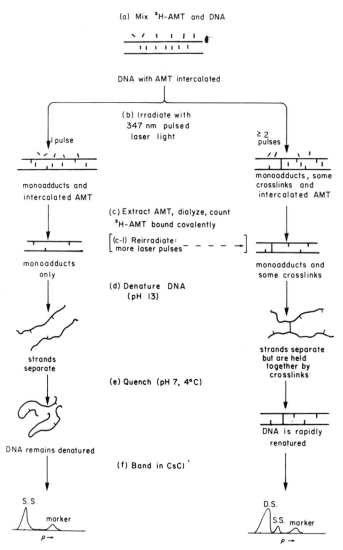

FIGURE 3. *Experimental Scheme for Pulsed Psoralen Photo-chemistry with DNA - a) Solution contained 50 μg of DNA and 5.6 μg of AMT per ml. in 0.01 M phosphate buffer (pH 7.0) and 1 mM EDTA. b) The pulse energy was 50 to 100 mjoule with a cross section of approximately 3 mm in diameter. d) The solution was brought to pH 13 with NaOH, allowed to stand for 10 minutes at room temperature (22°C), and then cooled to 4°C and neutralized (e) by addition of 0.2 M NaH$_2$PO$_4$. f) Samples analyzed in Beckman model E analytical ultracentrifuge at 42,000 rev/min. Abbreviations: SS, single stranded; DS, double stranded.*

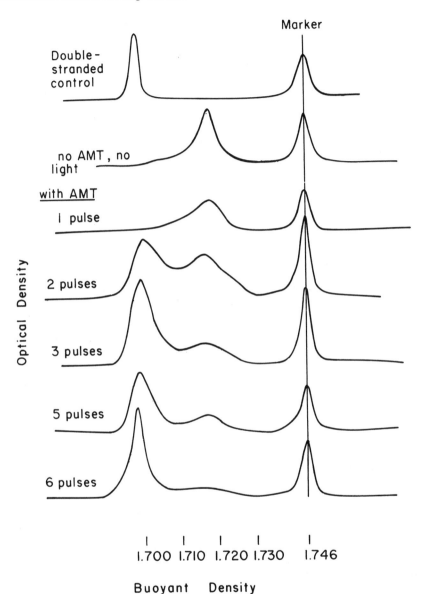

Marker

Double-stranded control

no AMT, no light

with AMT

1 pulse

2 pulses

3 pulses

5 pulses

6 pulses

Optical Density

1.700 1.710 1.720 1.730   1.746

Buoyant   Density

FIGURE 4. *Results of CsCl banding, where unreacted AMT was removed by extraction and dialysis after exposure to one laser pulse. This pure monoadduct sample was then further irradiated to produce the samples receiving two through six pulses, as in step c-1 of Fig. 3. The more leisurely decline of the denatured peak and the nearly constant buoyant density of the native peak are both due to the lack of new monoadduct being created after the first pulse.*

recovery of crosslink formation efficiency. The $t_{1/2}$ of this first order process is about 1 μs.

Molecular interpretation of this result awaits further experimentation. Possible models include the need for conformational relaxation of the DNA helix between the formation of the first cyclobutane bridge and the second cyclobutane bridge. An alternate model more photochemical in nature would hypothesize that it is an excited singlet state which is photochemically reactive during monoadduct formation and the decay of a long lived triplet is required to reach this reactive state.

III. UTILIZATION OF HIGH LASER INTENSITIES IN A KINETIC ANALYSIS OF PSORALEN INTERCALATION IN THE DNA HELIX

A. *Theoretical Considerations (19,28)*

The most elementary mechanistic representation of psoralen photoaddition to DNA includes three reactions. First, the drug must intercalate between the bases in the DNA helix; second, the intercalated molecule absorbs a photon and adds to the DNA; and third, free drug absorbs a photon and is destroyed as a reagent by photolysis.

$$P + S \underset{k_{-1}}{\overset{k_1}{\rightleftharpoons}} PS \tag{1}$$

$$PS + h\nu \xrightarrow{k_2} A \tag{2}$$

$$P + h\nu \xrightarrow{k_3} B \tag{3}$$

where P is free psoralen, S is binding site in the DNA, PS is psoralen bound to the DNA at an intercalation site, A is photoadduct to the DNA and B is photolysis product or breakdown product.

The laser pulse experiment discussed in the preceding section was performed in the strong binding domain where a large fraction of the psoralen (the AMT) in the solution was intercalated in the DNA prior to the irradiation. Let us now consider the weak binding limit in which a small fraction of the psoralen in solution is intercalated. This limit can be achieved either by using a psoralen derivative which binds weakly to the DNA or by using low concentrations of a strong binder. The mathematical assumption is that [P],[S]>>[PS] during the reaction.

Under these conditions, for low levels of non-covalent binding to DNA a steady state assumption on [PS] leads to the following equation:

$$\frac{\text{rate of photobreakdown}}{\text{rate of photoaddition}} = \frac{k_3}{k_2} \frac{K_D}{[S]} + \frac{k_3}{k_1} \frac{I}{[S]} = \frac{B}{A} \tag{4}$$

where K_D is the dissociation equilibrium constant, (k_{-1}/k_1), and I is the light intensity.

B. *The Potential Experiment*

These reactions can easily be studied using radioactive drug under conditions of vast excess of DNA reactive sites so that [S] is essentially constant. Under such conditions Eq. 4 is time independent and leads to the conclusion that the ratio of psoralen breakdown to psoralen addition to DNA is light intensity dependent. K_D is readily measured by equilibrium dialysis, and k_3 is the first order rate constant for psoralen photolysis in the absence of DNA. Thus the ratio B/A in the limit of I → 0 provides a measure of k_2 while the slope of the intensity dependent term provides a measure of k_1, the second order rate constant for drug intercalation into the DNA.

The constant, k_1, is especially interesting to measure for it is likely to be highly dependent upon chemical conditions including interacting proteins and tertiary structure. One of my major objectives is to measure this constant for DNA inside virus particles or in chromosomes in order to establish the degree of protection the DNA experiences from external drugs in these complex structures.

From values of k_1 which have been reported for other inter-calating drugs I have estimated that intensities of 10^3 to 10^4W/cm^2 will be required to measure k_1 by this method for purified DNA *in vitro*. Such peak intensities are easily achievable using pulsed lasers; however, the pulses introduce additional complications relating to polarization of the light and the rate of segmental diffusion of DNA in solution. Argon ion CW lasers can be focused to such intensities using the multiline, 351.1 and 363.8 nm, so by pumping the DNA solution or moving a closed capillary containing the solution past the focused laser beam the conditions for the experiment can be achieved. Because intercalation rates in virus particles are likely to be 100 times slower, model experiments have been initiated at intensities of 10 to 30 W/cm^2 using an unfocused beam from a Spectra-Physics Model 171 argon ion laser and T-7 bacterial virus.

IV.   CONCLUSION

I have outlined two specific applications to biochemistry which utilize psoralen photochemistry with nucleic acids and lasers.   There are many more applications which are obvious. DNA replication in *E. coli* occurs at a rate of one nucleotide added to the growing chain per millisecond.   Photofixation of the "growing fork" in milliseconds is feasible and might provide information on how far ahead of the growing chain the DNA helix is unwound, or on what initiates DNA synthesis on the "lagging" strand where DNA synthesis must be discontinuous.
    Similar kinetic questions can be asked of transcription, translation, messenger RNA processing, and ribosome allostery. In each of these cases the existence of a rapid photochemical probe of nucleic acid structure such as the psoralens coupled with lasers provides the biochemist with a valuable experimental tool.

REFERENCES

1.   Watson, J. D. and Crick, F. H. C., *Nature 171*, 964 (1953).
2.   Arnott, S., Hutchinson, F, Spencer, M. Wilkins, M. H. F, Fuller, W., and Langridge, R., *Nature 211*, 227 (1966).
3.   Meselson, M. and Stahl, F. W., *Proc. Nat. Acad. Sci., USA 44*, 671 (1958).
4.   Crothers, D. M., *Accts. Chem. Res. 2*, 225 (1969).
5.   Marmur, J. and Doty, P., *J. Mol. Biol. 5, 109* (1962).
6.   Geiduschek, E. P., *J. Mol. Biol. 4*, 467 (1962).
7.   Britten, R. J. and Kohne, D. E., *Science 161*, 529 (1968).
8.   Wetmur, J. G. and Davidson, N., *J. Mol. Biol. 31*, 349 (1968).
9.   Davis, R. W., Simon, M., and Davidson, N., *Methods Enzymel. 21D*, 413 (1971)
10.   Borer, P. N., Dengler, B., Tinoco, I., and Uhlenbeck, O. C., *J. Mol. Biol. 86*, 843 (1974)
11.   Inman, R. B., *J. Mol. Biol. 18*, 464 (1966).
12.   Cech, T. R., Wiesehahn, G. and Hearst, J. E., *Biochemistry 15*, 1865 (1976).
13.   Kim. S. H., Sussman, J. L., Suddath, F. L., Quigley, G. J., McPherson, A., Wang, A. H. J., Seeman, N. C. and Rich, A., *Proc. Nat. Acad. Sci. USA 71*, 4970 (1974).

14. Robertus, J. D., Ladner, J. E., Finch, J. T., Rhodes, D., Brown, R. S., Clark, B. F. C. and Klug, A., *Nature 250*, 546 (1974).

15. Cech, T. R. and Hearst, J. E., *Cell, 5,* 429 (1975).

16. Wollenzien, P. L., Youvan, D. C. and Hearst, J. E., *Proc. Nat. Acad. Sci., USA 75,* 1642 (1978).

17. Shen, C.-K. J. and Hearst, J. E., *Proc. Natl. Acad. Sci., USA 73,* 2649 (1976).

18. Shen, C.-K. J, Ikoku, A. and Hearst, J. E., *J. Mol. Biol.,* in press.

19. Isaacs, S. T., Shen, C.-K. J., Hearst, J. E. and Rapoport, H., *Biochemistry 16,* 1058 (1977).

20. Musajo, L. and Rodighiero, G., *Photochem. Photobiol. 11,* 27 (1970).

21. Cole, R. S., *Biochem. Biophys. Acta 217,* 30 (1970).

22. Hanson, C. V., Shen, C.-K. J. and Hearst, J. E., *Science 193,* 62 (1976).

23. Hallick, L. M., Yokota, H. A., Bartholomew, J. and Hearst, J. E., *J. Virology 27,* 127 (1978).

24. VanHolde, K. E., Sahasrabuddhe, C. G. and Shaw, B. R., *Nucleic Acids Research 1,* 1579 (1974).

25. Noll, M., *Cell 8,* 349 (1976).

26. Chatterjee, P. K. and Cantor, C. R., *Nucleic Acids Research, in press.*

27. Johnston, B. H., Johnson, M. A., Moore, C. B. and Hearst, J. E., *Science 197,* 906 (1977).

28. Hyde, J. E. and Hearst, J. E., *Biochemistry 17,* 1251 (1978).

Index